Crop Production:
Cereals and Legumes

Crop Production: Cereals and Legumes

by

BRIAN F. BLAND

Formerly Tutor in Agriculture at the University of Glasgow and Lecturer in Agriculture at the West of Scotland Agricultural College, Auchincruive, Ayr

1971

Academic Press: London and New York

ACADEMIC PRESS INC. (LONDON) LTD
Berkeley Square House,
Berkeley Square,
London, W1X 6BA

U.S. Edition published by

ACADEMIC PRESS INC.
111 Fifth Avenue,
New York, New York 10003

Library of Congress Catalog Card Number: 74–149693
ISBN: 0–12–104050–X

PRINTED IN GREAT BRITAIN BY
BUTLER & TANNER LTD, FROME AND LONDON

Preface

This book has been written primarily to meet the needs of students of Agriculture who cannot gain experience of all or even many arable crops in their practical year before commencing their studies.

The new Ordinary and Higher Diplomas in Agriculture are already under way in Scotland and are following in England and Wales. It is hoped that this account of cereal and legume production will find favour with entrants to these courses. Those qualifying will be our technicians and technologists for the 1970's, an age destined to demand high degrees of efficiency in the husbandries.

It is equally suitable for those who have embarked or are about to embark on courses similiar to the present National Diploma in Agriculture (NDA), since crop production in its widest sense represents about one-quarter of the syllabus.

Students entering the field of Agricultural Engineering, with a basic training in Engineering, may well appreciate this technological approach to the growing of crops. It is important for all our agricultural engineers to have sympathy with the structure and needs of agriculture, and their main contribution to the industry as a whole has, and will, come from the crop side.

It is hoped that students at university, reading Agriculture or Agricultural Science, will find the accounts of crop production informative and perhaps stimulating to further reading or research. In this connection references appear at the end of each chapter and although this practice has resulted in a little duplication in respect of authorities on general concepts, it was considered desirable in this form.

References are made to scientific literature of both official and commercial origin and the results of pertinent field experiments are included in the many tables to illustrate concepts or simply to present information in a highly condensed form.

Very few principles are left in our fast-changing agricultural industry and although they receive some attention, the main purpose has been to present an up-to-date technical account of the growing of cereals and legumes.

Variety recommendations are included for the sake of completeness and for these I wish to thank the National Institute of Agricultural

Botany, the three Scottish Agricultural Colleges and the Ministry of Agriculture in Northern Ireland. Although they are the most ephemeral parts of the book, readers are constantly reminded of the sources of up-to-date information on this topic.

Most farmers do not find time to indulge in much reading and they usually have access to numerous advisers, but it is hoped that those who seek information about cereals and legumes will not be disappointed with this volume.

The generous assistance received from many commercial companies, the *Farmers Weekly*, the *Farmer and Stockbreeder* and the National Institute of Agricultural Botany, who supplied many of the excellent photographic plates and explanatory diagrams is gratefully acknowledged.

Finally, the encouragement and help from many colleagues throughout Britain was greatly appreciated, and to Margaret Leckie, who typed the full manuscript, goes a special thank you.

BRIAN FOSTER BLAND

*RHM Agriculture,**
Throws Farm,
Stebbing,
Dunmow, Essex.
January, 1971

---

* Present address

Contents

PREFACE   v

1. Wheat

I.	Introduction	2
II.	Grain Quality and Uses	3
III.	Straw Characteristics	6
IV.	Types of Wheat	8
V.	Soil Type and Soil pH	10
VI.	Climate and Temperatures for Growth	11
VII.	Rotations and Disease Aspects	13
VIII.	Cultivations and Manuring	16
IX.	Seed Details	20
X.	Varieties	25
XI.	Details of Drilling, Time of Sowing and Early Spring Management	25
XII.	Chemical Weed Control	29
XIII.	Irrigation	34
XIV.	Harvest	34
XV.	Drying and Storage	38
XVI.	Disposal of Grain and Production Economics	45
XVII.	Production Economics	48
	Appendix	51
	References	60

2. Barley

I.	Introduction	64
II.	Botanical Classification	64
III.	Grain Uses	65
IV.	Grain Quality for Livestock Feeding	66
V.	Malting and Brewing	68
VI.	Requirements for Malting	68
VII.	Soil	71
VIII.	Climate	71
IX.	Rotations and Monoculture	73
X.	Manuring of Feeding Barley	76
XI.	Manuring of Malting Barley	78

XII. Cultivations and Seed-bed Preparations . . . 80
XIII. Drilling 85
XIV. Seed Rates 86
XV. Seed Dressings and Seed Treatment 86
XVI. Varieties/Cultivars 87
XVII. Management of the Established Crop 90
XVIII. Chemical Weed Control 91
XIX. Harvest 93
XX. Drying and Storage 96
XXI. Disposal of Grain, Production Economics and Yield . 105
Appendix 110
References 118

3. Oats

I. Introduction 122
II. Botanical Classification 123
III. Grain Production and Uses 127
IV. Grain Quality 128
V. Straw Yield and Quality 134
VI. Soil 137
VII. Climate 137
VIII. Place in Cropping Sequence 138
IX. Manuring 139
X. Cultivations and Seed-bed Preparation . . . 141
XI. Drilling 141
XII. Time of Sowing 143
XIII. Seeding Rate 144
XIV. Seed Dressing and Seed Treatment 145
XV. Varieties/Cultivars 147
XVI. Spring Management of Winter Sown Crops . . 150
XVII. Undersowing Oats with Grass and Clover Seeds . . 151
XVIII. Control of Insect Pest 151
XIX. Manganese Deficiency 153
XX. Chemical Weed Control 153
XXI. Harvest 155
XXII. Drying and Storage 157
XXIII. Production Economics 162
XXIV. Yield of Grain and Straw 164
Appendix 166
References 174

4. Rye

I. Introduction 178
II. Botanical Classification 179
III. Utilisation of Rye 181

IV. Grain Quality 182
V. Soils and Soil pH 185
VI. Climate and World Distribution 186
VII. Rye for Grain 187
VIII. Rye for Forage 203
Appendix 211
References 213

5. Maize

I. Introduction 217
II. Evolution and Botanical Classification 218
III. Groups or Types of Maize 220
IV. Maize in Britain 222
V. Maize Development in Other Parts of the World . . 224
VI. General Requirements of Climate and Soil . . . 225
VII. Fertiliser Requirements 232
VIII. Time of Application, Placement and Type of Fertilisers . 234
IX. Cultivations and Seeding 235
X. Seed Dressings 236
XI. Variety Classification 237
XII. Seed Rate, Plant Populations and Spacing . . . 239
XIII. Weed Control 242
XIV. Pest Control 244
XV. Diseases of Maize 245
XVI. Growth and Mid-season Management 247
XVII. Harvest and Storage 248
XVIII. Yield and Crop Quality 252
XIX. Production Economics, Gross Output and Gross Margins 254
Appendix 257
References 260

6. Beans

I. Introduction 265

PART 1. HORSE OR FIELD BEANS

II. Winter Beans 269
III. Practical Considerations in Pollination . . . 281
IV. Weed Control 283
V. Pest Control 285
VI. Spring Beans 287
VII. Harvest 293
VIII. Yield 297
IX. Grain Quality and Use of Field Beans in Livestock Feeding 298
X. Production Costs, Output and Gross Margins with Winter and Spring Beans 300
XI. Diseases of Field Beans 302

PART 2. FRENCH BEANS (KIDNEY)

XII. Introduction 303
XIII. Soils 306
XIV. Place in Cropping Sequence 306
XV. Cultivations 306
XVI. Manuring 307
XVII. Plant Spacing, Row Width and Seed Requirements . 309
XVIII. Sowing 310
XIX. Varieties 311
XX. Cultural Weed Control 312
XXI. Chemical Weed Control 312
XXII. Pest Control 314
XXIII. Irrigation 315
XXIV. Harvest 315
XXV. Yield and Quality 317
XXVI. Crop Value, Output and Gross Margin 319
Appendix 320
References 323

7. Peas

I. General Introduction 327
II. Taxonomy of Peas 330
III. Soils and Climate for Peas 332

PART 1. PEAS FOR HARVESTING DRY

IV. Introduction 333
V. Varieties 335
VI. Seed-bed Preparation 337
VII. Manuring 337
VIII. Seed Dressings 340
IX. Drilling and Spatial Arrangements 341
X. Time of Sowing 342
XI. Weed Control 343
XII. Control of Pests 347
XIII. Harvest 349
XIV. Warning and Information on Desiccants . . . 353
XV. Quality of Samples 353
XVI. Yield 356
XVII. Production Economics 356

PART 2. CONTRACT VINING PEAS FOR CANNING, QUICK-FREEZING AND DEHYDRATION

XVIII. Introduction 357
XIX. Major Production Areas 358
XX. Varieties 358

XXI. Seed-bed Preparation 361
XXII. Manuring 362
XXIII. Seed Rate 362
XXIV. Seed Dressing 364
XXV. Time of Sowing 364
XXVI. Chemical Weed Control 365
XXVII. Control of Pests 366
XXVIII. Harvest 366
XXIX. Yields with Vining Peas for Freezing and Canning . 373
XXX. Economic Aspects of Vining Peas 373

PART 3. PEAS FOR MARKETING GREEN— "PULLING PEAS"

XXXI. Introduction 375
XXXII. Production Areas 376
XXXIII. Sowing Dates 377
XXXIV. Varieties 378
XXXV. Harvesting and Marketing 378
XXXVI. Yield 378
XXXVII. Crop Protection 379
References 380
Addendum 382

8. Forage Legumes

PART 1. LUCERNE

I. Introduction 384
II. Early History and Spread of Lucerne Throughout the World 385
III. Nomenclature 386
IV. Taxonomy of Lucerne 388
V. Classification of Varieties or Strains of Lucerne . . 389
VI. British Evaluation of Varieties 391
VII. Acreage and Production Areas in Britain . . 392
VIII. Utilisation of the Lucerne Acreage . . . 393
IX. Soil Requirements 397
X. Climatic Requirements 400
XI. Lucerne Monoculture or Lucerne-grass Leys . . 404
XII. Establishment 407
XIII. Variety and Type of Seed 412
XIV. Management and Utilisation of the Crop . . 418
XV. Productivity 427
XVI. Quality of Herbage 429

PART 2. SAINFOIN

XVII. Introduction 431
XVIII. Botanical Classification 432
XIX. Types of Sainfoin 433
XX. Seed and Plant Morphology 433
XXI. Soil and Climatic Requirements 436
XXII. Fertilisers 436
XXIII. Seed 437
XXIV. Chemical Weed Control 439
XXV. Management and Utilisation 440
XXVI. Yield 441
XXVII. Quality of Herbage 443
References 444
AUTHOR INDEX 451
SUBJECT INDEX 457

CHAPTER 1

Wheat

I. Introduction 2
   A. Botanical Classification 2
II. Grain Quality and Uses 3
   A. Milling Characters 4
   B. Bread-making and Biscuit-making Quality 4
   C. Milling and Disposal of Products 5
III. Straw Characteristics 6
IV. Types of Wheat 8
   A. Vernalisation Requirement 9
   B. Winter versus Spring Varieties 9
V. Soil Type and Soil pH 10
VI. Climate and Temperatures for Growth 11
VII. Rotations and Disease Aspects 13
   A. The Place of Wheat in Cropping Sequence 13
VIII. Cultivations and Manuring 16
   A. Cultivations 16
   B. Manuring 18
IX. Seed Details 20
   A. Seeding Rates 20
   B. Seed Dressings and Seed Treatment 21
   C. Seed Quality and Approximate Costs 22
X. Varieties 25
XI. Details of Drilling, Time of Sowing and Early Spring Management . 25
   A. Drilling 25
   B. Time of Sowing 27
   C. Spring Management of Autumn Sown Crops 27
XII. Chemical Weed Control 29
   A. Autumn Spraying 29
   B. Spring Spraying of Autumn and Spring Sown Crops . . 30
XIII. Irrigation 34
XIV. Harvest 34
   A. Grain Sampling and Rapid Moisture Content Determinations . 35
   B. Types of Meters 36
XV. Drying and Storage 38
   A. Type of Drying Equipment and Installations 41
XVI. Disposal of Grain and Production Economics 45
   A. Government Support for Wheat 46
   B. Home-grown Cereals Authority 47

XVII. Production Economics   48
     A. Profitability, Grain and Straw Yields   49
     Appendix   51
     References   60

I. INTRODUCTION

The earliest cultivation of wheat in Britain took place before 2000 B.C. when Neolithic or New Stone Age immigrants from Europe practised shifting cultivation on the higher land above the marshes and forests in Wiltshire, Somerset, Gloucester, Dorset and the West Country (Percival, 1934). These early introductions and subsequent developments have been admirably portrayed by this author with some excellent photographs of ancient British wheats from places as far apart as Devonshire and Morayshire. In early times, wheat, barley, oats and rye were grown as single crops for bread-making and quite often a mixture of two of these cereals was used. By the middle of the eighteenth century it was estimated that 62% of the population in England and Wales ate bread made from wheat. Further changes from barley, oat and rye bread took place and Percival (1934) suggested that by 1825–1830 or even earlier the chief bread corn everywhere was wheat. This is not surprising since the flour from wheat can be made into loaves of bread which are more palatable and digestible and more universally acceptable compared with the other three cereals which tend to produce a coarse product.

A. Botanical Classification

Wheat-growing was extensively practised throughout Europe in prehistoric times and this cereal was of great importance in the ancient civilisations of Persia, Greece and Egypt (Percival, 1921). It spread to all the temperate countries of the world where it now plays a major part in the food supply of many of these nations and it is also widely cultivated in tropical and sub-tropical areas. The wheat plant probably originated from the fortuitous hybridisation of certain wild grasses (many points of similarity can be seen when lax-eared wheats are compared with couch-grass (*Agropyrum repens* Beauv.)) and Vavilov (1951), with his concept based on the variations in diversity of form, has suggested South-west Asia as the point of origin of the bread wheats. With such a long history of cultivations and the widespread nature of its occurrence, many distinct forms of wheat have evolved. They all belong to the genus *Triticum* of the family Gramineae, whether wild or cultivated, with a basic chromosome number of 7. The most commonly recognised species of wheat have been listed by Peterson (1965), who arranges them into diploid (14), tetraploid (28) and hexaploid (42) groups, the corresponding somatic number being given in brackets. Darlington and Wylie (1955) give a full list of the species belonging to the genus *Triticum* with references to all papers associated with chromosome study. This, together with the account of the cytogenetics and evolution by Riley (1965), may be helpful to the reader whose main interest is in plant breeding or genetics. Rivet wheats, *Triticum turgidum* (28), disappeared from British farming in the mid 1950's and all the remaining varieties left in cultivation and new additions belong to *Triticum vulgare* (42), the bread wheats, nowadays referred to as common wheat, *T. aestivum*. The cessation of Rivet wheat cultivation was due to a number of factors. Firstly, the straw was very tall and reed-like with a great tendency to lodge; secondly, the yield of grain was only moderate and lastly, the flour which its grain produced was not suitable for the manufacture of bread as it tended to be very dense and non-porous. It was, however, used on occasions for mixing with bread wheats and it did find an outlet in biscuit-making.

II. GRAIN QUALITY AND USES

British wheat is mainly used for animal feed or in the production of flour for bread-making, biscuit-making and household baking. In 1965 the estimated production of 3·6 million tons, according to *Farmers Weekly*, October of that year, was utilised as follows:

Human food	Sold for animal feed	Sold for seed	Farm retained
44%	48%	4%	4%

Greer (1968) showed that most of the wheat bought by flour millers was used to make bread flour and his estimates of tonnages used and proportions home produced are reproduced below, and probably correspond to the wheat milled in 1967.

Flour type	Annual wheat usage (million tons)	% Home-grown
Bread	3·6	20
Cake and biscuit	0·65	90
Household	0·7	50
Total	4·95	33

After Greer (1968).

A. Milling Characters

The best milling varieties of wheat are those from which the white endosperm (flour) can be easily and cleanly separated from the brown seed coat, known as bran. This is related to the texture of the grain and in general terms the miller prefers hard wheats which are smooth-skinned, as soft-textured grains often present separation difficulties. Milling assessment can easily be made by plant breeders on their new hybrids or potential varieties through the colour of the bran which they produce. Known varieties of very good or equally bad milling character can be used to formulate extremes in a colour gradient chart and then any new material can be assessed by comparision with those predetermined standards. The milling characteristics of any variety are genetically fixed and only minor modification can be made through climate and cultural techniques. Low rainfall, high levels of sunshine and on occasions the application of fertiliser nitrogen have been associated with flinty grains which tend to show up harder than the softer starchy types.

B. Bread-making and Biscuit-making Quality

A good "bread wheat" is one from which the flour will produce a dough which is elastic by nature and which can be baked into large loaves with an even texture. Biscuit-making on the other hand requires a dough which will prove to be plastic or sticky. These distinctions can easily be seen when the doughs from good bread and biscuit wheats are subjected to extensometer tests. An extensometer is a piece of apparatus consisting of a pair of jaws linked to an automatic recording device. Dough placed over the jaws is pulled apart and the tension required to do this and the extension produced are recorded simultaneously on a graph. This is usually referred to as an extensometer curve and the graphs produced by bread and biscuit wheats are quite distinct (see Fig. 1). The characteristics exhibited by the dough can be mainly attributed to the grain proteins termed gluten.

The amount of grain protein can easily be enhanced by the liberal use of nitrogenous fertilisers as one might expect, but the type of gluten cannot be altered. Thus it is not possible to change a poor bread-making or biscuit-making wheat by varying the cultural techniques. A feature of twentieth-century wheat-growing in Britain has been the correlation of high yield and low bread-making quality and of high bread-making quality and low to moderate yield amongst the varieties. Millers did little or nothing to change this deadlock as quality payments to farmers were notional and quality wheats were obtainable on the world market at reasonable prices. Canada (the world-famous Mannitoba Nos 1, 2 and 3 wheats) and to a lesser extent Russia, Australia and Argentina to name but a few, were all contributing to our National Bread. This is still true but the plant breeders have at last managed to link high grain yield with bread quality and in 1967 farmers saw the introduction of a wheat premium scheme. A pilot scheme involving 12,000 acres of the variety Maris Widgeon, an introduction from the Cambridge Plant Breeding Station, was announced by the Ranks-Hovis-McDougall

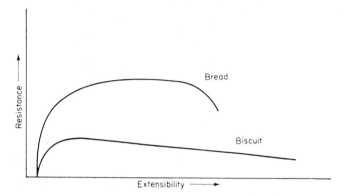

Fig. 1. Extensometer curves from the doughs of a bread wheat (SVENNO) and a biscuit wheat (PEKO). (After Thompson and Whitehouse, 1962.)

group of companies. This scheme was inaugurated to provide a guaranteed tonnage of high quality wheat for flour-milling and the farmers who contract to grow this variety receive a premium of 30 shillings per ton above the ruling milling price (*Farmer and Stockbreeder*, 28th June, 1967). With effect from the 1971 harvest, quality spring wheats will also command a premium, although in practice it may not be as high as that paid for Maris Widgeon.

C. Milling and Disposal of Products

Examination of the longitudinal cross-section of the wheat grain indicates three distinct parts; the central starchy core known as the endosperm, the embryo at the base and the skin or seed coat which encloses the endosperm and embryo (see Fig. 2). In the production of flour (endosperm) it is customary to remove the embryo and the skin. The grain is passed between varying sized rollers which crush it and remove successive layers of skin. At the same time these are separated by sieves and fans into bran, middlings and sharps. The bran, being the outside of the grain, is coarse and brown whereas the sharps and middlings are much paler in colour and are fine in texture since they represent the internal layers of the skin.

EMBRYO	SKIN or SEED COAT	ENDOSPERM
Wheat germ	Bran, Middlings, Sharps	Flour

Collectively known as wheat offals

Bran, middlings and sharps are important animal feeds and the germ is a valuable fraction. It can either be collected separately or it may be combined with the fine offals to produce a product known as wheat-feed.

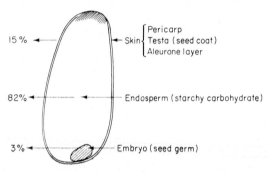

FIG. 2. Parts of a wheat grain (approximately twelve times normal size).

On average one-quarter of wheat weighing 500 lb will produce the following:

	lb	%
Flour	380	76·0
Bran	26	5·2
Middlings	50	10·0
Sharps	34	6·8
Loss	10	2·0

From McConnell (1883).

III. STRAW CHARACTERISTICS

In the past wheat straw has been used for many purposes, for example, thatching, covering clamps both outside and under cover and for packing bottles and fragile objects for transportation. This has resulted in the cultivation of longer strawed varieties which give a good yield of straw at the expense of the grain. Nowadays a small acreage of wheat is bindered to satisfy this demand. The higher costs of harvesting are well offset by an increased straw revenue in the region of £10 per acre. Many of the modern short varieties of wheat have straw which breaks up into small pieces when passed through the combine harvester. It is thought that the presence of pith in the centre may be a possible cause (see Fig. 3). As a result these varieties with short, thick-walled straw are not recommended for the quality straw trade and in practice they are often burned or ploughed-in following harvest.

The environment and some cultural techniques can markedly affect the yield of straw. High levels of fertility or large dressings of nitrogenous fertiliser will produce excessive quantities of straw with wheat. Under many circumstances this can lead to lodging, due to weakness and excessive length of straw and quite often this is accompanied by a concomitant drop in grain yield. Where spring grazing of winter wheat has been practised lower straw yields have been reported from all the experiments and this has been the net result of lower straw numbers and reduced straw height. The common belief that spring grazing will

(a)   (b)

Fig. 3. Cross-sections of wheat straw. (a) Typical of short, modern varieties with thick-walled straw. (b) Typical of older varieties with long, hollow straw.

result in a larger number of tillers is correct if applied to the vegetative tillers only as a reduction (10%) in ear-bearing straws has been recorded following spring grazing.

Recently a new chemical cycocel (C.C.C.)†, has been introduced to combat lodging in cereals. It reduces the crop height by shortening the internodes and its overall effect has been a reduction of approximately 4 inches in straw length. In the case of winter wheat 3 pints per acre of 40% W/V is recommended (2 pints for spring wheat) in 20–40 gallons of water for spraying after tillering and before the commencement of stem elongation (stages 3 to 5 on the Feekes–Large Scale, Fig. 10). It can be applied simultaneously with chemical weed control but growers should make certain that it is safe to mix cycocel with their choice of chemical for weed control. Using present-day prices an increase in grain yield of 1·5–2 cwt/acre with winter wheat and 1–1·5 cwt with spring wheat is necessary to cover the cost of this spraying. Experiments with this chemical have produced very variable results in terms of

† ((2-chloroethyl)—trimethyl ammonium chloride) = C.C.C.

increased crop yield. Cycocel has nearly always reduced the amount of lodging in wheat crops, but this has not always been followed by yield increases. However, since reduced lodging is almost certain, the level of fertiliser nitrogen could safely be raised.

IV. TYPES OF WHEAT

Varieties of wheat fall naturally into two main groups, those for establishment in the autumn or winter and the remainder for spring sowing. Occasionally a cultivar may be introduced which can be used for either winter or spring sowing as did Bersee in the early 1950's.

With considerable interchange of cultivars (cultivated varieties)

TABLE 1

Types of Wheat and their Associated Agronomic Characteristics

Prostrate	Semi-prostrate	Semi-erect	Erect
Scandinavian winter wheat	British, Dutch and German winter wheat	French and Belgian winter wheat	Most spring wheats
Very winter-hardy	Sufficiently winter-hardy	Winter-hardiness slightly suspect under extreme British conditions	Not reckoned to be winter-hardy
V.R.[a] large	V.R. medium	V.R. small	V.R. nil
Maturity: late	Maturity: medium/early	Maturity: early	—

[a] V.R. denotes the vernalisation requirement.

between European countries it is not surprising to find significant variation in the types of winter wheat which appear on our recommended list. (Details of these appear at the end of the chapter). Broadly speaking they may be grouped according to initial mode of growth and in Table 1 this is illustrated and a comparison with spring wheat is made.

A.   Vernalisation Requirement

Most of the recognised winter cultivars require to experience low temperature and growth during the short day–long night period of the year before they change from purely vegetative growth to reproductive growth. This is known as the vernalisation requirement and in order to produce a crop of grain these winter varieties need to be sown in the autumn or early winter otherwise there is a danger that they may remain in leaf form.

The classification suggested in Table 1 is not very precise and some cultivars may not fit in with this scheme. In spite of this it is worth consideration since cultural practices are quite often influenced by the type of variety to be used. Where Scandinavian wheats are chosen it is essential that they are sown early since their vernalisation requirement is large and their maturity rating late. At the other end of the scale, whilst it is not recommended that they should be sown late, it is possible to obtain good crops from the early French types with sowings as late as January or February. At that time of year one would have thought the natural choice would be a spring cultivar but even with this late sowing it is often possible to obtain better yields so long as either an early French or a dual purpose type is chosen. Occasionally dual-purpose varieties which can be used either for winter or spring sowing become available and Maris Ranger is one such cultivar.

B.   Winter versus Spring Varieties

Under comparable conditions winter varieties will outyield spring ones by at least 3 cwt/acre. On average the difference would be 5–6 cwt and many farmers have obtained 8–10 cwt/acre more from their winter sowings. Another big advantage of winter wheat over spring wheat is in time of maturity. Spring wheat is the latest ripening of the three main cereals and as one proceeds north the feasibility of successful cultivation diminishes fairly quickly. This is mainly on account of the later harvesting period and significantly lower temperature for growth. When the condition of grain is of prime importance then early harvesting under good conditions is essential. This applies particularly to the wheat crop, the majority of which is sold and leaves the farm. In order to qualify for economic support through the deficiency payment scheme, provided for in the Agriculture Act of 1947, the grain must be of good quality in sound condition and the sale of it must be certified by an authorized grain merchant.

V. SOIL TYPE AND SOIL pH

Up to the middle of the twentieth century successful wheat cultivation had always been associated with strong land which contained a significant clay or silt fraction. Light soils were regarded as being suitable only for oats and barley and in extreme cases rye. Today the heavier soils can be relied upon to produce high yielding crops of wheat and to carry a large proportion of this crop but wheat cultivation has spread to medium and light textured soils with a fair degree of success. Under such conditions the crop seldom reaches the level of yield of the stronger land but ploughing, subsequent cultivations and also the establishment are less exacting, quicker to put into effect and therefore

TABLE 2

Effect of pH and Additions of Lime on Wheat Yield

Results of a liming experiment on a medium soil at St Albans (1934–1950)				
Change in pH	4·3–4·8	4·8–5·3	5·3–5·8	5·8–6·3
Extra wheat yield (cwt/acre)	6·0	3·6	2·2	1·3
Actual yield at varying pH values (1946)				
Level of pH	4·3	4·7	5·3	5·8	6·5
Yield (cwt/acre)	3·8	10·2	15·4	17·7	18·4

are somewhat cheaper. Silty soils prove a little less arduous in their cultivation requirements than the heavy clays and on average produce some of the heaviest yielding wheat crops in Britain. Where cereals are to be grown on the highly organic Fen soils then the choice is nearly always wheat. With these reclaimed peats there is a significant annual release of mineralised nitrogen which is responsible for a large volume of vegetative growth with most plants. Under these conditions barley and oats have little chance of standing and wheat should be chosen in most cases. Short-strawed varieties of wheat are employed, fertiliser nitrogen applications are minimal and other cultural techniques should be used to cut down the volume of flag leaf in order to lower the disease risk and to reduce lodging. Peaty soils with their high buffering capacity due to the large organic matter content, are often associated with manganese deficiency in wheat. This, however, does not preclude wheat-growing on these soils as a foliar spray of a manganese salt is effective in curing the deficiency. The condition can arise either through a soil deficiency or it may be due to unavailability of manganese. Where it is known to occur then corrective measures become a standard practice. Some varieties can tolerate manganese deficiency but others will require to be sprayed with manganese sulphate at the young plant stage.

Much has been written about critical pH and optimum pH range for the various crops and the experimental data quoted by Gardner and Garner (1953) draws attention to this basic soil requirement (Table 2).

A single figure for critical pH is often quoted by agricultural writers. Watson and More (1962), Robinson (1958) and Gardner and Garner

TABLE 3

Relative Wheat Yields on Sands and Clays at Varying pH Values

Soil Type/pH	4·0	4·5	5·0	5·5	6·0	6·5	7·0	7·5
Sandy soil	10	70	94	99	100	99	—	—
Clay soil	—	—	50	67	80	89	95	100

(1953) suggest 5·1 as the value but this will depend to some extent on the soil type. The relationship between pH and crop yield varies on widely differing soil types, as illustrated in Table 3. La Potasse (1962) draws attention to this.

Although the actual figures may not be the same under British conditions the principle which the above data suggests will be the same. From the examples given it is clear that to obtain anything approaching the possible maximum yield a pH of at least 6 must first be obtained.

VI. CLIMATE AND TEMPERATURES FOR GROWTH

From the acreage statistics the following counties are the most important producers of wheat:

ENGLAND AND WALES	SCOTLAND	NORTHERN IRELAND
Cambridge	Aberdeen	Down
Essex	Angus	Londonderry
Lincoln	East and	
Norfolk	Midlothian	
Suffolk	Fife	
Yorkshire	Peebles	
	Perth	
	Ross	
	Roxburgh	

A brief glance at these counties will indicate that wheat is concentrated on the eastern side of the various countries where lower rainfall and higher sunshine are experienced. Wheat-growing is practised to a much smaller extent in the west of Britain mainly on account of the high rainfall. A wet autumn and winter prevents successful sowing in these areas and heavy winter rain can result in crop failures even after a good establishment. In these mainly western and northern districts with considerable rain falling in August and September, ripening is delayed and often impaired and under extreme conditions much of the grain can be seen sprouting in the stook. Saleable grain samples are difficult to obtain since the removal of large quantities of moisture leaves the grain thin and wrinkled and deficiency payments may be withheld as a result. Sunshine is one of the main factors in producing high grain yields of good quality. Recent fundamental research has shown that photosynthesis by glumes and awns (floral parts surrounding the grain) after ear emergence plays a significant part in grain formation and those areas with high sunshine figures will normally produce well-filled, bold grain which has properly matured. It also follows that the costs involved in artificially drying the grain in these areas will be minimal.

Wheat is a crop of the temperate zones of the world requiring the highest temperatures that many of these areas can supply.

TABLE 4

Average Germination and Growth Temperatures for Wheat

	Minimum	Optimum	Maximum
Germination	32–41°F	77–88°F	88–110°F
	0–5°C	25–31°C	31–43°C
Growth	41°F	84–85°F	108°F
	5°C	29°C	43°C

After Hall (1945).

Average minimum, optimum and maximum temperatures are given in Table 4 and although different temperatures may be required by the various parts of the wheat plant and variation probably occurs through-out the growing season, Hall's data are worthy of note.

The southern part of Britain experiences a temperature range nearest to the optimum and this in itself could explain the high yields obtained there. Other areas have temperatures considerably below the optimum and it follows that most of the wheat crops in these localities should be restricted to sea level or just above. It will be remembered that a rise in altitude means a lowering of the temperatures (1°F/300 ft) and a reduced growing season.

VII. ROTATIONS AND DISEASE ASPECTS

A. The Place of Wheat in Cropping Sequence

The phrase cropping sequence is chosen deliberately in preference to rotation as it describes the present-day attitude in arable farming. Gone are the days of complete adherence to strict rotations mainly as a result of the economic pressures on crops with high labour requirements and high capital investment. During the 1960's simplified farming systems have been sought for small to medium farms which require the minimum amount of labour and capital investment.

Winter wheat does extremely well after a bare fallow, provided the land is not bare in the previous July. Without a soil cover during this month wheat-bulb flies lay their eggs on the ground and the larvae emerge to cause serious damage to the young wheat plants in the autumn. Winter wheat after fallow however is only feasible in respect of a few thousand acres of heavy clay which are fallowed for cleaning purposes. High rents necessitate a crop to be taken from the land each year and the area destined for fallowing diminishes each year. Early harvested sugar beet and potatoes often leave the land in time for successful winter wheat drilling. Seed-beds are usually easy to prepare and there are normally sufficient reserves of nutrients left behind to carry the wheat through until the following spring.

Wheat following beans or red clover was traditional and although this had largely disappeared by the 1960's there is evidence that this practice may again become widespread as bean-growing continues to expand. Wheat grown after good crops of field beans usually produces high yields.

Winter wheat after ploughed up grassland can be successful, but carries with it some problems regarding pests such as frit fly, slugs, leatherjackets and wireworms. Early ploughing coupled with correct selection and timely use of insecticides can minimise the damage.

With such a large proportion of the arable land under cereals it is inevitable that some wheat should follow a cereal crop. Frequent cropping with wheat or wheat and barley magnifies the risk of crop reduction and failure through higher infections of the fungal diseases. The most notable of these being eye-spot (*Cercosporella herpotrichoides* Fron.) and take-all (*Ophiobolus graminis* (Sacc.) Sacc.), which attack the base of the plant causing widespread lodging. Thus if wheat is to follow wheat or barley on land which has been continually or closely cropped with cereals in the immediate past then an accurate assessment of the disease risk or potential should be made by the farmer or with the aid of his advisers. In recent years the National Agricultural Advisory

Service in the South-west Region has introduced a soil test to assess the likely incidence of take-all. This is done by growing young wheat plants in a sample of soil from the field in question and after 18–20 days the roots are examined for fungal attack. According to the percentage root

TABLE 5

Level of Take-all (Ophiobolus graminis) *in the Soil and Crop Recommendation*

Infection percentage	Recommendation
Below 10	Safe for all cereals
10–20	Safe for spring wheat or spring barley
20–30	Safe for spring barley
Above 30	Not safe for wheat or barley

Extract from "Keep Pace With Progress" No. 4. Farmer and Stockbreeder Publication. Edited by D. Barber.

FIG. 4. Direct drilling (no ploughing) into "Gramoxone W"—sprayed stubble or odd pasture. Photograph by courtesy of Howard Rotavator Co. Ltd.

infection with this take-all fungus so a scheme for suitability of the different cereals can be applied.

In an attempt to speed up winter wheat drilling after a cereal crop and to minimise the carry-over of disease, chemical ploughing has been

FIG. 5. Direct drilled wheat. A successful wheat crop without traditional ploughing photographed at Harbury, Gloucester, 1966. Photograph by courtesy of Plant Protection Ltd.

introduced. Stubble from the preceding crop is sprayed and drilling follows using a sod-seeder. The chemical used is based on paraquat (commercially known as Gramoxone W) and is applied 4 pints/acre in low to medium volume to stubble in late September to kill off any vegetation. Direct drilling of the wheat can then take place during the first or second week of October (Figs 4 and 5).

It is also possible for wheat to be successfully cultivated after oats, the latter being a crop which is not susceptible to the main races of eye-spot and take-all. Where continuous cereal cropping is envisaged oats may be considered as a "break-crop". Wheat after oats, however, is somewhat academic since the main concentrations of each are at opposite sides and ends of Britain, wheat in the south and east whereas oats are found mainly in the north and west.

Wheat crops which follow peas, either harvested green or dry, can produce some high yields as long as the land is not left bare in July otherwise damage from wheat bulb fly larvae can again be serious. Early harvested peas can be followed by quick-growing white turnips or mustard to give ground cover and they will be ploughed-in prior to cereal drilling. In the past wheat after peas was not popular since most pea crops left the land rather dirty (weeds and weed seeds). Nowadays with satisfactory chemical weed-control established for both crops, wheat after peas no longer presents any cultural problems.

VIII. CULTIVATIONS AND MANURING

A. Cultivations

It is often said that winter wheat does best on a stale furrow and this implies that a well-firmed seed-bed is most suitable. Ploughing should be done as early as possible and the normal depth would be in the region of 6 inches. This could be increased to 8 inches where large quantities of trash need to be properly buried. Details of the cultivations necessary following ploughing are impossible to define accurately since structure, texture and condition of the soil may vary greatly even between fields on the same farm. The type of seed-bed required for winter wheat can be described as one with a reasonable tilth in the top 2–3 inches, with a surface containing a high proportion of clods the largest of these being about the size of a man's hand. This is to prevent capping, a condition which can easily arise with heavy winter rain, when the soil surface runs together forming a crust. Under such circumstances, winter kill due to frost heaving may result and plant reductions either through water-logging or lack of oxygen can lead to poor brairds.

Where winter wheat follows the ploughing up of grassland then skim coulters will be required to bury the old sward properly so as to prevent re-growth between furrow slices.

A considerable amount of experimental work has been carried out on the subject of optimum seed-beds for the winter crop. The results have been somewhat disappointing to the purist as there were seldom

TABLE 6

Yield of Successive Crops of Winter Wheat Established by Traditional Methods Compared with Direct Drilling (cwt/acre)[a]

Harvest year and variety	Gramoxone W (pints/acre)	Units N top-dressing	Sprayed with Gramoxone W and direct drilled	Ploughed, cultivated and drilled
1962 Hybrid 46	12	0 30 50 70	55·4	58·3
1963 Hybrid 46	12	0 50 75 100	52·9	52·2
1964 Hybrid 46	8	0 40 80 120	40·5	37·1
1965 Cappelle-Desprez	4	0 40 80 120	36·0	26·3
1966 Cappelle-Desprez	4	0 40 80 120	24·3 34·2 37·8 35·5	31·0 41·0 38·1 35·8

[a] Average yield over N rates in the first few years.

A basal fertiliser dressing of 20 : 40 : 40 units of N : P : K was also applied.

From Guide to Experiments 1967, Jealott's Hill Research Station and Jeater (1967).

any significantly differing responses from the crop to any but the extremes of treatment. As long as the wheat was drilled at a reasonable time (October or November) and was covered by soil then a wide range of preparatory cultivations gave equal results when yields were considered.

At this point it is therefore necessary to consider the method already

briefly described as chemical ploughing, using a paraquat-based formulation coupled with "direct drilling" or "sod-seeding" as it is sometimes
termed. Under some circumstances the speedy land preparation obtained by a single spraying may be preferred to traditional ploughing
and cultivating which is much more time consuming. Experiments have
been in progress since 1962 to compare these two drastically differing
systems of cereal growing at Jealott's Hill Research Station (I.C.I. Ltd.).
An extract of the data appearing in their 1967 Guide to Experiments
is given in Table 6.

It would appear from the figures in Table 6 that there is little to
choose between these two widely diverging methods of establishment
when yield averages over five years are compared, thus confirming the
results of the earlier cultivation experiments. From the data relating to
varying nitrogen levels in this and other similar trials it is clear that
high yields can be obtained at low to medium nitrogen levels following
traditional lines. Where chemical spraying and direct drilling is employed,
higher applications of fertiliser nitrogen are required to maintain
yield. These figures would suggest that mineralisation of soil nitrogen is
restricted without conventional ploughing and cultivations and a
higher expenditure on fertiliser is necessary to maintain output when
the soil is not disturbed. Against this must be set a quicker and possibly
cheaper rate of establishment, and the likelihood of lower incidences
of the fungus diseases take-all and eye-spot (as in 1965) with this no-
ploughing regime. Perennial grass weeds are controlled but not killed by
paraqual sprays and it is therefore important to ensure that the land is
relatively free from grass weeds before embarking on a no-ploughing
system.

B. Manuring

With all crops it is essential to ensure that adequate supplies of
phosphate and potash are available during the first few weeks of growth.
Once observed it is not possible to correct properly any deficiency and
both of these major elements are required either in advance of drilling
or they may be combine-drilled with the seed. Combine-drilling is the
most economical way of applying these fertilisers, but with winter
wheat, timeliness of sowing is the number one priority and the faster
method of application using fertiliser spinners is more often preferred.
For average conditions 30 units (one unit is equal to 1·12 lb and is the
same as 1% on analysis) each of phosphate and potash will be sufficient.
Where soil analysis shows that the land is low in these essential nutrients,
or they are to be applied by spinner rather than combine-drilled, then
the dressing may be increased to 40–50 units/acre. These recommenda-

tions apply equally well to spring wheat, the only exception being in the fact that the nitrogen is applied at the same time for the spring sown crop.

If the soil is rich in nitrogen then 30 units/acre of fertiliser nitrogen would suffice, but under average conditions levels up to 60 units are considered economic rising to 80 units in the low rainfall areas. Previous cropping, local environment and to some extent cultural techniques can also influence the optimum level of this nutrient.

When the soil is likely to supply some nitrogen for the early growth

TABLE 7

Fertiliser Nitrogen Levels for Winter Wheat

Location, soil condition and/or previous cropping	Recommended level of nitrogenous top-dressing (units N/acre)
1. After heavily grazed grass-clover leys in medium-high rainfall areas	30–50
2. Fen peats	30–50
3. After potatoes or sugar beet grown with F.Y.M.	50–60
4. Average conditions in medium-high rainfall areas	40–60
5. Northern Ireland Ministry of Agriculture recommendation	60
6. Average conditions in low rainfall areas	60–80
7. Intensive grain growing in low rainfall areas	80–100
8. Continuous cereal growing	80–110
9. Continuous cereals—chemical ploughing	80–120

of a winter sown crop then it is unlikely that any autumn fertiliser nitrogen would be required. On the other hand if the present crop is the third or fourth consecutive cereal then 15–30 units of nitrogen will be required along with phosphate and potash to obtain a satisfactory braird.

Due to the diversity in British farming and the weather which these islands experience there is considerable variation in the optimum level of fertiliser nitrogen for the winter crop. Examples to illustrate the range of recommended fertiliser levels are given in Table 7.

The short, stiff-strawed varieties of wheat can stand high levels of fertiliser nitrogen whereas the taller ones used to produce quality straw will only tolerate moderate amounts. Of all the cereals winter wheat will give the highest response to this fertiliser and to obtain the best return the appropriate dressing should be applied at the correct time. Top-dressings applied in mid April will generally give the best return in additional grain yield in the dry areas, but this may be too early for crops in wetter districts. There, early applications of nitrogen will produce much more straw and thus increase the risk of lodging. Where large quantities of this fertiliser are used it will often pay to split the dressing, applying half during late March to early April and the rest about a month later.

As far as spring wheat is concerned up to 60 units of nitrogen can be economic and from the literature there is no consistent evidence to support splitting this dressing. It should be applied prior to drilling, combine-drilled with the seed or immediately the crop has braided.

The results from 114 experiments with winter wheat over 7 years showed that Cappelle Desprez would give an extra 6 cwt grain/acre when top dressed with 60 units of nitrogen (Lessells and Webber, 1965).

IX. SEED DETAILS

A. Seeding Rates

The amount of seed required for autumn wheat will vary between 1 and 2·5 cwt/acre. Early sowings need the least since the temperatures for germination are higher than those later on and a larger number of the seeds produce plants. As one goes north the autumn temperatures become significantly lower and hence to obtain the optimum number of plants in the braird it is necessary to sow larger quantities of seed.

Many experiments have been conducted to ascertain the optimum

TABLE 8

Seeding Rates for Winter Wheat (cwt/acre)

Time of sowing	South	North
Early (End Sep.–Mid Oct.)	1–1·25	1·5
Mid (Mid Oct.–Mid Nov.)	1·5	1·75
Late (Late Nov. and December)	1·75–2	2–2·25
Very late (January onwards)	2–2·25	2·25–2·5

seeding rate for winter wheat and these have shown that for early drilling under good conditions 140 lb/acre (1·25 cwt) appears optimal (Mundy & McClean, 1965).

Since much of the winter crop goes in under far from ideal conditions then it is logical to sow larger quantities of seed to obtain the required plant stand.

Varieties of spring wheat have neither the time nor the genetic potential for tillering like their winter counterparts and 1·75 cwt/acre will be required under most circumstances.

B. Seed Dressings and Seed Treatment

Most wheat seed purchased through the trade is dressed with proprietary compounds for the control of seed-borne fungal diseases or insecticides to ward off attacks by the various predators. In many cases a combined seed dressing is employed to cover both diseases and pests. The various recommended single dressings or combinations, given in

TABLE 9

Recommended Seed Dressings for Wheat

Seed-borne diseases	Wireworm	Seed-borne diseases and wireworms	Wireworms and wheat bulb fly on autumn sown crops	Seed-borne diseases, wire-worms and wheat bulb fly on autumn wheat
Organomercury either dry or liquid	BHC liquid[a]	Organomercury with BHC	Liquid aldrin or BHC (dry)[a, b]	Organomercury with BHC or dieldrin[b]

[a] Can be used as additional dressing where liquid organomercury has been applied.
[b] It is illegal to sow grain which has been treated with aldrin, dieldrin or heptachlor after 31st December.

Table 9, have been taken from the Ministry Booklet—Agricultural Chemicals Approval Scheme.

Hot water treatment of seed will control the seed-borne fungus loose smut (*Ustilago tritici*). This was the task of plant breeders, their agents or seeds merchants and it maintained loose smut susceptible varieties free from this fungus in the first year and helped to reduce infections in later years of multiplication. In recent years chemicals have been introduced as seed dressings which will control loose smut and Murganic RPB is one commercial formulation for this purpose.

B

C. Seed Quality and Approximate Costs

Quality and cost of seed is listed below indicating the range of prices available to farmers.

Terms such as pure line, pedigree or original stock although used quite frequently are purely descriptive terms and seldom give a good guide as to the quality of the seed or degree of control exercised in its production. In general terms the grades in Table 10 are listed in descending order of quality, but this does not mean that a farmer's own stock is necessarily poor seed. If he starts with one of the officially recognised grades of seed, is prepared to keep it free from admixtures of other varieties or other cereals, harvest it under good conditions at full maturity and take care in any artificial drying which may be done, then there is no reason why this should not be satisfactory for several years.

TABLE 10

Types of Cereal Seed

Type of seed	Approximate cost or value/cwt
1. Breeder's elite stocks—Basic Seed	£5–8
2. British certified seed	£5
3. Multiplication seed	£4–5
4. a. Field approved (England and Wales)	
b. Stock seed (Scotland)[a]	£3·50–4·50
c. Certified seed (Northern Ireland)	
5. Field certified seed (Scotland)[a]	£3–3·50
6. Commercial seed	£2·50–3
7. Once grown seed	£2–2·50
8. Farmer's own seed	£1·50–2

[a] Restricted to oats and barley only.

He must also grade the seed sample by taking out broken or small grains, weed seeds and trash and dress it with an appropriate seed dressing. Should there be any doubt about the vigour or germination then one of the official seed-testing stations will carry out the appropriate tests for a small nominal charge.

High quality seed must be purchased by those who are intending to produce seed stocks for sale and this is often best done in co-operation with recognised merchants in the seed trade. Contracts can be obtained and often the high cost of purchasing elite stocks for multiplication are borne partially or wholly by the contracting seedsman. Where crops are simply grown for commercial use then it is not necessary or economic to use the highest quality seed. There is a common belief that change of

FIG. 6. Yellow rust of wheat, caused by the fungus *Puccinia striformis*. N.I.A.B. photograph.

seed in itself can be responsible for higher yields since stocks may de-
generate if grown on the same farm for several years. Genetical degen-
eration of the self-pollinating cereals (wheat, barley and oats) is possible,
though highly improbable, over a short period of several years provided

FIG. 7. Loose smut of wheat, caused by the fungus *Ustilago nuda*. Controlled by (a)
choosing the least susceptible varieties; (b) using high quality seed; (c) using seed
which has been chemically or hot-water treated. N.I.A.B. photograph.

the original stock is good. It is more usual for seed to be discarded as the result of admixture or because it has been harvested under adverse conditions, resulting in poor vigour and germination.

X. VARIETIES

Unlike some other European countries where it is only possible to grow government-approved varieties, in Britain there is a long list of varieties (some of which carry official recommendation) from which a farmer may select one or more for growing. This freedom of choice is useful in many ways. Perhaps the most important factor lies in the fact that information can be obtained about many cultivars, in particular their field performance. This information although less scientific and less accurate can augment the results of National Trials and Tests should it be necessary to seek further evidence about some variety. When the farmer's choice is restricted to a Government Recommended List, with one very obviously high-yielding cultivar heading it, then there is a real danger that the majority of the crop acreage could be planted with it. Under these circumstances there is no insurance against substantial losses occurring if this particular variety should fall victim to a new race of a disease as did Rothwell Perdix to yellow rust (*Puccinia glumarum–syn. Puccinia striiformis*) in 1966–67 (see Fig. 6). This freedom of choice acts as a safety valve but it must be stressed that unless there is local evidence to the contrary, the best variety the farmer can choose will come from the officially recommended lists. For England and Wales this is issued by the National Institute of Agricultural Botany at Cambridge; for Scottish conditions there are general and regional recommendations from the three Colleges of Agriculture; for Northern Ireland the lists are prepared by the Ministry of Agriculture. Full details of these appear in the Appendix to this chapter. It must be pointed out that they are reviewed annually and farmers should make sure that the most up-to-date information is at their disposal when choosing cultivars. The right choice can lead to substantial profits whereas the wrong decision will often result in diminished financial returns. When hundreds of acres are involved this can represent significant sums when the farm accounts are analysed.

XI. DETAILS OF DRILLING, TIME OF SOWING AND EARLY SPRING MANAGEMENT

A. Drilling

Cereals are best established by drilling rather than broadcasting and harrowing, the latter method being resorted to only in cases of extreme

FIG. 8. Cereal mildew on wheat caused by the fungus *Erysiphe graminis*. Controlled by (a) choosing the least susceptible varieties; (b) avoiding high seed rates at very high nitrogen levels; (c) using fungicides as seed dressings, incorporated with ferti-lisers or as foliar sprays. N.I.A.B photograph.

difficulty in the late autumn and winter when the movement of tractors and drills on wet land is impossible. Force-feed or gravity-feed drills may be used with a wide range of coulter types varying from the old Suffolk coulter to the more modern single or double disc types. Drilling ensures a fairly uniform depth of seed placement and according to the early Norwegian experiments referred to in the Ministry's publication on Wheat (Robinson, 1958) this is a much more critical factor than many people have thought.

Depth of drilling	1 in	2 in	3 in	4 in
Yield (cwt/acre)	20·4	17·3	13·6	9·4

On occasions winter wheat has to be planted a little below optimum depth in order to obtain proper seed coverage but under normal conditions it should not go in deeper than 1·5 inches. Besides obtaining a better depth control with drilling compared with broadcasting there is also a better and more uniform spatial arrangement of the plants. Holliday (1963) has summarised the experimental data relating to row width and yield of cereals and with winter wheat the highest grain yields have been obtained at coulter spacings of 3–5 inches. Very little work in this connection has been done on spring wheat, but from the limited data it would appear that the same recommendations could be made. This evidence corroborates the old concept that cross-drilling increased yield. Instead of putting in 2 cwt seed/acre at 7 inch coulter spacing, 1 cwt would be drilled in each direction at right angles to produce higher grain yields. The differences involved do not merit recommendation of cross-drilling but simply point to the fact that maximum yield can only be achieved through narrow coulter spacings.

B. Time of Sowing

This has already been considered when the type of wheat was discussed earlier in the chapter and in summarising the position in general terms, the best time to sow winter wheat is in the month of October. Drilling after November has always resulted in lower yields and when late sowing is unavoidable it is important to choose early maturing varieties requiring the least vernalisation.

Spring wheat is by far the latest cereal to mature in most areas where it is grown and the aim should be to bring forward this date as much as possible. It should therefore be the first cereal to be planted in the spring. Following a particularly mild winter and early spring it is often possible to sow in late February but more often the earliest sowings occur in March.

C. Spring Management of Autumn Sown Crops

Where the land still carries surface clods or has run together to produce a surface pan, a spring harrowing followed by rolling will greatly benefit the crop. It is said and written that the harrowing of winter proud crops and those with an excess of plants will produce good results but in the author's opinion it is not worthwhile on either account. Where there is an excess of vegetative growth in the spring, i.e. winter proudness, it can be removed by grazing animals (preferably sheep) or a light topping with a mower. It is usually at a time of year when keep is scarce and the spring grazing of forward crops of winter cereals can produce significant contributions to livestock feeding. The principle involved must be a high stocking density for a short time and never the reverse otherwise the animals may remove the flowering initial and defoliate too severely many of the plants. Holliday (1956) has reviewed fodder production from winter sown cereals and its effect upon grain yield and readers who are especially interested in this technique may wish to consult this paper with its many references. Removal of herbage by cutting or grazing invariably results in a lowering of grain yield and to apply economics to this, one should set the probable loss in yield against the value of the grazing obtained. This can be assessed as live-weight gain, milk produced if dairy cows are involved or as daily maintenance charges. It would appear that a proportion of this loss in yield can be recovered by the application of a small additional top dressing of fertiliser nitrogen. Other benefits following grazing can include reduced lodging associated with less straw through a reduction both in straw height and straw numbers. This is of special interest because in many areas there is a belief that tillering is increased after grazing. Straw numbers were reduced by 10% in some experiments and an increase in the number of vegative tillers was recorded but these are unlikely to bear inflorescences (ears). Possible reductions in the incidence and spread of fungal diseases such as mildew and eye-spot may also result from these defoliations. Early rather than late grazing is recommended in order to minimise the reduction in yield.

Month of grazing	December	January	February	March	April	May
Probable yield Reduction (cwt/acre)	0–2	0–2	3–4	5–6	6–8	Over 8

Early grazing of cereals can lead to a high risk of hypomagnesaemic tetany and whilst animals are obtaining most of their keep from cereal crops, farmers must see that their diet contains sufficient magnesium. This is usually achieved by adding magnesium salts to any concentrate feed.

In respect of a normal crop, spring harrowing can often prove useful in the eradication of weed seedlings but chemical control has become more popular and more effective in recent years with the introduction of new, improved materials.

If the land is still rough in the spring then rolling can be of great benefit in root consolidation, in making available capillary moisture and for producing a flat surface to facilitate easy operations for the harvesting machinery.

XII. CHEMICAL WEED CONTROL

A. Autumn Spraying

(i) GENERAL WEEDS

If crops of winter wheat are being smothered by annual weeds (dicotyledonous) in the autumn and provided the land is dry enough to take tractors and spraying equipment, then it is possible to spray with some materials to eliminate this unwanted competition. DNOC, dinoseb-ammonium or dinoseb-amine formulations are recommended at up to 6, 1·125 and 1·5 lb active ingredient/acre respectively.

(ii) WILD OATS (*Avena fatua and Avena ludoviciana*, see Fig. 9)

Previously no chemical control was possible with this troublesome weed of cereals but Barban at 5 oz active ingredient (a.i.) per acre applied in low volume when the majority of wild oats have between 1 and 2·5 leaves, now offers some measure of control. This material can be applied either as a winter spray or in the spring but it is not recommended between mid January and the end of February. Barban can be used on most varieties of winter wheat but it must be remembered that *a few varieties are very susceptible to damage.* These include *Elite Lepeuple. Hybrid 46* and *Professor Marchal* and farmers must make sure that it is safe to spray the varieties which they are growing.

(iii) BLACKGRASS (*Alopecurus myosuroides*)

In recent years blackgrass has emerged as one of the most difficult weeds to eradicate from arable farms on heavy land and until 1966 only cultural means were available for its control (for details see Weed Control Handbook). The introduction of Prebane by Fisons Pest Control Ltd from Geigy (Switzerland), active ingredient 2-methylthio-4-ethylamino-6-tertiary butylamino-s-triazine, offers some control of this grass in early sown winter crops (Fisons Agricultural Technical Information, 1966 and 1967). This new herbicide will control autumn germinating blackgrass and also a range of broad-leaved weeds when

Fig. 9. Wild oat seeds (*Avena fatua*). Note the hairy grain base and large, heavily twisted awn attached to the middle.

sprayed at 4 lb/acre commercial product in low to medium volume and all the present varieties of winter wheat may be safely sprayed.

B.   Spring Spraying of Autumn and Spring Sown Crops

A large number of chemicals are now available for post-emergence weed control in winter and spring wheat crops. These are reviewed annually by the British Weed Control Council (Weed Control Handbook) and by the Ministry of Agriculture usually in January and it is essential that farmers consult the most up-to-date information before making their choice of weed-killer.

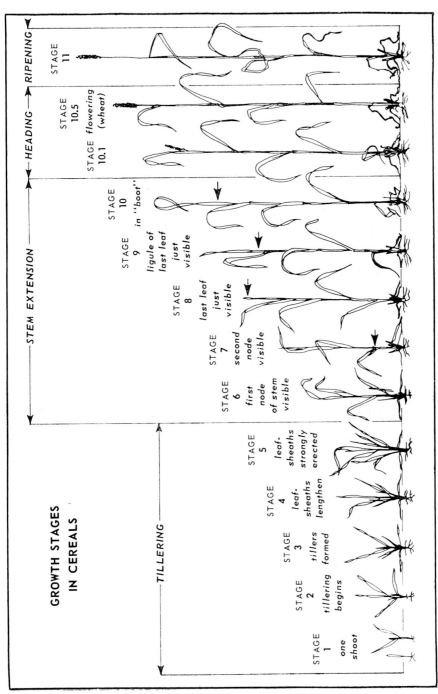

Fig. 10. The Feekes–Large scale. Reproduced from *Plant Pathology* **3**, (1954) by courtesy of H.M.S.O.

TABLE 11

Weed-killers that may be Used on Wheat in Spring (M.A.F.F. 1970)[a,b]

	Winter Wheat	Spring Wheat 3–4 leaves	5 leaves to 'joining'
I Weed-killers for the control of broad-leaved weeds			
(a) Cereals not undersown with clover			
MCPA	√	√[c]	√
2,4-D	√	NO	√
mecoprop	√	√	√
mecoprop + fenoprop	√	NO	√
mecoprop + 2,4-D	√	NO	√
dichlorprop	√	√	√
dichlorprop + 2,4-D	√	NO	√
dichlorprop + MCPA	√	√[d]	√
dicamba + mecoprop with or without MCPA	√	NO	√
2,3,6-TBA + MCPA	√	√[d]	√
2,3,6-TBA + dicamba + mecoprop + MCPA	√	NO	√
dicamba + benazolin + MCPA	√	NO	√
ioxynil + mecoprop	√	√	√
ioxynil + linuron	NO	√	√
ioxynil + dichlorprop + MCPA	√	√[d]	√
bromoxynil + ioxynil + dichlorprop	—	√	√
bromoxnil + MCPA	√	√	√
(b) Cereals undersown with clover			
MCPB + a little MCPA[e]	√	√	√
2,4-DB + a little MCPA[e]	√	NO	√
2,4-DB + a little 2,4-D or MCPB and MCPA[e]	√	NO	√
benazolin + 2,4-DB or MCPB and a little MCPA[e]	√	NO	√
dinoseb[f]	√	√	√

II Weed-killers for the control of wild oats and/or blackgrass in the spring.
  Tri-allate
    May be used in spring barley or spring wheat, either before or just after drilling.
    Thorough incorporation by cultivation immediately after application is essential.
  Barban
    May be applied to spring wheat and most varieties of spring barley (but not
    Proctor, Impala and some others) when the wild oats or blackgrass has 1–2
    leaves per plant.
  Methoprotryne + simazine[g]
    For use on late-drilled winter wheat between mid February and the end of
    March when the crop has at least 4 leaves. Only recommended for use in
    Eastern Counties and there not on free draining or highly organic soils.
  Metoxuron[g]
    May be used on some varieties of winter wheat and winter barley when the crop
    has at least 3 leaves.   *Notes on opposite page.*

It is important to recognise the three growth stages which are critical when deciding upon and using herbicides and to avoid any confusion these are given below with their equivalent growth stage number as shown in the Feekes–Large Scale (Fig. 10).

Cereal Growth Stages

Description	3-Leaf	5-Leaf	Jointing
Position on the Feekes–Large Scale	Stage 2	Stage 4	Stage 6

In Table 11, recommended chemicals are listed with an indication as to their suitability at the various growth stages of wheat. This is an extract from MAFF Short Term Leaflet No. 19, but the full data must be consulted to make sure that the best choice is made in relation to specific circumstances and weed problems. When farmers are not very familiar with herbicides and their use then it is worthwhile consulting the local advisory officers who can assist in the technical and economic aspects having seen at first hand the problem in question.

Where wheat crops contain a range of weed species which can be controlled by the herbicides MCPA or 2,4-D then these should be chosen in preference to the more complicated mixtures which often cost considerably more. When weed species are encountered which do not succumb to these two common chemicals then farmers must select from the wider range of material which has been developed in recent years for the control of the more difficult species. The following examples serve to indicate this principle and illustrate the appropriate herbicide when the stipulated weeds are encountered.

Where crops are undersown with grass and clover seed for the

---

[a] For trade names, see the Ministry list of approved products.
[b] For the correct dose, read the manufacturer's instructions.
[c] At half normal dose. Some deformity of the ear may follow application at this stage of growth; do not use on malting barley or seed crops.
[d] Some deformity of the ear should be expected; do not use on malting barley or seed crops. *Follow the manufacturer's instructions.*
[e] Clover should have its first trifoliate leaf before spraying.
[f] Clover should have two trifoliate leaves before spraying.
[g] Some broad-leaved weeds also controlled.

√ Indicates suitability.
NO = Not Suitable.

This extract from MAFF Short Term leaflet No. 19—1970 is reproduced by kind permission of the Ministry of Agriculture, Fisheries and Food.

establishment of leys then the herbicidal treatment requires modification to preserve the young clover seedlings. Here MCPB, 2,4-DB,

TABLE 12

Examples of Difficult Weeds and the Appropriate Herbicide for their Control

Mecoprop with or without Fenoprop	Mecoprop plus Ioxynil	Dichlorprop
Chickweed Cleavers Fumitory	Corn marigold Mayweed	Black bindweed Redshank Spurrey

Benazolin and Dinoseb may be used, bearing in mind the specific requirements of legume growth stage.

Full details of all the recommended herbicides may be found in the newest edition of the Weed Control Handbook, issued by the British Weed Control Council.

XIII. IRRIGATION

Wheat growing is concentrated in the drier areas of the South and East and in many years the limiting factor in respect of grain yield is available moisture. With winter sown crops irrigation would often show significant responses when applied in April or May and with spring sowings a little later in May or June. However, there are many factors which make the irrigation of cereals impractical. Firstly, the water is seldom available in these extremely dry areas of Britain. Secondly, there is a better response from crops with a much higher gross output financially and thirdly there are many practical difficulties in the application of the water. With yield increases of the order of 10–20% this could seldom, if ever, justify the high capital outlay necessary plus the direct and indirect charges associated with irrigation schemes.

XIV. HARVEST

Winter wheat is normally harvested from August to October, depending on the type of summer experienced and also the geographical location. There is a difference of up to 4–6 weeks in time of harvest as one moves from one end of the country to the other, ripening being, on average, up to 6 weeks later in the North. Spring wheat matures much later than winter wheat and later than the other cereals, and is quite often the last grain to be harvested on many farms.

Following a hot, dry summer grain may be combine-harvested under very good conditions and if the moisture does not exceed 14% then it can be stored in sacks or bulk without drying. Moisture tests can be carried out by farmers at harvest and these are often used to indicate the stage of ripeness or readiness for combining. It must be pointed out that in the absence of significant amounts of precipitation or dew then these figures will be a useful guide as to the stage of maturity. If on the other hand August is wet then the grain moisture content will almost certainly be higher than usual and this can mask the maturity stage. Although most of the wheat is cut by combine harvester, there is still a small, but significant acreage which is bindered to satisfy the demand for long straw. It is said to be binder-ripe when the grain is fairly firm, has a cheesy texture and does not exude any milky fluid when pressed. This stage is usually reached between 1 and 2 weeks before it can be combine-harvested. The actual binding should not take place until the morning dew has disappeared and the sheaves should be stooked as soon as possible. Once cut the grain will mature in the ear, the straw will dry out and the green material in the bottom of the sheaf (butt) will also lose much of its moisture. Readiness to stack can vary from a few days in dry weather to 2–3 weeks or even longer under adverse rainy conditions. In some areas, notably Perthshire, specially constructed barns were erected for sheaf drying.

Stacks should be built on dry sites close to the granary and losses may often be avoided by erecting them on loose or baled straw. Temporary covers in the form of tarpaulins or polythene sheets should be available during building and once complete the stacks must be properly thatched with straw. The stacks will sweat slightly at first but they will soon dry and the grain will harden off. Once this stage has been reached the remaining dangers include rooks and jackdaws which may damage the thatch. Vermin in the form of rats and mice can be a problem with stacks and constant vigil will be necessary.

A. Grain Sampling and Rapid Moisture Content Determinations

When grain is combine-harvested, it is essential to know the moisture content in order to decide whether or not artificial drying is needed. Rapid determinations can be made to an accuracy of approximately 1% moisture content and these figures are needed to determine the extent of drying necessary with moist grain.

Moisture content determinations are carried out on small samples representative of large bulks of grain. It is therefore very important that these are taken accurately at random and in sufficient quantities to give reliable representations. Where time permits, the moisture content of

each sample should be determined and this will demonstrate any pattern or variation within the bulk. Otherwise the samples should be thoroughly mixed and two or three sub-samples drawn and used to obtain an average figure. Where the grain is held in bulk stores a sampling spear will be invaluable to check on moisture content at varying depths.

Some moisture meters require the grain samples to be milled, others work on whole grains. Hand-operated coffee-grinder types are useful for a small amount of samples, but a small motor-driven mill will be necessary for large numbers. The fineness of grinding is not thought to be important, except in the case of the Acetylene Gas Meter, but it is essential that estimates of moisture content are made immediately the grain is ground to avoid any loss of moisture. Whole grain moisture meters are liable to give erroneous results if the surface of the grain is wet or unusually dry.

B. Types of Meters

Several different types of moisture meters are available and these have been described in some detail in M.A.F.F. Bulletin No. 149, Farm Grain Drying and Storage (1966), and also in H.-G.C.A. Technical Notes No. 5. It is proposed to deal briefly with each type in turn, indicating the levels of accuracy and ranges. For readers wishing to know more about any of these the two publications referred to above will prove useful and the manufacturers will always send details on request.

(i) INFRA-RED METER

This instrument works on the principle of driving off the moisture from a weighed sample of milled grain using an infra-red lamp. The loss in weight is calculated and the instruments are usually calibrated to read the percentage moisture content of the grain.

Time to complete one reading: 10–20 minutes
Accuracy: better than $\pm 1 \cdot 0\%$ m.c.
Range: virtually no limit
Power source: mains electricity.

(ii) RESISTANCE METERS

(a) *Using milled grain.* These portable meters are ideally suited for farm use and measure the resistance of a grain sample which is then converted to a direct reading of moisture content.

Time to complete one reading: 5–10 minutes (including grinding)
Accuracy: $\pm 1\cdot0$–$1\cdot5\%$ m.c.
Range: 10–22% m.c. (occasionally up to 25%)
Power source: battery.

(b) *Using whole grain.* Larger samples are needed (3–4 lb) and the resistance between two groups of electrodes in the instrument is measured by an electrical insulation test.

Time to complete one reading: 1–2 minutes
Accuracy: $\pm 1\cdot0$–$1\cdot5\%$ m.c.
Range: 13–25% m.c.
Power source: battery.

CAPACITANCE METERS

These whole grain meters measure the permittivity (dielectric constant) of samples and this value is directly correlated with moisture content.

Time to complete one reading: 1–2 minutes
Accuracy: $\pm 1\cdot0\%$ m.c. (with a good meter)
Range: 11–28% m.c.
Power source: battery.

ACETYLENE GAS METER

This meter works by measuring the pressure of acetylene which is produced when the moisture from a milled grain sample reacts with finely powdered calcium carbide.

Time to complete one reading: 10–15 minutes
Accuracy: $\pm 2\cdot0\%$ m.c.
Range: 3–26% m.c.
Power source: none required.

HAIR HYGROMETERS

These instruments measure the relative humidity of the air surrounding grain in sacks or bulk stores. The hair element, surrounded by a perforated tube, contracts or expands with changing humidity and the moisture content of the grain is read directly from a suitably calibrated dial scale.

Time to complete one reading: 30–45 minutes
Accuracy: $\pm 1\cdot5\%$ m.c.
Range: 0–30% m.c.
Power source: none required.

This type of meter is also calibrated to read the relative humidity of air (0–100%) and is fairly accurate in assessing grain moisture content provided that the results are not urgently required.

XV. DRYING AND STORAGE

On occasions wheat may be harvested under ideal conditions when the moisture content does not exceed 14% and then it can be stored

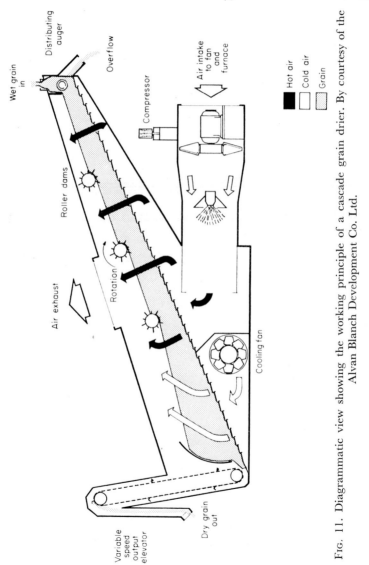

Fig. 11. Diagrammatic view showing the working principle of a cascade grain drier. By courtesy of the Alvan Blanch Development Co. Ltd.

without artificial drying. The vast majority of the grain is brought in at moistures ranging from 16–24% and in the main will require drying before storage. At 16% it may be stored for a considerable period in sacks but not in bulk. Grain up to 20% can be kept for a short period in open sacks but at this high moisture content a constant watch must be maintained for over-heating and the onset of mould. A moisture meter

Fig. 12. Alvan Blanch Cascade "27" Grain Drier. This photograph represents a wide range of driers on the cascade principle, with outputs ranging from 0·75 to over 6·5 tons/hour, based on a 5% moisture extraction at normal drying temperature. Photograph by courtesy of the Alvan Blanch Development Co. Ltd.

has become the stock-in-trade of the cereal farmer and many different kinds are available which will give fairly accurate assessments between 10 and 30%. At the beginning of harvest it is often worthwhile waiting for reductions in the moisture content before combining in order to reduce drying costs. But as time goes on and conditions deteriorate with a much shorter combine-period in a day, then the crops should be harvested as fast as possible once the surface moisture has disappeared.

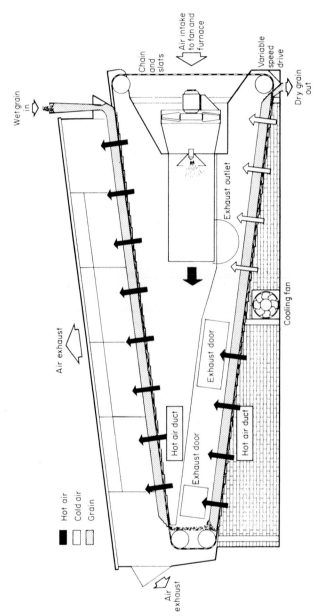

Fig. 13. Diagrammatic view showing the working principle of a double flow grain drier. By courtesy of the Alvan Blanch Development Co. Ltd.

Fig. 14. Electrically operated Double Flow Grain Drier, representing machines with an output range from 7 to 23 tons/hour. One of the main advantages with this type of drier is its own roof and access to a building is only required at the output/input end. Photograph by courtesy of the Alvan Blanch Development Co. Ltd.

A. Type of Drying Equipment and Installations

(i) IN-SACK OR PLATFORM DRIER

When the acreage of wheat is small and the combine harvester is a bagger-type, then it is possible to use the simplest form of drier, namely the in-sack or platform. Partially filled sacks of grain are placed over openings on a raised platform through which hot air is blown. It is probably best to dry the grain straight off the combine, leaving dressing and grading until later.

(ii) CONTINUOUS FLOW DRYING

Where tanker combines are used and the grain is handled in bulk then continuous flow driers are much more suited to the grain handling system as a whole. The principle involved with this type of drying is for the grain to pass slowly through an area which is heated with warm or hot air. Drying temperatures are always important and quite often they

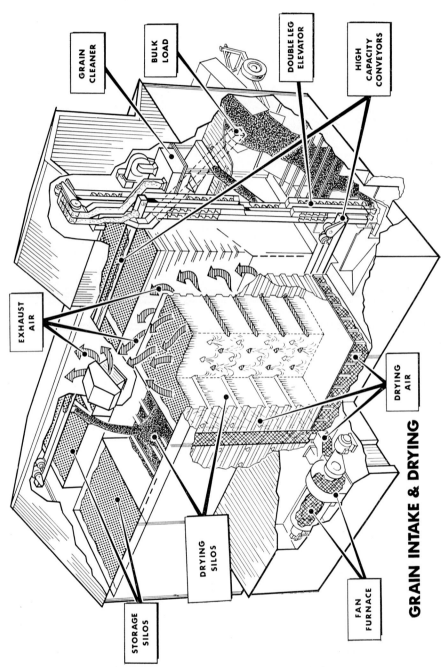

GRAIN CLEANER

BULK LOAD

DOUBLE LEG ELEVATOR

HIGH CAPACITY CONVEYORS

EXHAUST AIR

DRYING AIR

DRYING SILOS

STORAGE SILOS

FAN FURNACE

GRAIN INTAKE & DRYING

Fig. 15. The Jack Olding grain drying system. Reproduced by kind permission of Jack Olding and Co. Ltd, Scotland.

are critical when it is necessary to retain the germination potential of the grain. Maximum safe temperatures are listed below.

It is unwise to exceed these temperatures and where accurate assessment cannot be readily obtained it would be better to err on the low side. These continuous flow driers have capacities between 0·75 and 12 tons/hour removing up to 5–6% moisture each time the grain passes through.

The Cascade driers produced by Alvan Blanch Ltd come under this group of continuous flow types and the range of drying capacities is

TABLE 13

Maximum Safe Temperatures for Drying Wheat

Grain use		Drying temperature
Livestock Feed		180°F, 82°C
Milling Wheat		150°F, 65°C
Seed		
Moisture content	18%	150°F, 65°C
	20%	142°F, 61°C
	22%	134°F, 57°C
	24%	127°F, 53°C
	26%	120°F, 49°C
	28%	114°F, 46°C
	30%	110°F, 43°C

extensive. At the lower end of the scale, the smallest will deal with up to 0·75 ton/hour and 1·2, 2, 2·5, 3·5, 4·25, 4·75, 5·75 and 6·75 ton/hour models are available when restricted to a single flow basis. This company has also marketed a range of double flow continuous (Cascade) driers and with a second movement of grain back-tracking the first, the capacities have been markedly increased. These will cope with 10, 13·5, 16·67, 20 and 23 tons of grain/hour, extracting 5% moisture at operating temperatures between 150 and 160°F (65–71°C). Figures 11 and 12 illustrate the Cascade driers and Figs 13 and 14 illustrate the double flow types.

(iii) TRAY OR BATCH DRIERS

With these driers a set quantity of grain is placed on a tray and dried *in situ*, without any movement. Under these conditions, overheating and excessive temperatures can easily result and the safe maximum temperatures quoted earlier should all be reduced by 10°F for safety. The range of capacities is similar to the previous group, i.e. 0·75–12 tons/hour.

(iv) BULK OR BIN DRYING

With most of the grain cut with tanker combines and handled in bulk, grain drying in bulk has been the natural development. Self-emptying bins or containers, holding on average 15–25 tons, are used for drying and storing the grain. Most systems employing this principle also have in addition some bins which are simply used for grain storage, since they cost considerably less than those which can also be used for drying. Hot air is passed through the grain radially (Simplex driers, Figs 1 and 2 in Chapter 3) or at various levels through the bulk (as in the Jack Olding system illustrated in Fig. 15) or it may start at the bottom of the containers and move upwards through perforated floors. The last-mentioned is rather an old system, it is often referred to as in-bin ventilation and is somewhat outmoded. All these systems of drying are slow relative to the continuous-flow types, but speed is not so essential as greater quantities are involved. Bulk drying with the Jack Olding system is rated at 5% extracted from 40 tons in 20 hours, whereas the older in-bin ventilation systems would probably remove only $\frac{1}{2}$–1% in 24 hours. The energy is supplied through solid fuel stoves, by electricity or oil burners, each being connected to fans for propelling the hot air to and through the bulk grain.

(v) ON-THE-FLOOR DRYING

In recent years simple systems of grain drying have developed on farms, using existing buildings and with the minimum of capital invested. On-the-floor drying is one such method, requiring a dry building with a level concrete floor, a source of warm air, suitable air ducting and weld mesh/hessian channels below the grain. This is perhaps the simplest and cheapest grain drying system, which although developed for drying the fodder cereals, barley and oats, can also be used for limited quantities of wheat.

(vi) CONTRACT DRYING SERVICES

Under special circumstances or where only small quantities are involved, it may prove more economical or convenient to use contract drying services through feed or seed merchants. The costs involved in such an operation are extremely variable since they will depend on a number of factors, the most important of these being how much moisture needs extracting. Up-to-date prices can be obtained from current agricultural papers or through "Farm Mechanisation". At the time of writing, removal of 5% moisture could cost between £1 and 1·50/ton and for 10% the corresponding charge would be £2–3/ton. The cost of transporting the grain to and from the drier needs to be added to the above figures to arrive at the total amount involved.

With most systems of drying it is recommended that the grain pass through a pre-cleaner before entering the drier in order to remove any green material, dust and rubbish. Quite often this unwanted trash is very much wetter than the grain and it is most uneconomic to waste money drying it. When systems of grain drying and handling are being planned it is important to ensure that these pre-cleaners do not limit the throughput of grain otherwise there could be serious hold-ups in the process.

Advice on grain drying can be obtained through the various Agricultural Advisory Services and from the companies which manufacture and sell drying equipment. There is a wide range of official literature concerning this subject which has only been dealt with very briefly here and it would be wise to consult some of these publications before making major decisions. A list of publications appears in the Appendix.

XVI. DISPOSAL OF GRAIN AND PRODUCTION ECONOMICS

The vast majority of wheat grown in the United Kingdom is sold and is used mainly for human and animal feed. Two aspects of "quality" will govern the price and outlet of the grain. The first is known as the condition of the grain, referring to its physical state and freedom from various impurities. All wheat offered for sale should have been properly and evenly dried, and this will mean a moisture content of around 14%. Samples should contain plump well-filled grains, free from seconds (small thin grains well below average size) and admixtures of other cereals and other varieties of wheat, should the consignment be destined for milling or for seed purposes. At the same time there should not be any weed seeds present or any defective grains, i.e. those exhibiting mould growth, and smutted, sprouted or overheated grains. Dust and other trash should have been removed in the dressing process and the sample must be free from the harmful pieces of the fungus ergot (see Fig. 4, Chapter 4). All these factors describe and define the physical state or condition of the grain and they apply to all samples put up for sale.

The other aspects of quality are those which are genetically determined and will govern the method of utilisation. These refer to milling and baking characteristics already defined. Where farmers wish to produce quality grain for the milling trade then the choice of variety is important and they should consult the most up-to-date recommended lists. Contract schemes, reported earlier, may be entered into in order to benefit growers through the quality bonus payments but it is important that high yields are maintained in order to ensure profitable wheat crops.

Sales of grain should take place through authorised merchants so that farmers can obtain the appropriate government subsidy through the deficiency payment scheme.

A. Government Support for Wheat

Under the provisions of the 1947 Agriculture Act, wheat qualifies for a guaranteed price and this is fixed annually at the time of the Price Review. With effect from 1st July, 1964, and as a result of the 1964 Annual Review and Determination of Guarantees, a "Standard Quantity" for wheat was introduced. (An unfortunate choice of terminology since the "standard quantity" has been changed each year

TABLE 14

Cereals Deficiency Payments—Wheat

The agricultural departments in the United Kingdom announced the following particulars of deficiency payments for the cereal year 1968/69.
Wheat (guaranteed price 27s. 5d./cwt)

Accounting period	Average market price/cwt		Quantity qualifying for deficiency payment	Deficiency payment/ cwt		Advance payment cwt already made		Final payment cwt to be made	
	s.	d.	'ooo tons	s.	d.	s.	d.	s.	d.
1968									
1. July/Sep.	21	8·3	507	3	3·4	2	6		9·4
2. October	22	5·2	423	3	10·5	3	3		7·5
3. November	22	9·8	243	3	10·9	3	5		5·9
4. December	23	1·2	224	4	0·5	3	6		6·5
1969									
5. January	23	3·5	305	4	3·2	3	9		6·2
6. February	23	4·6	306	4	7·1	4	1		6·1
7. March	23	3·4	305	5	1·3	4	8		5·3
8. April	23	2·2	292	5	6·5	5	2		4·5
9. May/June	23	2·8	642	5	8·9	—		5	8·9
Year	22	10·5	3247	4	6·5	—		—	

A Home-grown Cereals Authority Levy of 2·6d./cwt will be deducted from the wheat deficiency payment for the May/June Accounting Period.

Sep. 4th, 1969

and in 1968 was dropped.) When production rose above this figure the amount of government subsidy, in the form of a deficiency payment was automatically reduced.

In order to encourage orderly marketing, the guaranteed price for wheat is converted to a rising scale of seasonal standard prices, which correspond to definite accounting periods within the harvest year. This is best explained through actual figures and the relevant data for 1968/69 appear in Table 14.

In conjunction with the standard quantity, target indicator prices are also used to assess the levels of deficiency payments.

TARGET INDICATOR PRICES

These are related to the proposed minimum import prices for wheat, allowing for handling costs, quality differences and the marketing conditions of the current wheat crop in the United Kingdom.

B. Home-grown Cereals Authority

The Home-grown Cereals Authority is a statutory body created by the Cereals Marketing Act, 1965. Its basic objective is to "improve the marketing of the home-grown cereals wheat, barley, oats and rye". The administrative costs of the authority are currently borne equally by the Exchequer and by cereal-growing farmers. It was agreed that home-grown cereals could compete more effectively with imported grain if supplies came forward throughout the year to match the demands of the market, if users had a greater assurance of continuity of supply and if more comprehensive market information was available. Should it be considered necessary, the authority would seek powers from the government ministers in respect of a trading plan. At the moment the non-trading powers and responsibilities as specified in the Act are:

1. To operate forward contract bonus schemes for home-grown cereals;
2. To operate bonus schemes in respect of deliveries of such cereals;
3. The making or guaranteeing of loans on forward contracts;
4. To improve market intelligence and statistical services;
5. Research and development.

The Home-grown Cereals Authority (H-G.C.A.) was obliged by the Cereals Marketing Act to introduce a forward contract bonus scheme for wheat and barley "as soon as practicable" and it was given permission to implement the other services.

The actual bonus payments for these forward contracts in respect of wheat and barley in 1967 were as follows (previous year given in brackets):

TYPE A CONTRACTS

1. Two calendar months, 8s./ton (10s.);
2. Three or more calendar months, 10s./ton (12s. 6d.).

TYPE B CONTRACTS

1. Two or more calendar months, 8s./ton (last year 2s. 6d. a ton less than Type A rates).

Type B contracts provide for deliveries to be made over a period of two consecutive months *at the call of the buyer*. Two other conditions apply to Type B contracts:

(a) A minimum tonnage of 25 tons (last year 100 tons);
(b) A fixed price to be stated in the contract.

Forms of Contract and Registration are obtainable from the authority and they must be signed by both grower and buyer.

FORWARD CONTRACT BONUS SCHEME 1969–70.

The contract bonus rate was 8s./ton and applied to deliveries at any time between October and June inclusive. Where October is the delivery month the contract must be made before the end of August. For all other deliveries, the delivery date cannot fall within the month in which the contract is made or the two following months.

XVII. PRODUCTION ECONOMICS

The specimen production cost shown on the opposite page has been based mainly on the Economic Reports of Anderson (1964) and Mathias (1965).

From the items of cost, it would appear that an acre of winter wheat will cost approximately £26 to grow, assuming that it is combine-harvested.

Winter Wheat—Combine-Harvested

Specimen Production Costs/Acre

Item	Cost
Labour (11 man hours at 32½p)	£3·57½
Tractor (7 hours at 25p)	1·75
Machinery depreciation and repairs	5·00
Contract services	0·60
Other fuel	0·40
Materials	
Seed	4·20
Fertilisers	4·15
Sundries	1·25
Rent	4·50
Total Direct Costs	25·42½
Share of general farm expenses	2·00
Adjustment for manurial residues	1·50
Gross cost	28·92½
Credit value of straw	3·00
Net cost	25·92½

A. Profitability, Grain and Straw Yields

(i) YIELD OF GRAIN

The average yield/acre has been increasing in recent years, and in respect of England and Wales the estimated figure for the three years 1964–66 was 32–33 cwt/acre. Using this estimated average and the specimen costs given, the profitability of wheat can be assessed.

Item	Value
32 cwt of grain at £22/ton	£35·20
Deficiency payment £5/ton	8·00
Total	43·20 per acre

On this basis a margin or profit of approximately £17/acre can be expected. Lower than average yields mean reduced profits and when yields exceed 32–33 cwt/acre the returns are correspondingly higher. Looking at the economic aspects in a slightly different way, the first ton of grain will cover the costs incurred, and sales above this could be classified as profit. Under these circumstances one ton of grain is said to be the break-even yield.

It can be seen that where average yields are experienced, about half of the profit comes from government support and without it a large number of farmers would not be able to show a significant return bearing in mind the large amount of capital invested in the growing, harvesting, drying and storage of this cereal.

The highest recorded yield of winter wheat on a field scale stands at 72 cwt/acre and although this is unique many farmers consistently achieve figures of over 2 tons/acre of dried grain. Growers should aim to produce 2·5–3 tons/acre and they ought not to be satisfied with yields below 2 tons/acre if they are farming good land in a recognised arable area. Yield categories for both winter and spring sown crops are suggested below.

	Winter wheat	Spring wheat
	(cwt/acre)	
Low	20–29	20–24
Medium	30–39	25–34
High	40–49	35–44
Exceptionally high	50 and over	45 and over

(ii) YIELD OF STRAW

Short strawed varieties suitable for combine-harvesting will produce about 1 ton of straw/acre which is mainly used for bedding. The special varieties with long straw, to be harvested by binder, are capable of giving 1·5–1·75 tons/acre.

Appendix

Recommended Varieties of Winter Wheat for England and Wales—1971†

The National Institute of Agricultural Botany tests all promising new varieties of cereals in field trials at many centres over a number of years. Special tests are also made to determine the effects of diseases and

Varieties classified for General use G, Provisional recommendation P	Cama	Joss Cambier	Maris Ranger	Champlein	West Desprez	Maris Widgeon	Capelle-Desprez	Maris Nimrod
				Recommended				Provisionally recommended
	G	G	G	G	G	G	G	PG
Agricultural Characters:								
Yield as % of control	106	106	103	102	98	97	97	114
Standing power	7½	8½	8½	8	7	7	6½	7
Shortness of straw	9	9	7½	8	7½	5	7	7½
Earliness of ripening	8½	9	8½	9	7½	7	7½	7½
Resistance to eyespot	6	8	8	3	8	8	8	8
Resistance to loose smut	7	3	2	7	3	2	2	7
Resistance to mildew	3	3	4	6½	6½	7	6	8
Resistance to yellow rust	2	6	2	3	8	6	4	5
Resistance to shedding	7	5½	6½	6	7	7	6½	6½
Resistance to sprouting	7½	5	8	6½	6½	6	7	7
Tolerance of manganese deficiency	4	5	4	7	4	4	4	6
1000 corn weight	7	6½	8	6½	8	7½	7½	7½
Latest safe sowing date	Mid Feb.	End Jan.	End Mar.	End Feb.	End Feb.	End Feb.	End Feb.	End Feb.
Quality of Grain: (greatly influenced by conditions of growth)								
Milling quality	7	5	4	5	7	9	5	4
Bread-making quality	5	3	3	5	8	8	5	2
Biscuit-making quality	2	4	7	4	3	1	4	7
Year first listed	1969	1968	1968	1962	1969	1964	1953	1971

A high figure indicates that the variety shows the character to a high degree.

† Farmers' Leaflet No. 8. Issued by the National Institute of Agricultural Botany, Huntingdon Road, Cambridge.

the quality of the produce as well as other characters which affect performance.

The control for yield comparisons is the mean of Cappelle-Desprez and Maris Ranger. Comparisons other than with the control are not strictly valid, and any differences of 3% or less should be treated with reserve.

No variety is recommended on less than three years' trial. The lists are reviewed each year and the varieties are classified as follows:

G   Recommended for General use
S   Recommended for Special use
P   Provisional Recommendation for a variety on which further trials are still in progress. Seed may not be available for commercial production
O   Becoming Outclassed by other recommended varieties

Varieties are listed in the tables in the order of expected yields as a percentage of the control, based on accumulated results in the last five years of N.I.A.B. trials. Numerical scales indicate the relative value for other characters, such as standing power, earliness of ripening, resistance to diseases and grain quality; these are based on a 0–9 scale with a high figure indicating that the variety shows the character to a high degree. But the scales in one table do not necessarily correspond to those used in another table. Special note should be taken of varietal differences in disease resistance as these may have an important effect on yield and grain quality. The figures indicating resistance to disease are based on the field reaction to established races of the organisms concerned. The appearance of new races may therefore alter the reaction of varieties.

VARIETIES RECOMMENDED FOR GENERAL USE

CAMA

Very high yield combined with stiff and very short straw. Very early. Rather low resistance to mildew. Susceptible to yellow rust.

Cappelle-Desprez × H.392/70. Gembloux, Belgium.

JOSS CAMBIER

Very high yield, very early and very short stiff straw. Rather low resistance to mildew.

(Cambier 194 × Tadepi) × Cappelle-Desprez. Cambier, France.

MARIS RANGER

High yield, very early and very stiff straw. Very low vernalisation requirement. Susceptible to yellow rust and loose smut.

Peko × Cappelle-Desprez.
Plant Breeding Institute,
Cambridge.

CHAMPLEIN

A very early variety combining high yield with short stiff straw. Good tolerance of manganese deficiency. Susceptible to yellow rust and eye-spot.

Yga-Blondeau × Tadepi.
Claude Benoist, France.

WEST DESPREZ

High milling and very high bread-making quality. Straw short and moderately stiff.

[(90 × Choisy) × Cappelle-Desprez] × Cappelle-Desprez.
Desprez, France.

MARIS WIDGEON

Very high milling and bread-making quality. Straw rather long. Susceptible to loose smut. Moderately resistant to the Septoria diseases.

Holdfast × Cappelle-Desprez.
Plant Breeding Institute,
Cambridge.

CAPPELLE-DESPREZ

Short moderately stiff straw. Susceptible to loose smut. Suitable for a wide range of conditions.

Hybride du Joncquois × Vilmorin 27.
Desprez, France.

NEW VARIETY PROVISIONALLY RECOMMENDED WHILE FURTHER TRIALS ARE STILL IN PROGRESS

(Seed may not be available for commercial production)

MARIS NIMROD

Outstanding yield. Straw moderately stiff. Grain of poor appearance.

[(CI.12633 × Cappelle-Desprez) × Hybrid 46] × Professeur Marchal.
Plant Breeding Institute,
Cambridge.

C

WINTER WHEAT VARIETIES FOR THE WEST

In some parts of the west the diseases mildew and Septoria are particularly prevalent, and some varieties have given comparative yields significantly different from the national mean.

Yields from trials centres in the west midlands and Wales expressed as a % of the mean of Cappelle-Desprez and Maris Ranger are as follows:

Maris Ranger	..	105	Maris Widgeon..	101
Cama	.. ..	104	West Desprez ..	95
Champlein	..	103	Cappelle-Desprez	95
Joss Cambier	..	102		
			Maris Nimrod ..	117

LATE SOWING OR WINTER WHEAT VARIETIES

Many winter wheats can be sown during early spring with a reasonable chance of success provided the field conditions are satisfactory. However, winter wheat sown too late may result in a complete crop failure, and the latest safe sowing date for varieties is shown in the table opposite. Although sowing after these dates can sometimes be successful, it is safer to use a spring wheat.

Recommended Varieties of Spring Wheat for England and Wales—1971†

The control for yield comparisons is the mean of Kolibri and Rothwell Sprite. Comparisons other than with the control are not strictly valid, and any differences of 3% or less should be treated with reserve.

	Recommended						Provisionally recommended		Becoming outclassed
Varieties classified for General use G, Special use S, Provisional recommendation P Becoming Outclassed O	Kolibri	Cardinal	Sirius	Maris Ensign	Rothwell Sprite	Troll	Kleiber	Maris Dove	Janus
	G	G	G	G	G	S	PG	PG	O
Agricultural Characters:									
Yield as % of control	104	104	100	97	96	95	110	103	94
Standing power	6½	6½	8	7	7	5½	6½	7½	7
Shortness of straw	6	5½	8½	6	6½	6	6½	6½	6½
Earliness of ripening	6½	6	7½	5½	5½	6½	6	6	7
Resistance to loose smut	7	9	6	8	9	3	7	9	8
Resistance to mildew	7	8	6	7	5	6	7	8	6
Resistance to yellow rust	4	7	4	4	4	6	4	6	4
Resistance to shedding	5	6	4½	5½	6	6	6	6	4½
Resistance to sprouting	7	6	6	7½	6	5	7	6	7
Tolerance of manganese deficiency	6	6	4	7	8	4	6	8	4
1000 corn weight	6	4½	4	6	5	4	5½	5	5
Quality of Grain: (greatly influenced by conditions of growth)									
Milling quality	7	6	9	5	6	8	6	7	8
Bread-making quality	6	5	7	5	6	7	7	5	5
Biscuit-making quality	1	2	1	4	4	3	1	2	3
Year first listed	1968	1970	1970	1968	1967	1968	1971	1971	1968

A high figure indicates that the variety shows the character to a high degree.

† Farmers' Leaflet No. 8. Issued by the National Institute of Agricultural Botany, Huntingdon Road, Cambridge.

VARIETIES RECOMMENDED FOR GENERAL USE

KOLIBRI

High yield; grain of moderately high milling and bread-making quality.

(Heines 2174 × Peko) × Koga II.
von Lochow-Petkus, Germany.

CARDINAL

High yield, and good all round disease resistance.

Koga II × Alba.
von Lochow-Petkus, Germany.

SIRIUS

Combines high yield with very early maturity, very short and stiff straw, and very high milling and high bread-making quality. Liable to shedding when over-ripe.

Bavarian variety × Probat.
von Rümker, Germany.

MARIS ENSIGN

Moderately high yield with stiff straw.

Teutonen × Cappelle-Desprez.
Plant Breeding Institute, Cambridge.

ROTHWELL SPRITE

Moderately high yield with stiff straw. Good tolerance of manganese deficiency.

Hd. 12 × Cappelle-Desprez × Fasan.
M.G.H., Holland.

VARIETY RECOMMENDED FOR SPECIAL USE

TROLL

For use where grain of high milling and bread-making quality is required. Moderately high yield.

Ring × (Pondus × Karn).
Weibull, Sweden.

NEW VARIETIES PROVISIONALLY RECOMMENDED WHILE FURTHER TRIALS ARE STILL IN PROGRESS

(Seed may not be available for commercial production)

KLEIBER

Very high yield; grain of moderately high milling and bread-making quality.

(Heines 2174 × Peko) × Koga II.
von Lochow-Petkus, Germany.

MARIS DOVE

Combines high yield with very stiff straw and good tolerance of manganese deficiency.

Koga II × H.8810/47.
Plant Breeding Institute,
Cambridge.

VARIETY BECOMING OUTCLASSED

JANUS

Stiff strawed, early maturing. Liable to shedding when over-ripe.

[(v. Rümker's Dickkopf × Erli) × (Erli × Hope)] × Dickkopf.
von Rümker, Germany.

1970 Varieties of Cereals for Scotland

Recommendations in this leaflet are based on trials and experience of the Scottish Agricultural Colleges.

In case of doubt, or for information about varieties not on the list, farmers should consult their Agricultural Adviser.

BRITISH CEREAL SEED SCHEME
(SCOTLAND)

Certified seed, Multiplication seed or Field Approved seed of the following recommended varieties will be available through normal commercial channels for sowing in 1970:

Wheat—Champlein, Maris Ranger, Cappelle.

		Trial Yields as % of Control Variety
WINTER WHEAT		
G	Joss Cambier	107
G	Champlein	104
G	Maris Ranger	103
G	Cappelle-Desprez	100
S	N. 59	
SPRING WHEAT		
G	Kolibri	107
G	Kloka	100

WINTER WHEAT

JOSS CAMBIER

Early ripening with straw shorter and stiffer than Cappelle. Mildew susceptible. Similar eyespot tolerance to Cappelle. Grain shows a tendency to shed when ripe.

CHAMPLEIN

An early ripening variety with straw shorter and stiffer than Cappelle. Tends to shed when ripe. Susceptible to yellow rust and eyespot.

MARIS RANGER

Short stiff straw, early ripening. Tolerance to eyespot similar to Cappelle.

CAPPELLE-DESPREZ

A dependable variety over a wide range of conditions. Ripens moderately early and has short, stiff straw. Good resistance to sprouting. Susceptible to yellow rust but shows high tolerance of eyespot.

N. 59

Recommended primarily for use where long straw is wanted but nevertheless gives satisfactory grain yields. Resistance to lodging is good and ripens reasonably early.

SPRING WHEAT

KOLIBRI

Several days later than Kloka but higher yielding.

KLOKA

Early ripening with short, stiff straw. Liable to shedding.

Varieties of Wheat for Northern Ireland†

No local trials are carried out in respect of winter wheat or spring wheat and data from the National Institute of Agricultural Botany are supplied as an indication of possible performance in Northern Ireland. Details as per England and Wales List.

MINISTRY OF AGRICULTURE
DUNDONALD HOUSE
BELFAST BT4 3SB

Ministry Publications Related to Grain Drying and Storage

1. Buildings for Grain Drying and Storage. F.E.F. 10.
2. Bulk Grain Driers. M.L. 6.
3. Farm Grain Drying and Storage. Bulletin No. 149.
4. In-Sack Grain Drying. F.M.L. 12.
5. Preservation of Grain Quality During Drying and Storage. S.T.L. 24.
6. Farm Seed Cleaning and Grading Machinery. M.L. 11.
7. Grain Driers: Continuous Flow and Batch Machines For Grain in Bulk. F.M.L. 4.
8. Grain Silos. F.E.F. 9.
9. Drying Grain On The Floor. S.T.L. 24.

References

Agricultural Chemicals Approval Scheme. (Insecticides, Fungicides and Herbicides.) List of Approved Products (1967). For Farmers and Growers. Ministry of Agriculture, Fisheries and Food. H.M.S.O.

Anderson, J. L. (1964). "Wheat Production in the East of Scotland (1963)". Edinburgh School of Agriculture Economic Report No. 85.

Barber, D. "Keep Pace With Progress". No. 4. Farmer and Stockbreeder Publication.

Bell, G. D. H. (1948). "Cultivated Plants of the Farm". Cambridge University Press, London.

Darlington, C. D. and Wylie, A. P. (1955). "Chromosome Atlas of Flowering Plants", 2nd edition. Allen & Unwin Ltd, London.

Fisons. Agricultural Technical Information. No. 7 Autumn 1966 and No. 9 Autumn 1967.

† Northern Ireland Ministry of Agriculture Leaflet No. 96—Varieties of Cereals.

Farmer and Stockbreeder. June, 1967.

Gardner, H. W. and Garner, H. V. (1953). "The Use of Lime in British Agriculture". Farmer and Stockbreeder Publications Ltd. E. and F. N. Spon Ltd, London.

Greer, E. M. (1968). "The Wheat Variety Maris Widgeon, Ceres". Vol 1. H.-G.C.A.

Hall, Sir A. D. (1945). "The Soil". John Murray, London.

Holliday, R. (1956). "Fodder Production From Winter-Sown Cereals And Its Effect Upon Grain Yield". *Field Crop Abstracts* **9**, 3, 1–13.

Holliday, R. (1963). "The Effect of Row Width on the Yield of Cereals." *Field Crop Abstracts* **16**, 2, 71–81.

Jealott's Hill Research Station. *1967 Guide to Experiments.* I.C.I. Ltd, Agriculture Division, Bracknell, Berkshire.

Jeater, R. S. L. (1967). 8th Brit. Weed Control Conf. Proc. Vol. 3. 19, 874–883.

Kent, N. L. (1964). "Technology of Cereals, with Special Reference to Wheat". Pergamon Press, London.

La Potasse. (1962), No. 297, p. 146.

Lessells, W. J. and Webber, J. (1965). *Expl. Husb.* **12,** 74–88.

Mathias, K. (1965). "Winter Wheat 1963–64". University of Manchester, Dept. of Agriculture, Economics Bulletin No. 116/EC. 61.

McConnell, P. (1883). "The Agricultural Notebook". Re-written and edited by H. I. Moore, 14th edition, 1962. Ilffe Books Ltd for *Farmer and Stockbreeder.*

M.A.F.F. Short Term Leaflet No. 19, "Choosing Selective Weed-killers for use on Cereals in the Spring".

Mundy, E. J. and McClean, S. P. (1965). *Expl. Husb.* **12,** 143–163 and 164–172.

Percival, J. (1921). "The Wheat Plant. A Monograph". Duckworth and Co., London.

Percival, J. (1934; re-issued 1943). "Wheat in Great Britain". Published by the author, Leighton, Shinfield, Reading, England.

Peterson, R. F. (1965). "Wheat, Botany, Cultivation and Utilisation". Grampian Press Ltd., London, Interscience Publishers Inc., New York.

Riley, R. (1965). "Cytogenetics And The Evolution of Wheat. Crop Plant Evolution". (Sir J. Hutchinson, ed.) Cambridge University Press, London.

Robinson, D. H. (1958). "Wheat". Ministry of Agriculture, Fisheries and Food, H.M.S.O.

Smith, H. P. (1964). "Farm Machinery and Equipment" (5th edition). McGraw Hill, New York.

Thompson, J. B. and Whitehouse, R. N. H. (1962). *Euphytica* **11**, 2, 181, 196.

Vavilov, N. I. (1951). "The Origin, Variation, Immunity and Breeding of Cultivated Plants". Translated by K. Starr-Chester. Chronica Botanica Co., Walthan, Mass., U.S.A.

Watson, J. A. S. and More, J. A. (1962). "Agriculture, the Science and

Practice of British Farming". First Published 1924, 11th edition 1962.
    Oliver and Boyd, London.
Weed Control Handbook, Issued by the British Weed Control Council
    (E. K. Woodford and S. A. Evans, eds). Blackwell Scientific Publications,
    Oxford.

CHAPTER 2

Barley

I.	Introduction	64
II.	Botanical Classification	64
III.	Grain Uses	65
IV.	Grain Quality for Livestock Feeding	66
V.	Malting and Brewing	68
VI.	Requirements for Malting	68
	A. Malting Barley Characteristics	68
VII.	Soil	71
VIII.	Climate	71
IX.	Rotations and Monoculture	73
	A. Position of Winter and Spring Barley in Cropping Sequence	73
X.	Manuring of Feeding Barley	76
XI.	Manuring of Malting Barley	78
XII.	Cultivations and Seed-bed Preparations	80
	A. Ploughing	80
	B. Final Seed-bed Preparations	82
XIII.	Drilling	85
XIV.	Seed Rates	86
XV.	Seed Dressings and Seed Treatment	86
	A. Type of Seed	87
XVI.	Varieties/Cultivars	87
	A. For Winter Sowing	87
	B. Varieties for Spring Sowing	89
XVII.	Management of the Established Crop	90
	A. Spring Management of Winter Sown Crops	90
	B. Undersowing Barley with Grass and Clover Seeds	91
	C. Control of Insect Pests	91
XVIII.	Chemical Weed Control	91
XIX.	Harvest	93
	A. Feeding Barley	93
	B. Malting Barley	95
XX.	Drying and Storage	96
	A. Drying	96
	B. Grain Chilling and the Storage of Cool Grain in Bulk	97
	C. Bulk Grain Aeration	100
	D. Storage of High Moisture Grain in Sealed Silos and Containers	100
	E. Acid Treatment for the Storage of Moist Grain for Animal Feed	104
XXI.	Disposal of Grain, Production Economics and Yield	105
	A. Disposal of Grain	105

B. Government Support for Barley 106
C. Production Economics 106
D. Yields 107
E. Profit Margins 108
Appendix 110
References 118

I. INTRODUCTION

From discoveries in Mesopotamia, Beaven (1947) estimated that cereal production began some 20,000 years ago and it could possibly have been earlier. According to Fraser (1912), who assessed cereal antiquity through mythology, barley cultivation preceded that of wheat and Vavilov (1926), using his concept of variations in the diversity of evolutionary forms, suggested two centres of origin for barley, namely North Africa (Abyssinia) and East Asia. Barley, reputed to be the gift of Ceres, was probably the first plant to be cultivated and this is not surprising since many wild forms were known to have been used for human consumption whilst the populations in these areas of origin were completely nomadic (Körnicke, 1895; Beaven, 1947). With such a long history of cultivation coupled with the possibility of two centres of origin, barley has developed a much wider ecological range than any other cereal. This range has been described by Hunter (1952) in the introductory chapter to his book "The Barley Crop", indicating cultivation from inside the Arctic Circle to tropical India where it is grown to heights over 15,000 feet. Barley growing in Britain goes back at least as far as the Iron Age and probably earlier and although historians are not precise about the dates of first cultivation they are fairly confident that it originated on the upland chalk soils. This assumption is reasonable when one considers how intolerant present-day varieties are to acid conditions.

II. BOTANICAL CLASSIFICATION

Barley belongs to the genus Hordeum and according to De Candolle (1882), the immediate progenitor of the present cultivars is *Hordeum spontaneum* Koch., which very closely resembles current two-row types. The wild barley grass indigenous to Britain, *Hordeum murinum* (sometimes termed wall-barley) is similar in many respects to *Hordeum spontaneum* and *Hordeum agriocrithon* Aberg., found wild in Tibet, is thought to be concerned with the origin of six-row cultivated forms.

Classification of the various forms of this cereal has received much attention in the past. Linnaeus (1753), Körnicke (1882), Körnicke and Werner (1885), Voss (1885), Rimpau (1891), Atterberg (1899) and

Beaven (1902) have all produced schemes but fail to agree on many points. However, it is now fairly widely accepted that all cultivated forms belong to the single species *Hordeum sativum* Jess. Their previous specific classification has been listed by Gill and Vear (1958) and for those who may be interested in phylogeny and taxonomy the early attempts at grouping should be consulted along with "The Cromosome Atlas of Flowering Plants" by Darlington and Wylie (1955).

Cultivated barleys (*Hordeum sativum*) will cross-fertilise readily with each other; they are diploids with fourteen chromosomes. Six-row and two-row barleys have been grown in Britain and both have representatives on the current recommended list of varieties although in recent years the two-row varieties have been more numerous. The so-called four-row or beer barleys (strictly speaking a lax six-row), previously grown under upland conditions in Wales and Scotland, have disappeared from cultivation except in some of the Western Isles (e.g. Tiree). Although rather coarse, they could be grown on poorer soils and at higher elevations than any other varieties.

III.　GRAIN　USES

In the distant past barley was largely used directly for human consumption, but once wheat was firmly established as the major source of flour most of the production was used for malting or livestock feeding. Before the Second World War, over half the produce of the barley

Pre-1939	%	1958	%	1965	%
Malting	56	Malting	25	Human food and	
Distilling	8	Distilling	6	drink	16
Other industrial		Other industrial		Sold	
use	3	use	2	Feed	39
	—		—	Seed	3
	67		33	Export	1
Seed	8	Seed	5	Farm retained	41
	—	Export	3		
	75		—		
			41		
Farm retained or		Farm retained or			
sold for feed	25	sold for feed	59		
Total	100	Total	100	Total	100
Estimated pro-					
duction 800,000 tons		3,000,000 tons		7,500,000 tons	

crop went for malting but with a very large increase in the amount of land devoted to this crop after the war the percentage dropped to 25% by 1958 and to 16% by 1965. In practical terms the position changed from more than one sample in two accepted for malting, to one in four and to one in six as the acreage and production rose (see above). Pre-1939 the premiums obtained for malting samples were important and sought-after as production was aimed at a quality product for this market. Today when maltsters can be much more choosy and with intensification of production requiring the highest output per acre (usually associated with maximum yield) malting barley production is often more fortuitous than calculated.

Barley grain with its high starch content and relatively low percentage fibre or husk is admirably suited for the feeding of pigs with their single stomach and simple digestive system and for a long time this was the main outlet for feeding samples. In recent years feeding experiments carried out at the Rowett Research Institute, Aberdeen, both with beef and dairy stock, have demonstrated that barley feeding can be both safe and profitable and a large proportion of their concentrate ration is now made up of this cereal.

IV. GRAIN QUALITY FOR LIVESTOCK FEEDING

Samples of grain destined for livestock feed should be free from dust, weed seeds and trash. They should be well filled with a plump appearance and free from fungal or bacterial growth. The best feeding grain is that which also has a low husk percentage and a high protein content which is reflected by a N content of approximately 2%. When grain is to be sold for livestock feeding, then under most circumstances it should be dried to 14% moisture in order to maintain its condition during storage. When the barley produced is to be home-fed to stock then storage as wet grain straight from the combine may be both attractive and economic. The essential feature of this method is to obtain as quickly as possible an atmosphere surrounding the grain which is free from oxygen (anaerobic conditions) which means properly sealed containers. (For further details see under methods of storage.) Moist grain is more easily prepared for animal rations and although there is some degree of fermentation and an associated characteristic odour, there are no difficulties in feeding it. In fact, animal intake often appears higher than with the conventionally dried product, although when dry matter intake is calculated then little difference is observed.

(1 qr) 4 cwt quality barley is germinated on floors or in drums

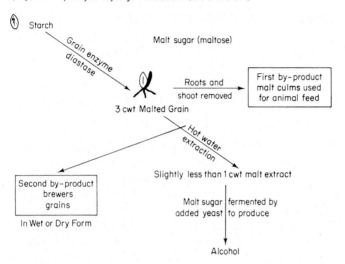

Predicted extract of malt using grain size and N Content

Amount of malt extract from
4 cwt of barley (1 qr)   =   110 – (11 x N Content of the grain)
                                           + (0·22 x 1000 grain weight)
                                           (Dry weight of malt in lbs)

FIG. 1. Malting and brewing.

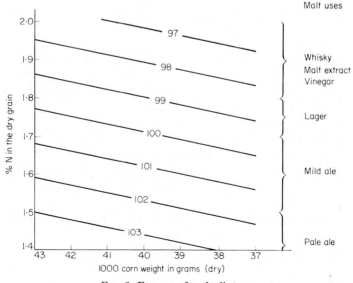

FIG. 2. Extract of malt, lb/qtr.

V. MALTING AND BREWING

The malting process is essentially the conversion of the starch to sugar in the grain, by inducing germination. Brewing, which follows, is a yeast fermentation when the sugars are converted to alcohol. These processes are outlined in Fig. 1 and the amount of malt which can be obtained from a quarter (4 cwt) of barley is shown in Fig. 2.

VI. REQUIREMENTS FOR MALTING

Samples of barley purchased for malting are bought mainly on hand evaluation, but there is an increasing trend towards laboratory backing to determine accurately the amount of protein which the grain contains. Thompson (1966) has pointed out that, whereas in the past barley buyers could rely on a low nitrogen content with the light-coloured samples, this is not true of many of the new varieties. He has enumerated the characteristics which the brewing industry and maltsters consider important in the production of new malting varieties and although they are recommendations from the standpoint of the industrial user to the plant breeder, it is always interesting to see what "pre- and post-growing associates" have in store for future farming. Many factors, when integrated, constitute malting quality and these have been listed below. They do not appear in order of importance and are briefly discussed following enumeration.

A. Malting Barley Characteristics

1. Large grain.
2. Well filled grain with starchy cross-section.
3. Thin husk.
4. Good colour.
5. Low moisture content, evenly dried.
6. Even sized sample.
7. Recognised degree of wrinkle.
8. Correctly threshed or combined.
9. Free from dirt, weed-seeds, other cereals, diseased and broken grains.
10. Free from green immature grain.
11. All grain to be of the same variety.
12. The variety to be a recognised malting type.
13. High total germinative capacity.
14. High germinative energy.
15. Low total N content.

16. Low permanently soluble N content (PSN).
17. Grain dormancy.
18. Time to malt.
19. Satisfactory diastatic power.
20. High cold water extract (CWE).

1. *Large grains* are preferred on two accounts. Firstly, the samples of malted barley which result from them are bold and attractive in appearance and secondly the proportion of inner carbohydrate to outer seed coat is likely to be higher than with small grains (see graph of Malt Extract, Fig. 2).

2. *Well filled grains* also contribute towards the two factors in 1. above and will ensure a high extract of malt and at the same time they should exhibit a white starchy appearance when cut. By comparison feeding samples produce a flinty, translucent cross-section, usually indicative of high N content (Rayns, 1959).

3. *A thin husked* grain automatically ensures a high proportion of starch and under most circumstances will lead to higher yields of malt following the germination process. This is largely a varietal characteristic with the present six-row winter cultivars Senta and Dea exhibiting a very thick husk resulting in a "feed" classification.

4. *Grain colour*, in the past has been associated with protein content, dark grains were considered high in N and light-coloured samples were assumed to be low in N. This is not true to the same extent today and attractive samples with a pale appearance may be as high as $1 \cdot 7\%$ N. Grain colour without association with protein content is important in itself when high quality malt is being produced for pale ale and the lighter coloured samples are preferred.

5. *Moisture content* is important since the majority of grain purchased for malting will have to be stored until the following year when it is used. When farm drying is necessary this should be done carefully in order to maintain the germination capacity and a strict watch on drying temperatures (related to the original moisture content as combined) must be maintained. A slow, even drying process to reduce the moisture content to 14% should be the aim.

6. *Size.* Before offering samples of barley for sale it may be necessary to remove the small grains, usually referred to as "seconds". This is relatively unimportant compared with other characteristics but on a hand evaluation it could result in the securing of malting price, an additional £2–4 per ton compared with feeding barley.

7 and 8. Quality samples of barley are finely wrinkled (Rayns, 1959), and should not be overthreshed to break the ends of the grain.

This could lead to a reduced germinative capacity and also it gives a point of entry to fungus moulds which become active once the grain has been water-soaked and heated to initiate the malting process via germination.

9 and 10. In selecting samples for malting, potential buyers look to see that extraneous material such as dirt, free awns and weed seeds has been removed. The presence of other cereals or broken and diseased grains is highly undesirable and should part of the crop be lodged resulting in green immature grains due to second growth, this portion should not be included in the bulk.

11 and 12. Varieties offered for malting should be recognised "malting-types". Much work goes into the testing and evaluation of cultivars for this purpose and it is important that farmers choose those which are suitable (see Appendix Recommended List of Cereals, England and Wales). Bulk grain should contain only *one* of these since "time to malt" is a varietal characteristic and variation within a batch is undesirable.

13 and 14. The malting process is essentially the controlled germination of the grain requiring a quick, even germination of as much of the grain as possible.

15 and 16. *Nitrogen content of the grain* and solubility of this protein are two of the most important characteristics of the grain and alone can decide whether the sample is suitable for malting or whether it should be used for feeding. The total N content of the grain is the concern of the barley grower and it should be kept as low as possible when producing quality malting barley. From the equation used to predict extract and the graph derived from it, the need for low nitrogen grain is apparent since it affects the amount and uses of extract of malt obtained.

17, 18, 19 and 20. These four grain characteristics are the concern of the maltster. Dormancy is that period after harvest when grain will not germinate and in order to eliminate risks on this account malting does not usually take place until the turn of the year. Should a variety exhibit an unusually long dormancy period then it will not be recommended for malting.

The time which variety takes to malt (sometimes termed modification) may be affected slightly by the physical environment during growth but is mainly an inherited characteristic. Maltsters in the main prefer those varieties which are quick to germinate and grow in order to maximise the throughput of a given malt-producing unit.

The barley grain should contain a sufficient quantity and quality of the enzyme diastase in order to effect the change from starch to malt sugar during germination. This is referred to in brewing circles as the diastatic power (D.P.) of the grain.

Finally, the amount of malt extracted after germination by a cold water extraction process (C.W.E.) gives a good indication of the overall capacity of a variety or a batch of grain and this factor is used extensively by research scientists and the trade in comparative studies.

VII. SOIL

The first requirement in the production of any crop is to see that soil conditions are as close to the optimum as possible. Barley prefers well-drained soils, light to medium in texture with a high pH. As a result large areas of the Downs and Wolds are to be seen under this cereal. When fertility is high and weather conditions are favourable high yielding crops of good quality are obtained from these areas. With continued government support for lime since 1936 it means that much of the ploughable land of Britain will have a pH sufficiently high to support barley-growing. Rayns (1959) suggests that it is safe to grow barley as low as pH 6·2, but states that a pH of 6·5 would be safer due to the extreme variability of many soils. When pH values are recorded below 6·0 it would be wise to lime specifically for this crop and it should be worked into the top-soil in advance of sowing. Since the grain yield with barley is likely to be higher than with oats and due to its better feeding value (starch equivalent 70 compared to 60 for oats) it has replaced the traditional oat crop on many dairy and stock farms and it has therefore spread to a much wider range of soil types. In the past barley was not an important cereal on the fens where annual release of N from the soil resulted in grain of high N content and conditions which favoured lodging. With shorter, stiffer-strawed varieties and a much larger market for "feeding barley" it has become an important crop here and also on clays and silts. As long as the pH is over 6 and the soil is in reasonable conditions there is no reason why this crop cannot be grown on most soils in Britain, one notable exception being the wet soils associated with upland conditions.

VIII. CLIMATE

Cereal growing of any kind is easier in the drier Southern and Eastern regions of Britain where moderate rainfall and high sunshine hours are recorded. Only on the very light soils, which tend to "burn-up" during the summer, does higher than average rainfall produce significantly better yields. When barley is grown in the wetter areas of Britain it does best when the rainfall is below normal and when sunshine hours are higher than usual. Malting barley is seldom considered since the grain produced in these areas is much darker in colour with a high nitrogen content, and in any case large quantities are required for cattle feeding. In the South, barley is successfully cultivated to an altitude of approximately 1000 feet on the Cotswolds and in Yorkshire it is found between 400 and 800 feet on the Wolds. It can be seen at 650 feet (a.s.l.) in the West of Scotland and although in some years ripening is much delayed, barley may be found up to 1000 feet in the Clyde area. In the North of Scotland successful cultivation every year of growing is rarely seen above about 700 feet.

Kramer *et al.* (1952) reviewed the effect of climate on this crop through data from Zealand and Groningen. Their results suggest that low rainfall in April and early May and cool weather in May, at ear emergence and at anthesis, is required for high yields. High rainfall in the previous winter appeared detrimental and warm dry weather was required during ripening. Sunshine was considered important for malting barley and a little rain in July appears to improve the quality, a feature also suggested by Rayns (1959). In respect of these and other aspects of climate the reader may wish to consult the review article by Smith (1967) entitled "The Effect of Weather on the Growth of Barley". From the acreage statistics the following counties are the most important producers of barley, but it must be remembered that cultivation is widespread throughout Britain.

England	Wales	Scotland	Northern Ireland
York	Pembroke	Aberdeen	Down
Lincoln		Fife	Antrim
Norfolk		Angus	Londonderry
Suffolk		Perth	
Essex		Berwick	
Hampshire		East Lothian	
Devon		Roxburgh	
Cambridge			

The counties have been placed in order of acreage with the highest at the top of the lists. Temperatures required for germination and growth appear below and as with wheat, they indicate that the Southern part of Britain again has the most favourable range.

Spring barley, however, has a much shorter growing season than the other cereals and grain yield does not suffer unduly when this is extended through lower than optimum temperatures.

Germination and Growth Temperatures for Barley

	Minimum	Optimum	Maximum
Germination	40°F	77–80°F	100–110°F
	4°C	52–57°C	38–43°C
Growth	41°F	84°F	100°F
	5°C	29°C	38°C

After Hall (1945).

IX. ROTATIONS AND MONOCULTURE

A. Position of Winter and Spring Barley in Cropping Sequence

Winter barley is often sown after early harvested sugar beet on the lighter soils, since seed-beds can often be easily and quickly prepared for sowing in October and November. It may, however, be more convenient to sow it after short term leys which are broken-up during the late summer and it is also possible for this crop to succeed peas or beans. With such a large acreage of arable land in cereals, many crops of winter barley will go in after spring cereals, but it would be unwise to grow winter barley following winter barley due to the increased disease risks involved.

Spring barley may follow almost any other crop provided the land is not in too high a state of fertility otherwise widespread lodging can result. Under systems of cereal monoculture or close cereal cropping spring barley is the most commonly chosen crop since it appears least affected by disease. Although a yield depression has nearly always been recorded, with intensive cereal growing it has been the least with spring barley and with such a short growing season cultural weed-control, disease control, timely ploughing and cultivation can easily be achieved. The longest sequence of barley with accurate yield determinations was recorded at Rothamsted Experimental Station, Hertfordshire from Hoosfield which has been continuously cropped since 1852 (with the exception of three years). The original manurial treatments were still being carried out 100 years later; a few of these showing 10-year means or yields from individual years between 1852 and 1966 are given in Table 1. It will be noted that where adequate mineral fertilisers are returned to the land or via the application of farmyard manure the yield of barley can easily be maintained and in most cases it is likely to greatly exceed that recorded at the beginning of the experiment.

In the past, continuous cereal growing on a farm scale has not been

TABLE 1

Yields of Barley on Hoosfield, Rothamsted Experimental Station (cwt/acre) 1852–1966[a]

Manurial treatment[b]	1852–1861	1862–1871	1872–1881	1882–1891	1892–1901	1902–1911	1912–1921	1922–1931	1932–1941	1942–1951	1952–1961	1852–1962 Means Grain	Straw
None	11·4	8·8	6·8	6·2	5·3	5·2	6·5	3·7	6·9	9·3	7·4	7·0	7·9
P	13·9	11·8	9·0	9·6	7·1	9·3	10·2	6·7	11·6	11·7	10·6	10·1	9·8
P, K	12·3	10·9	7·5	7·4	6·2	6·9	7·3	4·8	10·9	15·0	8·9	8·8	10·1
N	17·0	15·8	13·2	12·0	8·8	10·7	11·2	5·4	10·4	11·5	10·8	11·6	13·2
N, P	22·9	24·7	20·4	18·0	15·5	16·1	16·2	11·8	18·8	15·9	16·8	17·9	19·4
N, P, K	21·7	23·1	19·8	15·0	14·1	15·2	16·1	10·9	19·5	14·2	19·8	17·2	21·9
N[1]	—	16·0	14·2	14·0	11·6	12·6	12·3	7·1	12·1	13·0	12·9	12·4	15·6
N[1], P	—	24·0	21·1	21·0	19·4	20·3	19·7	14·7	22·0	18·5	20·2	19·9	22·5
N[1], P, K, Na, Mg, Si	—	25·5	23·6	22·1	20·2	21·4	17·9	14·1	23·1	22·5	23·3	21·1	24·6
Castor meal (Nitrogen) + P	23·9	23·7	21·2	19·1	17·1	18·3	14·6	14·8	21·3	20·9	19·5	19·5	21·4
FYM	22·7	26·7	25·7	23·7	22·9	23·6	18·6	15·0	26·1	26·7	28·0	23·6	28·6

Manurial treatment[b]	1963		1964				1965				1966			
			Plumage Archer		Maris Badger		Plumage Archer		Maris Badger		Plumage Archer		Maris Badger	
	Grain	Straw	Grain	Straw	Grain	Straw	Grain	Straw	Grain	Straw	Grain	Straw	Grain	Straw
None	3·6	3·6	5·3	2·2	7·4	2·5	9·1	6·5	7·4	2·9	10·2	5·1	10·0	4·3
P	5·9	4·1	9·1	3·8	11·3	4·5	11·6	6·8	11·3	4·8	13·2	7·0	10·7	3·8
P, K	6·5	4·2	10·4	6·1	10·2	5·6	13·9	10·0	9·2	7·1	10·6	5·6	8·4	4·4
N	5·4	6·8	7·5	5·1	11·5	5·9	12·5	10·1	14·4	9·7	16·7	11·3	17·7	11·6
N, P	15·8	15·1	17·4	10·6	24·4	16·0	23·2	17·2	27·7	26·1	26·7	18·0	24·7	18·9
N, P, K	20·2	19·0	24·6	18·9	36·5	25·7	31·0	31·2	37·5	39·9	34·3	21·9	40·5	32·3
N¹	5·3	7·1	7·8	6·3	10·4	8·2	13·0	11·8	12·2	12·4	12·1	11·5	11·0	12·1
N¹, P	17·4	17·6	20·1	14·5	28·7	20·2	26·6	24·5	34·1	36·6	30·2	22·2	34·0	29·6
N¹, P, K, Na, Mg, Si	25·5	19·7	21·3	14·4	40·0	28·0	31·2	28·1	35·3	40·3	32·8	22·7	45·1	36·8
Castor meal (Nitrogen) + P	17·9	12·7	21·3	12·8	38·1	22·9	25·5	19·6	37·3	28·2	29·6	17·8	40·3	24·4
FYM	29·8	25·8	33·2	22·0	37·0	22·9	36·7	38·6	44·9	42·8	39·8	25·2	37·4	26·3

[a] Continuous barley since 1852 except 1912, 1933 and 1943 when the land was bare-fallowed.

[b] Fertilizer Levels

	lb/acre	
N	43	(Sulphate of ammonia)
P₂O₅	65	(Superphosphate)
K₂O	98	(Sulphate of potash)
N¹	43	(Nitrate of soda)
Na	100	(Sulphate of soda)
Mg	100	(Sulphate of magnesia)
Si	400	(Silicate of soda)

FYM = 14 tons.

practised to any great extent but there have been a few pioneers who have come very close to it with their systems. On heavy land near Sawbridgeworth (Hertfordshire) Mr John Prout and his son Mr W. A. Prout took six or seven consecutive corn crops and followed this sequence by a break of red clover, beans or fallow. This system which lasted for 45 years is described by J. Prout (1881) and W. A. Prout and Voelcker (1905) and records show that there were no reductions in barley or wheat yields during this long period.

On light land in Oxfordshire, Mr F. P. Chamberlain grew barley and occasional wheat crops from 1909 until 1952 without stock being kept on the farm. The success of this continuous cereal growing on a light soil has been ascribed to the undersowing of the cereals with short term ryegrass–trefoil mixtures which helped to maintain fertility and at the same time controlled the fungus Take-all, Sylvester (1947). Recent results of continuous barley growing on a 200-acre farm in the sand land of Nottinghamshire have been described by Daw (1963). Yields were 7% lower than comparable farms on normal crop rotation, but fixed costs were 27% lower per acre which resulted in a much higher net farm income. Bullen (1964) reviewed the problems of intensive cereal production and concluded that almost continuous cropping of spring barley was possible in many areas provided that due regard was paid to crop nutrition, weed-control and husbandry standards. This would seem reasonable in the light of the farming evidence and from the extensive yield data from Rothamsted.

Results from a twelve year continuous barley cropping experiment reported by Mundy (1969) suggested that yields can be maintained by the application of 75–100 units of nitrogen and although pests and diseases are not serious, perennial grass weeds appeared the greatest hazard to a continuous cereal farming system.

X. MANURING OF FEEDING BARLEY

With ever-rising production costs it is essential for farmers to aim for maximum yield. This will result in a high gross output per acre and is nearly always associated with the best profit margins. There is no conflict between yield and quality where feeding barleys are concerned since a high protein content in the grain is desirable and reasonable quantities of fertiliser nitrogen will be required to obtain good yields. In Table 2 a number of examples of the manurial requirements of winter and spring barley grown for stock feed have been given and unless farmers have local evidence to the contrary it would be unwise to greatly exceed these recommended rates.

TABLE 2

Fertiliser Levels for Feeding Barley

	Units of fertilisers per acre[a]		
	N	P_2O_5	K_2O
Winter sown			
(a) Following well manured root and potato crops, and ploughed-up grass-clover leys			
At seeding	—	30	30
Spring top dressing	30	—	—
(b) In low rainfall area with normal crop rotation			
At seeding	15	40	40
Spring top dressing	35–55	—	—
(c) In high rainfall area with normal crop rotation			
At seeding	15	40	40
Spring top dressing	25–45	—	—
(d) Following a spring cereal, in a mainly cereal sequence			
At seeding	25	50	50
Spring top dressing	30–55	—	—
Spring sown			
(a) Conditions of high fertility, or grown in areas of very high rainfall	0–30	0–30	0–30
(b) In low rainfall area with normal crop rotation	50	40	40
(c) In high rainfall area with normal crop rotation	40	40	40
(d) Undersown crops (grass and clover)	20–30	40–60	40–60
(e) Continuous cereals, mostly spring sown barley, with ploughing and normal cultivations	60–80	40–50	40–50
(f) Continuous cereals and "chemical" ploughing in low rainfall areas	80–100	40–50	40–50

[a] 1 Unit = 1·12 lb N, P_2O_5 or K_2O.
1 Unit is therefore equivalent to 1% or analysis/cwt.

Where possible, the short, stiffer-strawed varieties should be chosen in order to exploit grain yield to the full by application of inorganic fertiliser. If the longer strawed varieties are grown (where straw is considered an important part of the crop) then the levels of fertiliser N suggested in Table 2 should be reduced by at least 10–20 units in most cases.

Winter barleys require a little nitrogen in the seed bed when the contribution from the soil is negligible or very small and the phosphate and potash must always be applied before or at seeding to obtain a full and vigorous braird. The majority of the nitrogen should be applied as a top dressing in the spring and in respect of the South and East this should go on in March or early April. In other areas it is usual to apply the top dressing in mid to late April in order to avoid excessive straw production which is sometimes the case when nitrogen is applied early. In respect of the spring sown crop, under most circumstances all the fertiliser should be applied before or at the time of sowing. The growing season is so short that all the nutrients need to be close at hand immediately following germination so that growth can go ahead unimpeded. However, where large quantities of fertiliser N are required (e.g. (e) and (f) in Table 2) it may be necessary to split the nitrogen dressing, half going on with the seed and half to be applied as an early top dressing.

Where farmers are in doubt about the capabilities of their land (which may just have been acquired) then it will often pay to apply the average levels of phosphate and potash to the barley and half of the nitrogen. Once the crop is well established it will be evident if more nitrogen is required and the deficiency can be rectified by a timely top dressing. Spring barley crops are frequently undersown with grass-clover seed mixtures and here a reduction in the quantity of nitrogen fertiliser should be considered. In order to obtain a satisfactory "take" or "establishment" a little of the potential grain yield should be sacrificed by withholding some nitrogen so that the cereal crop has the best chance of standing. Undersown cereals which lodge badly are the most difficult to harvest on account of the large volume of green material which grows through them. Early lodging will also mean a poor establishment of the grass and clover seedlings usually through greatly reduced light penetration. Besides reducing the nitrogen it is often beneficial to apply larger quantities of phosphate and potash which also helps the establishment of the herbage species, particularly the clover. It would be wise to adopt this modification of a standard manurial practice particularly when phosphate and potash soil reserves are low.

XI. MANURING OF MALTING BARLEY

Samples of barley which are suitable for malting will command a quality premium in the market and from past experience this has been in the region of an extra £2–4 per ton compared with feeding barley prices. With average yields for barley presently around 27–30 cwt/acre this will mean an additional revenue of £3–6 per acre should malting quality be obtained without a corresponding reduction in yield of grain. However, in order to satisfy the requirements of the maltster and in particular the low % N in the grain, the amount of fertiliser nitrogen which can safely be recommended will have to be reduced in order to achieve this "quality" product. Instead of using 60 units of nitrogen and over, the maximum dressing will have to be in the region of 30–40 units with similar levels of phosphate and potash. The fertiliser nitrogen applied must be responsible for a significant increase in grain yield and then its effect on the amount of grain protein will be minimal.

The effects of increasing levels of seed-bed nitrogen on the yield

TABLE 3

The Effect of Fertiliser Nitrogen on the Quality of Spring Barley

Treatment	% N in the dry matter of the grain				
	1950[a]	1951[b]	1952	1953	Average
Control (no nitrogen)	1·49	1·43	1·57	1·44	1·48
20 units N/acre	1·57	1·36	1·55	1·41	1·47
40 units N/acre	1·66	1·33	1·58	1·48	1·51
60 units N/acre	1·81	1·41	1·67	1·57	1·62
Average	1·63	1·38	1·59	1·48	—

[a] 1950 results are means from the two varieties, Spratt-Archer and Kenia.
[b] 1951–53 results are means from the three varieties, Spratt-Archer, Kenia and Proctor.

and quality of spring barley, grown on light land, were reported by Jones (1956). Average yearly results relating to grain quality are reproduced in Table 3 from the two or three varieties tested.

Mean figures have been calculated from the original data and it can be seen that yearly variations in average nitrogen content are slightly greater than those produced by the varying fertiliser levels. Season, therefore, plays an important part in producing quality grain and as long as fertiliser levels do not exceed 40 units N/acre, then on the lighter soils there is a good chance of obtaining malting quality. It is interesting to note that in three years out of four the first 20 units of nitrogen were responsible for a drop in the nitrogen content of the grain and 40 units of nitrogen only slightly increased the grain protein content compared with the control.

The time of application of a nitrogen top dressing to winter barley is important in this connection and the results obtained (at the Norfolk Agricultural Station) from a light soil in 1952, using the variety Pioneer are worthy of note and have been reproduced in Table 4.

TABLE 4

Nitrogen Top Dressings on Winter Barley

Treatment	Yield of grain (cwt/acre)	Nitrogen content of grain (%)
Control (No Nitrogen)	12·8	1·33
40 Units N early[a]	24·6	1·41
40 Units N mid[b]	24·2	1·49
40 Units N late[c]	18·4	i·66
13⅓ Units N early, mid and late	24·0	1·43
20 Units N early and mid	24·6	1·39
20 Units N early and late	21·2	1·44
20 Units N mid and late	23·0	1·55

[a] Early: 4th March.
[b] Mid: 2nd April.
[c] Late: 7th May.

The application of 40 units of nitrogen in March or April resulted in substantial increases in grain yield and although the nitrogen content of the grain increased, it did not rise above 1·5% and would therefore be acceptable for malting purposes. Late top dressings of nitrogen did not produce the same increase in grain yield and more went in elevating the grain protein as can be seen from the data. Thus in order to produce economic returns from malting barley, moderate levels of nitrogen should be applied early.

XII. CULTIVATIONS AND SEED-BED PREPARATIONS

A. Ploughing

For winter barley ploughing will normally take place in September or immediately following the harvest of the previous crop. On light soils it may even be left until October just prior to seed-bed working but wherever possible earlier turning of the soil on medium to heavy land is recommended. Where stubble from the previous crop is to be turned in together with weed-seedlings which have germinated as a result of stubble-cleaning or early autumn cleaning, then it may be necessary to plough to a depth of six inches. Barley is a relatively shallow-rooting cereal and without a great deal of material to be buried the depth of ploughing might well be reduced to four inches. Where close cereal cropping is practised it is often advisable to vary the

TABLE 5

Chemical Ploughing and Direct Drilling Compared with Ploughing and Cultivating for Spring Barley

Year and variety	Gramoxone W (pints/acre)	Seed-bed fertiliser (units N/acre)	Grain yields (cwt/acre)		
			Direct drilled after Gramoxone W spray		Ploughed, cultivated and drilled
			Early	Late	
1963	12	0	19·8	21·6	32·1
Rika		50	36·5	34·2	43·1
		75	39·3	40·8	42·9
		100	39·3	44·2	40·4
			26th April	27th April	27th April
1964	8	0	36·5	32·1	40·7
Rika		50	47·0	40·9	42·5
		75	49·2	40·0	41·3
		100	48·5	39·0	37·7
			14th Feb.	15th April	15th April
1965	6	0	27·7	28·6	46·9
Impala		50	41·7	47·6	52·5
		75	45·3	50·0	52·4
		100	47·0	52·4	50·7
			10th Feb.	31st March	31st March
1966	4	0	23·5	23·0	33·7
Impala		50	36·8	38·2	45·3
		75	42·2	41·9	47·4
		100	46·6	45·8	49·7
			8th March	30th March	30th March

Source of data: Guide to Experiments 1967: Jealott's Hill Research Station and Jeater (1967).

ploughing depth to something below the normal once in three years in order to avoid the build-up of the condition known as a plough pan. This can also be avoided by an occasional sub-soiling when deemed necessary and on occasions may prove quicker in the long run than a deeper ploughing.

With regard to the spring sown crops then much of what has already been said in respect of winter barley applies equally well. Medium to heavy soils should be ploughed early before the turn of the year in order that the weather, especially the action of snow and frost, may play its part in producing a fine seed-bed. Early ploughing on the light soils may also be useful in easing the burden of spring work, however it is not essential and spring ploughing will often do just as well following a mild, open winter. In the distant past two and sometimes three ploughings were practised in order to produce the very best of seed-beds but as Rayns (1959) points out, there is no experimental evidence to support the view that the second and third ploughings produced higher yields. The reverse tends to be the case today when any cultivation other than the barest essential is closely scrutinised in order to reduce production costs and, as with wheat, there have been attempts at growing barley following barley without the traditional ploughing and subsequent cultivations.

The results obtained at Jealott's Hill Experimental Station over the four-year period 1963–1966, are reproduced in Table 5 and indicate that spring barley does not fare as well as winter wheat under the system of spraying and direct drilling. Where fertiliser N is withheld and comparing yields at the same sowing date, the traditional ploughing and cultivating produced approximately 12 cwt more grain per acre than the spraying and direct drilling technique. Yields with this newer method of growing can be maintained by applying large quantities of fertiliser N, in the region of 75–100 units/acre, and under some circumstances may be justifiable. However, it still looks as though the traditional fine seed-bed for barley will produce high yields at relatively low costs and will be preferred under normal circumstances. The initial braird and mature plants resulting from direct seeding can be seen in Figs 3 and 4 and although the establishment appears satisfactory the crop of spring barley is not as good as the direct-drilled winter wheat in Fig. 5, Chapter 1.

B. Final Seed-bed Preparations

Assuming the conventional approach to seed-beds by ploughing and cultivating, spring barley will respond to a good tilth at sowing. After ploughing to 4–6 inches, the land should be left as flat as possible and this could mean the "ploughing-back" of the furrows to eliminate these depressions. Cultivation or heavy harrowing will follow to the lowest depths required and where perennial or stoloniferous weeds are present special equipment may be brought into use to remove them at this stage. Medium to light harrowing just prior to drilling will normally

Fig. 3. Emergence of barley on plots sprayed with Gramoxone W a fortnight before drilling. Photograph taken on the 8th April, 1964, at Frensham, Surrey. Reproduced by courtesy of Plant Protection Ltd.

Fig. 4. Barley drilled directly into stubble which had been sprayed with Gramo-xone W. Photograph taken on 15th July, 1964, near Winchester. Reproduced by courtesy of Plant Protection Ltd.

be required to produce a fine seed-bed in the top 2 inches of the soil. On the heavier soils rolling may be necessary to break any clods which remain, but this is not always successful and farmers should satisfy themselves that the implement is doing what is required otherwise it may simply add to the production cost without producing significant improvement. The object in seed-bed preparation is to produce the right degree of tilth with the minimum operations and soil moisture conservation in many areas must always be kept in mind. Seed-beds for barley should end up shallow, flat and even, containing adequate moisture for germination and where possible free from the difficult weeds which are not eliminated at a later stage by chemical means.

In respect of winter sowings the only difference lies in the fact that the soil surface is left in a rough condition in order to prevent surface capping and losses which could result from excess moisture and from the action of hard frosts.

XIII. DRILLING

The optimum depth of drilling is in the region of 1–1·5 inches. Placing the seed much deeper can result in a lower field germination, an irregular emergence and plants which come through the ground in a feeble condition having used up most of the nutrients from the seed before they reach the light. Drilling closer to the surface than the optimum can lead to poor erratic brairding since moisture levels are

	Early	Mid	Late
Winter Barley	End September–early October	Mid-October–November	December onwards
Spring Barley	February	March–beginning April	Late April–May

usually very low in this region and full germination may be held up until rain falls. This very shallow drilling more often leads to higher incidences and degree of pest attack. Once the larger birds have located a coulter row they seldom leave until several yards have been removed.

The time of sowing is an important consideration since it can have a marked effect on grain yield and in general terms the earlier the drilling date once the land is prepared and the conditions are right, the greater are the chances of reaching peak production. According to Bullen (1959), late sown crops tend to have a higher N content and thus where the quality market is the aim early drilling is doubly important. For those not familiar with cereal sowing the small table indicating early, mid-season and late sowing times may prove useful.

XIV. SEED RATES

In respect of winter barley 1–1·25, 1·5–1·75 and 2 cwt/acre are required for early, mid and late sowings respectively to obtain a satisfactory plant population.

Recommended seed rates for spring barley in the past have been of the order of 2 cwt/acre but it must be remembered that many of the older varieties had large grains which automatically gave rise to high seeding weights. Most of the new cultivars have slightly smaller grains and at the same time exhibit a much greater tillering capacity and thus the seed rates are much lower in present-day farming. Boyd (1952)

TABLE 6

Effect of Seed Rate on Grain Size and N Content with Spring Barley

Seed rate cwt/acre	1000 corn weight (g)	% N in grain dry matter
1	40·6	1·71
1·5	39·8	1·69
2	38·3	1·66
2·5	37·8	1·65

After Jackson and Page (1957)

in summarising experimental results from seed rate investigations indicated that 2 bushels/acre (1 cwt) was optimal for barley and Jackson and Page (1957), working in the East Midlands of England, could not produce evidence to support increasing spring barley seed rates above 1 cwt/acre. Two aspects of grain quality, namely size and nitrogen content, are influenced by seed rate and the trends can be judged from the figures in Table 6.

Under most circumstances it would appear that 1 cwt of seed is adequate; however, if sowing takes place particularly early in the spring when ground temperatures are still low, or there is evidence or a history of pest or disease incidence, then farmers would be wise to err on the safe side and put on 1·25 cwt/acre.

XV. SEED DRESSINGS AND SEED TREATMENT

As with wheat, most barley seed purchased through the trade is dressed with proprietary compounds for the control of seed-borne fungal diseases and in respect of barley these include Covered Smut, Leaf Stripe and Net Blotch. Organo-mercury compounds are applied at 2–2·5 ounces/bushel of seed and if this has not been carried out by the merchants or should farmers be using their own grain as seed then it can be done on the farm. Where liquid seed dressings are preferred it would be advisable for farmers to have their seed treated at authorised seed dressing premises since many of these chemicals are not on sale to the general public. Organo-mercury dressings incorporating the insecticide gamma-BHC should be chosen when there are likely to be problems with soil-inhabiting insect pests.

Loose Smut (Fig. 7, Chapter 1) is particularly important in barley. It is an embryo infection which is initiated at anthesis (flowering) and the control measures involve hot water or chemical treatment (Murganic RPB) of valuable seed stocks. Cereal mildew is also a serious fungal disease of barley most years and this can now be controlled by the new chemical seed dressing, Milstem.

A. Type of Seed

A similar range of seed quality as was described for wheat can be obtained for barley and in general terms it should be well filled, not too closely threshed, free from weeds and approaching 100% germinative capacity. For crops grown for commercial purposes, the cheapest seed which is adequate in all respects should be chosen, whilst producers of seed through the various official schemes, should choose seed of high quality.

XVI. VARIETIES/CULTIVARS

A. For Winter Sowing

From the discussions earlier on grain quality and uses it was seen that barley goes either for industrial purposes or for feeding and varieties have been produced with these outlets in view. Recommended lists are published annually by the various testing organisations in Britain and details from these appear in the Appendix to this chapter. Senta is an early maturing six-row barley representing the feeding group and Maris Otter is a good quality malting barley which should be chosen where the industrial outlet is the aim. These cultivars are considered to be the best at the time of writing but farmers must consult the most up-to-date recommendations if they are to take advantage of the highest yield potential through varietal choice. Vernalisation requirements can vary considerably with winter barleys and this agronomic character expressed through "winter hardiness" and "suitability for late sowings—early February" receives adequate attention on the recommended list for England and Wales.

In recent years the fungus disease Rhynchosporium (illustrated in Fig. 5) has become serious on barley grown in the South-west of

Fig. 5. Leaf blotch (known also as Rhynchosporium) of barley caused by the fungus *Rhynchosporium secalis*. N.I.A.B. photograph.

England and in this area varietal choice should be restricted to those which offer a reasonable degree of field resistance.

In the past a number of farmers in the South of Britain have "chanced" the winter sowing of spring varieties and on many occasions they have obtained satisfactory crops. This was a reflection on the available winter varieties which at that time were poor by the standards set by the spring ones. Today there are several useful winter varieties which could satisfy either the feeding or malting trade and this practice of gambling on a spring variety surviving the winter need not be continued.

The Northern Ireland Ministry of Agriculture use the variety recommendations produced by the National Institute of Agricultural Botany but in Scotland winter barley, although grown a little in the East and the South-west, does not enjoy official support from the three Colleges of Agriculture. Several varieties have been tried by the North College and to date none have proved completely satisfactory. Trials in the West have shown that the three cultivars Pioneer, Dea and Maris Otter can yield well if they survive the winter but with a history of one crop in four a complete failure and a further one in four a partial failure their general recommendation would be foolish. In those areas of the West of Scotland which enjoy significantly better winter weather due to the Gulf Stream, and if the soil is light and the aspect good then winter barley crops can be successful.

B. Varieties for Spring Sowing

With such a large amount of our arable land under barley, most of which is spring sown, there is a tremendous interest in and need for several good varieties. This is reflected in some of the recommended lists where there appears a much wider range or choice of cultivars compared with other cereals. Full details of these appear in the Appendix, where field characteristics, resistance to disease and comparative yields may be studied. Table 7 simply lists the current recommended varieties for all the regions for readers to obtain an overall picture and to see which are common to each area. The variety Zephyr features on all three lists and at the time of writing is probably the most widely grown variety suitable for malting. Other malting varieties for England and Wales include Imber and Gerkra and for feeding Midas, Vada and Deba Abed have very useful characteristics.

The 1962 cereal census in Northern Ireland indicated that 80% of the barley acreage was taken up by the four varieties Rika, Proctor, Pallas and Ymer but only Pallas still features on Leaflet 96—Varieties of Cereals.

In Scotland Golden Promise, Zephyr and Ymer are the three most important varieties and it is interesting to note that Ymer has been widely grown for nearly twenty years and can also be found in cultivation in North-east England. Two reasons are suggested for its continued success. The first lies in the fact that it was accepted for industrial use and secondly, it has seldom given rise to complaint in the field as have

TABLE 7

Recommended Varieties of Spring Barley

England and Wales	Scotland	Northern Ireland
Midas	Midas	Pallas
Sultan	Sultan	Bamba
Julia	Zephyr	Mentor
Zephyr	Ymer	Zephyr
Imber	Golden Promise	Ruby
Gerkra		Union
Inis		
Vada		
Impala		
Deba Abed		
Proctor		
Berac		
Hassan		
Lofa Abed		

so many of the newer varieties. In many areas of Scotland there has been a swing back to Ymer from some of the more recent introductions because of its greater dependability.

In future years naked barleys will become important. Some are being tested at the moment and appear promising since without a husk they have a higher starch and protein equivalent than present normal varieties.

XVII. MANAGEMENT OF THE ESTABLISHED CROP
A. Spring Management of Winter Sown Crops

In the review article by Holliday (1956), "Spring Grazing of Winter Sown Cereals", wheat is said to be more resistant to the adverse effects of defoliation than barley and the other two cereals, probably on account of the smaller amount of grazable herbage produced. Although there is little experimental data referring to barley, the practice of spring grazing under the right conditions can be beneficial when the vegetative growth is excessive. However, the damage caused by poaching, together with the reduction in grain yield which nearly always follows, markedly reduce the occasion on which it is both possible and desirable.

With much of the crop associated with light land then light harrowing followed by rolling can be beneficial in reducing weed-seedlings and in soil consolidation which is so important for root hold and moisture conservation and utilisation. Where chemical weed control has become a standard practice then harrowing may be omitted.

B.   Undersowing Barley with Grass and Clover Seeds

When winter sown barley crops have to be undersown with "seeds" then this should be done in the early Spring before the cereal crop has produced a full ground cover. On occasions a light harrowing may be necessary beforehand and the grass and clover seeds can either be drilled or broadcast over the crop. Rolling should follow broadcasting to bury the small seeds just below the surface and this operation will also help to conserve and bring up moisture from lower regions.

Undersowing of spring sown crops can be done either at the time of sowing or immediately they are established and in the wetter areas the latter is often preferred. In spite of spring barley having the shortest growing season of all the cereals, when rainfall is high the grass and clover can grow rapidly and often attain the same height as the cereal, thus making combine harvesting a difficult and slow operation.

C.   Control of Insect Pests

When barley or any other cereal follows the ploughing up of grassland then insect pests may present problems. Wireworms and particularly leatherjackets can be troublesome and where control by DDT is contemplated, it should be remembered that several varieties of barley are susceptible to injury by this insecticide.

Information concerning the reaction of barley varieties to DDT can be obtained from the N.I.A.B. or through the local Agricultural Advisory Services. Aldrin may be used to control leatherjackets on spring and winter barley crops where the varieties are DDT-sensitive.

XVIII.   CHEMICAL WEED CONTROL

Chemical weed control is invariably a spring operation on most barley crops and as in the case of wheat, a considerable range of suitable materials exists. This is constantly being reviewed and it is therefore important that farmers consult the most up-to-date information on this important topic. The successful choice and timely application of the appropriate chemical will lead to clean, high yielding crops which are

TABLE 8

Weed-killers that may be Used on Barley in Spring[a, b]

	Winter Barley	Spring Barley	
	Crop fully tillered but before the "jointing" stage	3–4 leaves	5 leaves to "jointing"
I Weed-killers for the control of broad-leaved weeds			
(a) Cereals not undersown with clover			
MCPA	√	√[c]	√
2,4-D	√	NO	√
mecoprop	√	√	√
mecoprop + fenoprop	√	NO	√
mecoprop + 2,4-D	√	NO	√
dichlorprop	√	√	√
dichlorprop + 2,4-D	√	NO	√
dichlorprop + MCPA	√	√[d]	√
dicamba + mecoprop with or without MCPA	√	NO	√
2,3,6-TBA + MCPA	√	√[d]	√
2,3,6-TBA + dicamba + mecoprop + MCPA	√	NO	√
dicamba + benazolin + MCPA	√	NO	√
ioxynil + mecoprop	√	√	√
ioxynil + linuron	NO	√	√
ioxynil + dichlorprop + MCPA	√	√[d]	√
bromoxynil + ioxynil + dichlorprop	—	√	√
bromoxynil + MCPA	√	√	√
(b) Cereals undersown with clover			
MCPB + a little MCPA[e]	√	√	√
2,4-DB + a little MCPA[e]	√	√	√
2,4-DB + a little 2,4-DB or MCPB and MCPA[e]	√	NO	√
benazolin + 2,4-DB or MCPB and a little MCPA[e]	√	√	√
dinoseb[f]	√	√	√

II Weed-killers for the control of wild oats and/or black grass in the spring.
   Tri-allate.
      May be used in spring barley or spring wheat, either before or just after drilling. Thorough incorporation by cultivation immediately after application is essential.
   Barban.
      May be applied to spring wheat and most varieties of spring barley (but not Proctor, Impala and some others) when the wild oats or blackgrass has 1–2 leaves per plant.
   Metoxuron[g]
      May be used on some varieties of winter wheat and winter barley when the crop has at least 3 leaves.

Notes on opposite page

easily harvested and with little time and labour needed in grain dressing and cleaning. Table 8, based on the Ministry of Agriculture's Short Term Leaflet No. 19 (1970), indicates the range of herbicides presently available and recommended.

Special emphasis has been placed on the stage of growth of the cereal crop, since it can often exclude a number of otherwise suitable weed-killers particularly early in the life of a crop. When farmers are not familiar with herbicides then prior consultation with the local agri-cultural adviser is always worthwhile. With such a wide range of materials available the principle involved is simply to choose the cheapest which will be effective in its task—easily said but not always easily achieved. To help in this connection, the weed spectrum found on most farms is listed in Table 9 and the "best buy" as far as a herbi-cide is concerned is indicated for straight sown barley crop and those which have been undersown.

XIX. HARVEST
A. Feeding Barley

Once the barley crop starts to ripen there is a constant translocation of plant foods from the vegetative parts to the seed. The assimilation or build up of carbohydrates goes on before and after ear emergence in nearly all the parts of the plant and an interesting piece of research

% Contribution to grain dry matter from the various parts of barley plant

26% Ears	$\begin{cases} 10\% \text{ Awns} \\ 16\% \text{ Glumes, etc.} \end{cases}$
59% Flag leaf, sheath and peduncle	
15% Shoot below the flag leaf	

After Watson *et al.* (1958)

a For trade names, see the list M.A.F.F. of Approved Products.
b For the correct dose, read the manufacturer's instructions.
c At half normal dose. Some deformity of the ear may follow application at this stage of growth; do not use on malting barley or seed crops.
d Some deformity of the ear should be expected; do not use on malting barley or seed crops. *Follow the manufacturer's instructions.*
e Clover should have its first trifoliate leaf before spraying.
f Clover should have two trifoliate leaves before spraying.
g Some broad-leaved weeds also controlled.
√ Indicates suitability.   NO = Not suitable.
This extract from M.A.F.F. Short Term Leaflet No. 19, 1970 is reproduced by kind permission of the Ministry of Agriculture, Fisheries and Food.

TABLE 9

Some Common Farm Weeds Listed Under the Appropriate Herbicide for their Control

1. *Barley crops not undersown*

MCPA or 2,4-D	Mecoprop	Dichlorprop	Mecoprop + Ioxynil	Bromoxynil + MCPA
Perennial Bindweed	Chickweed	Redshank	Corn Marigold[a]	Black Bindweed
Corn Buttercup	Cleavers	Spurrey	Mayweed	Groundsel
Creeping Buttercup	Cranesbill			Knotgrass
Charlock	Fumitory			
Docks				
Fat Hen				
Hempnettle				
Annual Nettle				
Perennial Nettle				
Pennycress				
Poppy				
Runch				
Scarlet Pimpernel				
Shepherd's Purse				
Annual Sowthistle				
Perennial Sowthistle				
Creeping Thistle				

[a] At the seedling stage.

2. *Barley crops undersown with grass and clover*

MCPB + a little MCPA	2,4-DB + a little MCPA or 2,4-D	Benazolin + 2,4-DB + a little MCPA	Dinoseb
Charlock	Black Bindweed	Chickweed	Cleavers
Corn Buttercup	Creeping Butter-		Corn Marigold
Docks	cup		Cranesbill
Fat Hen	Docks		Hempnettle
Fumitory	Fat Hen		Knotgrass
Pennycress	Groundsel		Mayweed
Perennial Bindweed	Annual Nettle		Redshank
Poppy	Perennial Nettle		Shepherd's ⎧ Dinoseb is also
Runch			Needle ⎪ recommended
Scarlet Pimpernel			Speedwell ⎬ to control
Shepherd's Purse			⎪ these weeds in
Annual Sowthistle			⎪ barley not
Perennial Sowthistle			⎩ undersown
Creeping Thistle			Spurrey

by Watson *et al.* (1958) showed the relative importance of each in the production of dry matter in barley grain throughout the season.

When the movement of plant foods to the ear ceases, the crop is said to be ripe. The leaves, stem and ears will have changed to a pale yellow colour and the ear bends over or "necks" until the awns point to the ground. In the absence of heavy dews and rain the moisture content of the grain will drop as ripeness approaches and farmers will often use a moisture meter around harvest time to ascertain the exact figures. With feeding barley which is to be stored dry, it is worthwhile allowing the grain to dry off as much as possible in the field to minimise the expense of artificial drying. In recent years silos have been introduced for the storage of wet grain when the crop is to be fed to stock on or near the farm where it has been grown. The principle involved is the speedy production and maintenance of a carbon dioxide atmosphere surrounding the grain so that it does not deteriorate through mould growths. Further details and a discussion of this method appear later under the section on grain storage and for this method of "moist storage" or when drying is contemplated, the barley crop should be combined as fast as possible once it is ripe.

B. Malting Barley

As far as malting barley is concerned a period of maturation follows ripening, before it is fit for industrial use. In the past when much of the crop was cut by binder, stooked, stacked and then threshed the grain slowly matured or "mellowed" into a good sample over a long period. Most of the barley is combine-harvested today and thus the maturation period is of necessity much condensed. According to Rayns (1959) the changes from a steely to starchy sample (mellowing) can be aided in the field by intermittent periods of sunshine, dew and light rain and this author recognises the fact that good maturation is difficult to obtain when the accent is on speed and labour efficiency. In the past it has often been said that a crop for malting should be left for 10–14 days after it is considered to be ripe. However, since there is a real danger of losing much of the crop should it become very windy and loss of condition or quality can follow heavy rainstorms, the maturity period in the field has been greatly reduced but it is still important. The second phase in the process of maturation takes place after harvest whilst the grain is in store and this is simply getting over the period of dormancy when germination is inhibited or restricted. In combine harvesting or the threshing of bindered crops it is important to see that correct drum speeds and concave settings are employed so that the grain is not threshed too closely. Samples of grain showing a little awn are much preferred to those which have been tightly threshed giving a clipped appearance.

XX. DRYING AND STORAGE

Following a hot, dry summer barley may come off the combine relatively dry and can be stored in small quantities for considerable periods without the aid of artificial drying. Maximum moisture contents for safe storage at ambient temperatures up to but not exceeding

TABLE 10

Maximum Grain Moisture % for Safe Storage (65°F, 18°C)

Storage period	Bulk storage (without turning or cooling)	Grain in open sacks with free access to air Grain in bulk which is cooled or turned
Up to 4 weeks	17	18
Up to 20 weeks	15	17
Up to 28 weeks	15	16
Longer than 28 weeks	14	15

From Farm Facts: *Farmer and Stockbreeder*, 3rd October, 1967.

65°F (18°C) are given in Table 10 and are related to method of storage or the amount of grain.

When moisture contents are higher than those indicated in Table 10 it will be necessary to consider artificial drying, cooling, sealed-storage or acid treatment of the grain in order that it may be kept in good condition until used or sold.

A. Drying

The various methods of grain drying have been briefly reviewed in the chapter on wheat and what has already been said there applies equally well in principle to barley. Malting samples of grain, however, should be dried at fairly low temperatures so that the nature of the grain and its germinative capacity are not impaired. In the past maltsters were very suspicious of farm dried samples in case the drying temperature used had been excessive. Where good quality barley has been harvested these should never exceed 120°F (49°C) and when moisture contents are between 25 and 30% then the maximum would be 110°F (43°C) and only 90–100°F (32–38°C) when exceptionally wet.

ON-THE-FLOOR DRYING

With such a large acreage of barley grown for feeding on the farms where it is produced, much attention has been paid in recent years to simple and cheap systems of "on-the-floor" drying. These can be recommended when grain moisture does not normally exceed 20–22% or when used in conjunction with conventional drying systems which are working at full capacity. Grain is placed on flat concrete floors to depths ranging from 3 to 8 feet (increasing depth beyond 8 feet often produces problems relating to air flow and building reinforcement and can seldom be recommended) and the building will need to be 12–14 feet high to facilitate proper working of elevating and conveying equipment. Hot air conveyed through a main duct and then through laterals placed every 3–4·5 feet on the floor, is forced through the depth of grain and sufficient ventilator space is required (2 sq. ft/1000 cu. ft of air/min) through which the moisture laden air can escape. Further details of this method of drying and in particular the fan capacities required for various quantities of grain, can be obtained from M.A.F.F. Short Term Leaflet No. 22 "Drying Grain On The Floor" and from articles by Lovelidge (1965), Greig and Boyce (1966) and Finn-Kelcey (1966) which appeared in *Farm Mechanisation Journal*.

The various other systems of grain drying outlined in the chapter on wheat are equally useful for the barley crop and readers should consult these when the installation of a drying plant is contemplated. However, there are other storage methods which are particularly suited to fodder grain and it is appropriate at this stage to consider them in some detail.

B. Grain Chilling and the Storage of Cool Grain in Bulk

When barley is harvested at moisture contents of 17% and above it may be kept in good condition without drying if the temperature of the bulk grain is maintained at a reasonably low level by blowing chilled air through it. Maximum temperatures for storage are related to moisture content, storage period and whether it is important to preserve the germinative capacity of the grain and approximate figures are given in Table 11 as a guide.

It is not economic to store grain which has over 20% moisture for long periods by chilling and for short term storage the moisture content should not exceed 22%. Rapid temperature drop (on the quick-freeze principle) is neither desirable nor economic and the reduction of temperature in large bulks of grain can be done over periods of 3–7 days. This is obtained by connecting chiller-units to sealed storage bins which have been insulated against direct solar radiation or the chilled air may be passed through heaped grain which is stored on the floor in the building. Once ventilation has ceased grain temperatures will rise slowly and thus a series of short periods of chilling will be needed at intervals of say 2–3 weeks in order to maintain satisfactory conditions.

TABLE 11

Maximum Temperatures for the Storage of Chilled Grain

% Moisture content of grain	Short term (up to 3 months) No deterioration		Long term (3–8 months) No deterioration		Free from moulds but germination impaired	
	°F	°C	°F	°C	°F	°C
17	55	13	51	10	55	13
18	52	11	48	9	52	11
19	48	9	45	7	48	9
20	45	7	42	$5\frac{1}{2}$	45	7
21	43	6	—		—	
22	40	4	—		—	

When the chiller units are attached to enclosed bins then a greater degree of efficiency can be obtained by re-circulating the exhaust air back through the chiller units, especially when it is at a lower temperature than the ambient air for considerable periods during storage.

The approximate chilling capacities of the equipment presently available range from 10 to 300 tons per 24 hours, and during a season these would cope with 200 to 6000 tons respectively. Further details concerning these chiller units together with the prices can be obtained either from *Farm Mechanisation* (May 1966) or the appropriate manufacturer. At first sight they appear to be costly items of capital expenditure to be added to costs of buildings and/or storage containers. However, it has been shown that less energy and air are required to cool grain, compared with drying it, and therefore the running costs of chilling systems would be lower.

The total requirements are given in respect of drying grain, whereas with chilling, further refrigerations at 2–3 week intervals will be necessary and so the saving in cost is not as large as one first imagines. The decision to buy and install chilling equipment cannot be taken

Requirement	To dry 1 ton grain (21% moisture) to 15% moisture	To chill 1 ton grain (21% moisture) from 77°F (25°C) to 41°F (5°C)
Energy	130,000 B.T.U.'s	40,000 B.T.U'S
Air	1·3 million cubic feet	60,000 cubic feet

After Burrell (1964).

quickly or without ample consideration and in order to facilitate the correct answer, the merits and disadvantages of grain chilling have been listed below in Table 12.

TABLE 12

Merits and Disadvantages with Grain Chilling Systems

Merits	Disadvantages
1. Capital costs lower than comparable continuous drying equipment.	1. Capital costs are higher than comparable in-bin drying units.
2. Running costs cheaper than with dried grain systems.	2. Although grain quality can be preserved by chilling, the trade is apprehensive about accepting it, since immediate use is imperative.
3. Maintenance of feeding value, milling or malting qualities possible.	3. As a result of No. 2 when the grain is to be sold it normally means drying and thus increased costs are incurred.
4. Rolling or bruising for stockfeed is easy and the grain does not smell "fermented" as with sealed-tower moist storage.	4. Insulated bins are required otherwise grain will heat up fairly quickly at the periphery.
5. Chiller units may be useful for other crop produce (e.g. vegetables, potatoes and fruit).	5. "Refrigeration" carried out at night will ensure maximum chilling and lowest cost since incoming air is then at its lowest temperature, but a close watch must be kept on air humidity.
6. Insect pests are not troublesome.	6. Constant vigil is required in respect of rise in temperature of the grain. Thermostats and automatic chilling are usually necessary.
7. Chiller units can be used to hold grain for a short period until conventional drying systems are unloaded.	7. Cannot be recommended for areas of high humidity.

C. Bulk Grain Aeration

Working at the Pest Infestation Laboratory and under commercial conditions Burges and Burrell (1964) studied the properties of bulk grain and the heat production and activity of granivorous insects in order to make recommendations regarding storage and aeration of grain in quantity. Much of their data relates to barley and more specifically to malting samples and they indicate successful conditions for the storage of a 6000 tons bulk, namely aeration from 56 to 94 hours at 0·16 cu. ft/bushel. These authors concluded that cooling by in-store aeration was reliable and labour saving and would eliminate the troubles which could arise in grain stores due to the activities of granivorous insects. Some economies in the amount of drying or the cooling of grain in driers are thought possible when aeration is implemented, but it can only be practised successfully in the autumn when the relative humidity of the air is below 75% and in the winter below 80%. On this basis and as a result of average weather conditions, the British Isles can be divided into three categories, good, satisfactory or doubtful for aeration.

Good	S. England, E. England, Central England
Satisfactory	N. England, Parts of S.W. England, Wales, Central Scotland, Most of Ireland
Doubtful	Extreme S.W. England, Southern Scotland, Extreme N. and W. Scotland and W. Ireland

The areas listed as doubtful for bulk grain aeration coincide with those unsuitable for grain chilling referred to on p. 99 under the heading disadvantages.

D. Storage of High Moisture Grain in Sealed Silos and Containers

In recent years sealed silos (Fig. 6) and flexible synthetic containers (Butyl rubber) supported by weld-mesh frames have been introduced for the storage of high moisture feeding grains. Surveys carried out by the N.A.A.S. and published in their Technical Report No. 16, showed that it was possible to successfully store grain in sealed silos at moisture contents from 15–26% although a maximum moisture content of 24% is recommended to eliminate some of the problems associated with the mechanical emptying. The essential feature is to keep the system air-tight when filling, during storage and also whilst emptying so that the grain does not lose its condition and feed value through moulds. In relation to the sealed silos, although filling can be done over several weeks, the first load should be large and proper sealing is essential between fillings. When the silos are air-tight, the

FIG. 6. Photograph of sealed silos for moist grain storage and grass conservation. Photograph by courtesy of Boythorpe Crop Stores.

oxygen in the air surrounding the grain is quickly used up and the atmosphere will contain 80% or more carbon dioxide, thus inhibiting the aerobic micro-organisms. Slight fermentation of the grain occurs giving it a characteristic "fruity" odour, not unlike that associated with some silages, but there appears no significant loss in feeding value and it is palatable to stock. Animal intake appears higher with moist grain, compared with similar dried material, but this is more apparent than real since the dry matter percentage will be significantly lower.

Grain is normally unloaded from the sealed silos from the bottom using motor powered augers connected to rolling or bruising machinery by grain tubes or inclined augers, with a view to a completely mechanised feeding system.

Little critical information is available on the synthetic rubber containers, but in respect of the sealed silos many details have been obtained from the N.A.A.S. survey of 150 farms between 1964 and 1966 where these have been installed. Emptying difficulties were associated with a high proportion of them and possible causes are listed below for the benefit of those who are about to embark on this method of grain storage or who have to advise on its various aspects. These difficulties of grain removal are mainly associated with "bridging", a condition where the grains bind together to inhibit or seriously restrict the working of the extracting mechanism (usually sealed augers).

1. Bridging and therefore extraction difficulties increase with increasing grain moisture content. (% of farms affected by bridging is indicated in the survey below.)

	Over 26% moisture grain	21–26% moisture grain	Below 21% moisture grain
Slight bridging	$31\frac{1}{2}$%	29%	14%
Serious bridging	$31\frac{1}{2}$%	9%	7%

2. These difficulties are more likely to occur towards the end of the emptying period when there is a greater chance of air entry and when ambient temperatures are rising.

3. Earlier bridging is almost certainly associated with

(a) Loss of air-tight conditions, leading to moulds.

(b) Dirty grain with presence of large quantities of awn, straw or trash.

(c) Increased moisture brought about by condensation on the cooler, shaded side of silos.

(d) The presence of larger than normal quantities of myceliated yeasts, which can grow under circumstances of very low oxygen tension and have often been recorded in large quantities with very wet grain.

When "bridging" occurs or when there has been a breakdown of the unloading mechanism, the silo will have to be entered usually via the filling hatch and the trouble dealt with.

WARNING

The air surrounding the grain in properly sealed, moist grain silos is practically devoid of oxygen and human beings quickly become unconscious through lack of oxygen in this atmosphere and CAN DIE WITHIN MINUTES. When entering these silos, the person concerned should wear a safety belt attached via a life-line to a colleague outside, and fresh air should be blown in before they are allowed to enter. In the case of extreme difficulty and danger it would be wise to call in the local fire brigade officers who are able to enter freely with breathing apparatus. A further danger exists with bridged grain as it may collapse, burying the person who has gone in to inspect the problem.

A further word of warning is necessary in respect of a possible microbial hazard and this is often associated with the last grain out of the silo. During the last stages of emptying, the air contains dust and microbes (actinomyces and thermophilic fungi) which when inhaled by humans gives rise to the disease known as "farmer's lung" and in livestock a non-infectious sickness may result. It is therefore recommended that the grain be used up by the middle of March and that the last loads out should be removed and dried. When emptying the silo or associating with the grain, especially in confined circumstances, workers should be properly masked.

Having mentioned all the difficulties and possible hazards with moist grain storage the reader may have concluded that there are too many risks involved. It must however be strongly emphasised that the great majority of farmers who have opted to store their feeding barley in this manner are satisfied with it and although some difficulties may be experienced from time to time, these are never insuperable. The main considerations are economic since the installation of one or more of the costly sealed silos will of necessity incur further expenditure on unloading augers, conveying augers, control units and electric motors to power them. Comparison with other systems must be made through financial estimates and budgeting before decisions are reached but the main criteria will always be economic ones related to the goal of fully mechanised feeding with minimum labour requirement.

MYCOTOXICOSIS

It is known that cereal grains can harbour many fungi which are capable, under some circumstances, of producing mycotoxins, i.e. substances which are poisonous when ingested by cattle, pigs and poultry or for that matter by humans. Harrison (1967) has listed these and suggests that mycotoxicosis may well become an important consideration with some intensified methods of stock rearing and feeding. Mould growth is certainly associated with moist grain storage where air has effected an entry and *Penicillia* and *Aspergilli* have been isolated from grain and forage at the N.A.A.S. Centre, Shardlow. When in doubt about the safety of feeding affected grain, samples should be taken to the local advisory mycologists.

E.   Acid Treatment for the Storage of Moist Grain for Animal Feed

The most recently introduced storage method is the application of acid to grain immediately it is harvested. Propionic acid is recommended and a mixture of propionic and formic acid also appears to be satisfactory in preserving moist grain. Wheat, barley and oats treated in this way must all go for stockfeeding. Samples of these three cereals which have been acid treated will not germinate and thus the method cannot be used on seed or barley which is destined for malting. Although propionic acid is permitted by the Ministry of Agriculture, Fisheries and Food as a preservative in a limited number of foods, the amount needed in moist grain storage is higher than the permitted level. Since there is every danger of infringing the Preservatives in Foodstuffs Regulations, acid treatment for grain storage cannot be used on wheat, barley or oats which go for human consumption in any way.

Grain straight from the combine is passed through one of the specially designed applicators (see Figs 6 and 7, Chapter 3) which have been previously set to supply the correct quantity of acid. This will depend firstly on the moisture content of the cereal and how fast the augers or machinery can move the grain. In respect of the moisture content, B.P. Chemicals (U.K.) Ltd make the following recommendations in their 1968 Propcorn users' manual.

These figures are used to calculate the required flow rate of acid in gallons per hour with varying auger throughputs. The acid treatment is said to kill bacteria and fungus moulds on the surface of the grain and to inhibit the development of them during storage. From the limited evidence in 1967, reported in the Arable Farmer, April 1968 and from farm observations and experiments in 1968 and 1969, this newest method of grain storage appears to be quite satisfactory and can be claimed as another "breakthrough" in new techniques during the 1960's. It is important that the recommended levels of acid, determined by the grain moisture content, be distributed evenly through the grain and once treated it should be augered and not blown into the store otherwise some of the preservative may be lost causing moulds and over-heating. When in doubt thermometers or temperature measuring devices can be employed to check on bulk grain temperatures whilst in store. Grain properly treated should not show any significant rise over ambient temperatures and if this is the case, the whole process should be examined for errors in moisture estimating and acid application.

The special applicators for distributing the acid onto the grain will cost approximately £250–300. Contractors' prices presently range from £1·25 to £2/ton of grain treated and the positions within this range will be determined by the grain moisture content and thus the amount of acid needed. Individual farmers or members of a group owning their own applicator will be able to treat their grain at a lower cost than these contractors' prices and when one considers that no special

Acid Treatment Rate per Ton of Grain
(Trade name, Propcorn; chemical name, Propionic acid)

Grain moisture content (%)	By weight (%)	Gallons
15	0·50	1·12
16	0·56	1·26
17	0·62	1·39
18	0·68	1·53
19	0·74	1·66
20	0·80	1·80
21	0·84	1·88
22	0·88	1·97
23	0·92	2·06
24	0·96	2·15
25	1·00	2·24
26	1·05	2·36
27	1·10	2·47
28	1·15	2·58
29	1·20	2·69
30	1·25	2·81

Extract from BP Chemicals (UK) Ltd., Propcorn Users Manual, June, 1968.

silo or container is necessary, this method of preserving moist grain becomes economically attractive.

From the feeding trials and observations on commercial farms to date, it has been shown that animals find the acid-treated grain palatable and there have been no adverse reports from either a nutritional or health aspect.

XXI. DISPOSAL OF GRAIN, PRODUCTION ECONOMICS AND YIELD

A. Disposal of Grain

In the mid 1960's approximately 40% of the production was farm retained and the remainder sold for malting, feeding, seed or export. The bulk of the barley leaving farms will have been dried and is handled by corn merchants for the malting or feeding trades or seed-merchants for next year's sowing. Samples offered for sale at corn exchanges or markets should be representative of the bulk, which should be clean, dry grain free from diseases and extraneous material. Where grain is offered for malting, the sample will be examined by experienced buyers who usually assess whether it is flinty or starchy by taking cross-sections of the grain with hand cutters. Where laboratory facilities are available N-analyses may also be done by the trade to verify what can be seen. Should an analytical service be available to farmers it may well be worthwhile for them to pay for this (% N) determination to be done. They would then be placed in a better position for the bargaining which inevitably follows.

B. Government Support for Barley

Government support for this cereal, through the 1947 Agriculture Act, takes place in the form of a Deficiency Payment, made on an acreage basis, regardless of whether the crop is actually sold. As a result of the 1964 Annual Review and Determination of Guarantees, the deficiency payments are worked out each year in relation to a standard quantity and target indicator prices, in a manner similar to that for the wheat crop. In attempting to regulate the flow of barley from farm to market, the Government has varied the standard price, with a reduced figure at and immediately after harvest and an increasing one for sales later on during the winter. This is designed to deter farmers from flooding the market at harvest and to assist towards the cost of drying plants through higher prices for later sold grain. Forward contracts are available for barley which is to be sold through the scheme introduced by the Home Grown Cereals Authority and where significant quantities of grain are involved and forward planning possible, then additional revenue may be obtained. In order to claim and obtain these acreage deficiency payments, farmers must comply with the regulations and see that the necessary forms of application are sent in at the right time.

C. Production Economics

The following specimen production cost has been based mainly on the Economic Reports by Martin (1962), Black (1963) and Nixon (1966).

The specimen production cost based on the economic reports on barley growing in East Scotland, Yorkshire, Cornwall and Devon indicates a net cost per acre in the region of £22 and this is related to barley growing on a fairly large scale. It is therefore unlikely that much reduction could be achieved even with very large scale production of this crop and where only a few acres are involved the net cost could

Spring Barley Combine-Harvested (Specimen Production Cost Per Acre)

Item	Cost
Labour (10 man-hours at 32½p)	£ 3·25
Tractor (6 hours at 25p)	1·50
Machinery Depreciation and Repairs	5·00
Contract Services	0·50
Fuel (other than for tractor running)	2·40
Materials: Seed	0·45
Fertiliser	3·00
Sundries (Sprays)	0·90
Rent	3·50
Total Direct Costs	20·50
Adjustment for manurial residues	1·50
Share of General Farm Expenses (Overheads)	2·00
Gross Cost	24·00
Credit Value of Straw	2·00
Net Cost/Acre	22·00

easily go up by £4–6 per acre. Macpherson's data (1969) relating to the West of Scotland indicated a total cost of £29–30 per acre without adjustment for straw values.

D. Yields

Statistical accounts of yields do not differentiate between autumn or spring sowings or whether it is for malting or feeding and thus the average will comprise several facets of barley growing. The official Ministry figures for England and Wales, however, reveal a marked and steady improvement in the output per acre with nearly twice the yield being produced now compared with the ten-year period 1920–29.

Year/Period	10-year average 1920–29	10-year average 1955–64	5-year average 1961–65
Yield cwt/acre	15·6	25·8	27·6

Year/Period	1963–64	1964–65	1965–66	1966–67	1967–68	1968–69	1969–70
Yield cwt/acre	28·3	29·4	29·9	28·2[a]	29·9[a]	27·7[a]	28·9[a]

[a] Subject to review by the Ministry of Agriculture, Fisheries and Food.

With ever increasing yields, further gains become more difficult to achieve without some major technological breakthrough.

In the past malting barleys were nearly always associated with lower yields compared with varieties for feeding and up to the introduction of the variety Proctor (bred by Dr G. D. H. Bell at P.B.I., Cambridge) they naturally fell into these two groups.

For a time varieties like Proctor, Freja and Ymer formed a dual-purpose group, being higher yielding than the malting varieties and

| Malting | Feeding | Dual Purpose | Varieties in 1970 | |
			Feeding	Malting
Spratt Archer	Rika	Proctor	Midas	Imber
Earl	Herta	Freja	Vada	Gerkra
Beorna	7–9 qtrs	Ymer	Deba Abed	Julia
5–6·5 qtrs^a		7–8·5 qtrs	9–11 qtrs	Sultan
				Zephyr
				9–11 qtrs

^{a} 1 qtr = 4 cwt.

under many circumstances they were equal to the high yields obtained from the purely feeding types. Proctor and Ymer became and are still popular with farmers since the yield obtained was always satisfactory and thus the financial output assured even if malting premiums were not obtained. Some indication of the relative performances of each of the groups can be seen from the yields quoted in quarters and the present level of output from feeding and malting barleys has also been included.

E.   Profit Margins

The combination of low production costs and high gross output will lead to maximum profit per acre and under most circumstances this is obtained from the highest yielding crops. Production costs rapidly become out of date, prices and the level of government support change from time to time and one hopes for higher yields in the future, but it is still worthwhile applying average costs to average yields to give a measure of the profit margin.

Item	Value
28 cwt. at £22/ton	£30·80
Deficiency payment	4·20
Total Output	35·00
Net Cost	22·00
Profit Margin/acre†	£13·00

† In order to simplify the calculations of profit margin the Home-Grown Cereals Authority levy and the Barley Incentive Scheme payment have been omitted.

Under a given set of circumstances the cost of production will be fairly constant and thus the profit margin is related simply to yield and price obtained. In recent years when average barley prices have been in the region of £22/ton it has meant that with low costs the first ton of yield will cover the production costs and the profit margin has come from grain in excess of this together with the appropriate deficiency payment. Where the cost of barley growing rises to £28 per acre then the profit margin is reduced to approximately £7 per acre.

Appendix
Recommended Varieties of Winter Barley for
England and Wales—1971†

The control for yield comparisons is the mean of Senta and Maris Otter.

	Recommended		Provisionally recommended
Varieties classified for General use, G Provisional recommendation, P	Senta G	Maris Otter G	Mirra PG
Agricultural Characters:			
Yield as % of control	105	95	110
Standing power	5	6	$5\frac{1}{2}$
Shortness of straw	$3\frac{1}{2}$	5	$2\frac{1}{2}$
Earliness of ripening	8	$7\frac{1}{2}$	8
Resistance to loose smut	6	4	2
Resistance to mildew	2	3	2
Resistance to Rhynchosporium	8	2	8
Resistance to yellow rust	7	4	5
Resistance to brown rust	7	5	6
Winter hardiness	7	$6\frac{1}{2}$	7
Latest safe sowing date	End Feb.	Mid Mar.	End Feb.
Malting grade (greatly influenced by conditions of growth)	1	7[a]	1
Year first listed	1967	1965	1971

A high figure indicates that the variety show the character to a high degree.
[a] Approved by the Institute of Brewing as a malting barley.

Attention is drawn to the note on varieties for the north and south-west.

VARIETIES RECOMMENDED FOR GENERAL USE

SENTA

A very early, high yielding 6-row feeding barley with good winter hardiness. Rather high husk content; very good Rhynchosporium resistance.

Weihenstephaner Stamm × Dea. Engelen, Germany.

†Farmers' Leaflet No. 8. Issued by the National Institute of Agricultural Botany, Huntingdon Road, Cambridge.

MARIS OTTER

High malting quality and early matur-   Proctor × Pioneer.
ity. Infection with Rhynchosporium in   Plant Breeding Institute,
the south-west may be severe.   Cambridge.

NEW VARIETIES PROVISIONALLY RECOMMENDED
WHILE FURTHER TRIALS ARE STILL IN PROGRESS
(Seed may not be available for commercial production)

MIRRA

A very early, very high yielding 6-row   G.109 × Herfordia.
barley with good winter hardiness; very   von Borries-Eckendorf,
good Rhynchosporium resistance but   Germany.
susceptible to loose smut.

Recommended Varieties of Spring Barley for
England and Wales—1971†

The control for yield comparisons is the mean of Proctor and Zephyr. Comparisons other than with the control are not strictly valid, and any differences of 3% or less should be treated with reserve.

† Farmers' Leaflet No. 8. Issued by the National Institute of Agricultural Botany, Huntingdon Road, Cambridge.

Varieties classified for General use G, Special use S, Provisional recommendation P, Becoming Outclassed O	Recommended						Provisionally recommended						Becoming outclassed
	Sultan	Julia	Zephyr	Vada	Proctor	Deba Abed	Berac	Hassan	Lofa Abed	Midas	Gerkra	Imber	Impala
	G	G	G	G	G	S	PG	PG	PG	PG	PG	PG	O
Agricultural Characters:													
Yield as % control	106	105	103	98	97	96	106	106	106	105	101	99	97
Standing power	6½	7	7	5½	5½	9	7½	6½	6	8½	6	6½	8
Shortness of straw	6	6½	6½	6½	6	7½	7	7	6½	8	5½	6	6
Earliness of ripening	7	7	7½	7	5½	5½	7½	7½	6	6	7½	6½	7½
Resistance to loose smut	2	4	3	3	8	7	2	4	5	4	8	5	3
Resistance to mildew	3	5	3	6	2½	6	4	3	6	4	4	4	5
Resistance to Rhynchosporium	2	2	2	3	5	1	2	3½	2	2	2	3½	3
Resistance to yellow rust	4	4	5	3	6	3	1	5	5	4	3	3	2
Resistance to brown rust	1	4½	3	5	3	2	4	4½	4½	1	3½	3	2
Malting grade (greatly influenced by conditions of growth)	4[a]	4	5[a]	2	6[a]	2	6	6	2	3	8	9	2
Year first listed	1968	1968	1969	1953	1953	1965	1971	1971	1971	1970	1970	1970	1965

A high figure indicates that the variety shows the character to a high degree.

[a] Approved by the Institute of Brewing as malting varieties. New varieties may not qualify until results of commercial malting and brewing are available.

Attention is drawn to the note on varieties for the North and South-west.

VARIETIES RECOMMENDED FOR GENERAL USE

SULTAN

High yield. Susceptible to loose smut and brown rust, and no longer very resistant to mildew.

Balder × [Agio × (Kenia × Arabian variety)].
Cebeco, Holland.

JULIA

High yield and fairly stiff straw. Moderately good resistance to most diseases.

Delta × Wisa.
Cebeco, Holland.

ZEPHYR

High yield with good malting quality and fairly stiff straw. Liable to ear loss when ripe.

H.2149 × Carlsberg.
M.G.H., Holland.

VADA

A feeding barley with good mildew and moderately good brown rust resistance; suitable for a wide range of conditions.

Hordeum laevigatum × Gull.
Wageningen, Holland.

PROCTOR

Recommended where a traditional malting barley is required, and for the south-west where Rhynchosporium is prevalent. Late maturing.

Kenia × Plumage-Archer.
Plant Breeding Institute, Cambridge.

VARIETY RECOMMENDED FOR SPECIAL USE

DEBA ABED

A feeding barley with straw very short and very stiff but liable to "brackle". Late maturing. Infection with Rhynchosporium in the south-west may be severe.

Abed Denso × Weihenstephaner II.
Abed, Denmark.

NEW VARIETIES PROVISIONALLY RECOMMENDED WHILE FURTHER TRIALS ARE STILL IN PROGRESS

BERAC

Combines high yield with stiff straw and good malting quality.

Balder × Erica.
C.I.V., Holland.

HASSAN

Combines high yield with moderately good brown rust resistance and good malting quality.

W

Delta × [Agio × (Keania × Arabian variety)]. Cebeco, Holland.

LOFA ABED

High yield and good resistance to most diseases. Rather late maturing.

Proctor × Minerva. Abed, Denmark.

MIDAS

High yield with very short and very stiff straw. Rather late maturing. Susceptible to brown rust. Requires good growing conditions.

[(Proctor × Wong) × M.R.A.] × gamma ray mutant from Maythorpe. Miln, Chester.

GERKRA

Very high malting quality with early maturity.

Volla × Proctor. Kraai, Holland.

IMBER

Very high malting quality. Earlier and stiffer than Proctor.

Elsa × Proctor. Guinness, Warminster.

VARIETY BECOMING OUTCLASSED

IMPALA

A feeding barley with stiff straw. Liable to ear loss when ripe.

(Wisa × Balder) × Nordstaat Stamm. M.G.H., Holland.

BARLEY VARIETIES FOR SPECIAL AREAS

THE NORTH

These notes refer particularly to Cumberland, Durham, Lancashire, Northumberland, Westmorland and Yorkshire North Riding.

Winter barley has not been an important crop in this region, but earliness in ripening makes it of potential value.

Senta and Mirra are winter hardy feeding varieties which may be useful if bird damage before harvest is not a limiting factor. Maris Otter may be hardy enough in milder areas but susceptibility to Rhynchosporium makes it unsuitable for areas near the coast or subject to hill mists.

Spring barley. The yields from trials centres in the above counties, expressed as a % of the mean of Proctor and Zephyr, are as follows:

Julia	106	Midas	109
Sultan	104	Hassan	103
Zephyr	104	Berac	102
Deba Abed	102	Lofa Abed	102
Vada	102	Gerkra	101
Proctor	96	Imber	99

Stiff strawed varieties are particularly useful but may be prone to ear loss.

The varieties Deba Abed, Lofa Abed, Midas and Proctor are relatively later maturing in the north.

THE SOUTH-WEST

Trials have shown that special care is required in the choice of varieties for the south-west. In particular Rhynchosporium has been especially severe in Cornwall, Devon, Dorset, Somerset and in other coastal regions.

The spring variety Deba Abed and the winter variety Maris Otter, which are given low figures for resistance to Rhynchosporium in this leaflet, should be avoided in these areas. All spring varieties sown in the autumn are particularly susceptible.

No European two-row barley is known to have high resistance to Rhynchosporium, but Proctor, Imber and Hassan are less susceptible than most other spring varieties. Among the winter barleys all the six-row varieties tested have shown high resistance.

Yields of spring barleys from trials centres in Cornwall, Devon, Dorset and Somerset, expressed as a % of the mean of Proctor and Zephyr, are as follows:

Zephyr	103	Proctor	97
Julia	102	Vada	95
Sultan	102		

Data for the provisionally recommended varieties Berac, Hassan, Lofa Abed, Midas, Gerkra and Imber are still too limited for inclusion.

1970 Varieties of Cereals for Scotland†

Recommendations in this leaflet are based on trials and experience of the Scottish Agricultural Colleges.

In case of doubt, or for information about varieties not on the list, farmers should consult their Agricultural Adviser.

† Issued by the three Scottish Agricultural Colleges.

BRITISH CEREAL SEED SCHEME
(SCOTLAND)

Certified seed, Multiplication seed or Field Approved seed of the following recommended varieties will be available through normal commercial channels for sowing in 1970:

Barley—Sultan, Zephyr, Ymer, Golden Promise.

SPRING BARLEY

S (N)	MIDAS	105
G	SULTAN	103
G	ZEPHYR	102
G	YMER	100
S	GOLDEN PROMISE	98

SPRING BARLEY

MIDAS

A variety with very short straw and good resistance to lodging which ripens up to 7 days later than Ymer. Has an upright growth habit similar to that of Golden Promise. Produces small grain.

SULTAN

On average slightly later than Ymer in maturity with slightly longer but rather stronger straw. Reasonably good for malting. Mildew resistant† but very susceptible to loose smut.

ZEPHYR

Short straw of good standing power. Grain large and of good malting quality, provided the grain is not split, the variety being rather prone to this defect. Liable to shed in exposed conditions.

YMER

Consistent yield over wide range of conditions. Moderately early; short straw of average strength.

GOLDEN PROMISE

An early ripening variety with exceptionally short straw and very good resistance to lodging. Very susceptible to mildew. Small grain, of good malting quality and appearance, which resists splitting. Not recommended for fields infested with couch grass because it lacks competitive ability, or for soils liable to drought. Very resistant to wind loss. High yield under suitable conditions.

† No longer mildew resistant.

Government of Northern Ireland Ministry of Agriculture: Varieties of Cereals†

Winter Barley

Details as given in the N.I.A.B. Farmers' Leaflet No. 8.

Spring Barley

Recommended List of Spring Barleys, 1967

Agricultural characters	Pallas	Banba	Mentor	Zephyr	Ruby	Union
Grain yield as percentage of Pallas	100	103	101	100	97	94
Standing power	7	7	7	$7\frac{1}{2}$	8	7
Shortness of straw	8	8	8	$7\frac{1}{2}$	$7\frac{1}{2}$	7
Resistance to loose smut	8	8	5	5	7	4
Resistance to mildew	4	$3\frac{1}{2}$	6	7	5	$7\frac{1}{2}$
Earliness of ripening	$7\frac{1}{2}$	$7\frac{1}{2}$	$7\frac{1}{2}$	8	$7\frac{1}{2}$	$8\frac{1}{2}$

PALLAS (G)

An X-ray mutant from Bonus. This variety, although better suited to fertile conditions, has proved a dependable yielder over a wide range of fertility and soils. It has short straw, good resistance to lodging and is rarely affected by loose smut.

BANBA (G)

Banba is similar to Pallas in length, stiffness of straw and time of ripening. This variety produces a lot of leaf and provides good ground cover, which is an advantage where weeds are present. Resistance to loose smut is good.

MENTOR (G)

Mentor is also similar to Pallas in length, stiffness of straw and earliness of ripening. It has a lower resistance to loose smut and higher resistance to mildew than Pallas. Grain yield is only slightly higher than that of Pallas.

ZEPHYR (N)

Zephyr ripens slightly earlier than Pallas. Its straw is slightly longer and slightly stiffer than that of Pallas. Zephyr is less resistant to loose smut and more resistant to mildew than Pallas.

RUBY (N)

This variety gives a slightly lower grain yield and has slightly longer

†Leaflet No. 96.

E

straw than Pallas. Its resistance to mildew is slightly greater than that of Pallas, and its resistance to loose smut slightly less. Ripening time is similar to that of Pallas. The main advantage of Ruby is its stiff straw. This makes it easier to harvest under high fertility conditions.

UNION (s)

Union ripens about four days earlier than Pallas. It has rather long but stiff straw. Resistance to mildew is good but it is rather susceptible to loose smut. Yield of grain is lower than that of Pallas. Union is recommended for late sowing or where early harvesting is desirable.

References

Arable Farmer. (1968). pp. 22–3. Farming Press Ltd, Ipswich.

Atterberg, A. (1899). "Die Varietäten und Formen der Gersten". *J. Landwirt.* **XVII,** 1–44.

Beaven, E. S. (1902). *J. Inst. Brew., London.* **VIII,** No. 5.

Beaven, E. S. (1947). "Barley: Fifty Years of Observation and Experiment". Duckworth, London.

Black, C. J. (1963). "Barley Production Today, 1962". University of Leeds, Department of Agriculture, Economic Section.

Boyd, J. (1952). "The Effect of Seed-rate on Yield of Cereals". *Emp. J. Exp. Agr.* **20,** 78, 115–22.

Bullen, E. R. (1959). *Exp. Husb.* **4,** 28.

Bullen, E. R. (1964). "Some Problems in Intensive Cereal Production". *Outlook on Agriculture* **4,** no. 2, 64–71. I.C.I. Publication.

Burges, H. D. and Burrell, N. J. (1964). "Cooling Bulk Grain in the British Climate to Control Storage Insects and Improve Keeping Quality". *J. Sci. Fd Agric.* **15,** 32–50.

Burrell, N. J. (1964). "Hot News on Cold Air". *Farmer and Stockbreeder,* 3rd November, 1964.

Darlington, C. D. and Wylie, A. P. (1955). "Chromosome Atlas of Flowering Plants". Allen and Unwin Ltd, London.

Daw, M. E. (1963). *Fm Mgmt Notes* **29,** 7.

De Candolle, A. (1882). "The Origin of Cultivated Plants". Engl. Edit. 1912. D. Appleton, New York & London.

Finn-Kelcey, P. (1966). "On-Floor Grain Drying". *Farm Mechanisation,* March 1966. Farm Journals Ltd, London.

Fraser, J. G. (1912). "The Golden Bough" (third edition). Vol. 1, pp. 131–2.

Gill, N. T. and Vear, K. C. (1958). "Agricultural Botany". Duckworth, London.

Greig, D. J. and Boyce, D. S. (1966). "The Control of On-Floor Grain Drying and Storage Systems". *Farm Mechanisation,* April 1966, p. 39. Farm Journals Ltd, London.

Hall, Sir A. D. (1945). "The Soil. An Introduction to the Scientific Study of the Growth of Crops" (fifth edition). John Murray, London.

Harrison, J. (1967). "Mycotoxicosis". N.A.A.S. Quarterly Review No. 78, Winter 1967, pp. 78–85. M.A.F.F., London.

Holliday, R. (1956). *Field Crop Abstracts* **9**, 3, 1–13. "Fodder Production From Winter-sown Cereals and its Effect Upon Grain Yield".

Hunter, H. (1952). "The Barley Crop". Crosby Lockwood and Son, Ltd, London.

Jackson, H. and Page, J. B. (1957). "Seed Rates for Spring Barley and Winter Wheat". *Exp. Husb.* **2**, 1–17.

Jeater, R. S. L. (1967). 8th Brit. Weed Control Conf. Proc., Vol. 3, no. 19, pp. 874–883.

Jones, P. J. (1956). "A Comparison of Three Varieties of Barley at Three Levels of Manuring". *Exp. Husb.* **1**, 10–17.

Körnicke, F. (1882). "Die Saatgerste, in der Zeitschrift für das gesammte Branwesen". Bonn Univ.

Körnicke, F. (1895). "Die hauptsächlichsten Formen der Saatgerste". Universitäts-Buchdrückerei von Carl Georgi, Bonn.

Körnicke, F. and Werner, H. (1885). "Handbuch des Getreidebanes". Verlag von Paul Parey, Berlin.

Kramer, C., Post, J. J. and Wilten, W. (1952). "Malting Barley and Climate". *Meded. Natn. Com. Brouwgerst.* **18**, 149.

Linnaeus, C. L. (1753). "Hordeum". *Species Plantarum* **1**, 84–85.

Lovelidge, B. 1965. "Grain On The Floor". *Farm Mechanisation,* Jan. 1965. Farm Journals Ltd, London.

Macpherson, J. F. (1969). West of Scotland Agricultural College Economics Dept. Notes for Advisers. No. 9.

Martin, P. C. (1962). "Barley Production in the East of Scotland 1961". Edinburgh School of Agriculture Economics, Report No. 77.

Mundy, E. J. (1969). *Exp. Husb.* **18**, 91.

N.A.A.S. Technical Report No. 16, September 1966. "Filling and Emptying Airtight High Moisture Grain Silos". Report prepared by N.A.A.S. Liaison Unit at N.I.A.E., Wrest Park, Silsoe, Beds. M.A.F.F.

Nixon, B. R. (1966). "Barley Production in Devon and Cornwall in 1964". University of Exeter, Department Agricultural Economics. Farmers' Report No. 2. April 1968.

Propcorn-Users' Manual. June 1968. B.P. Chemicals (UK) Ltd.

Prout, J. (1881). "Profitable Clay Farming under a Just System of Tenant Right". London.

Prout, W. A. and Voelcker, J. A. (1905). *J. Roy. Agr. Soc.* **66**, 35.

Rayns, F. (1959). "Barley". M.A.F.F., London.

Rimpau, W. (1891). *Landwirt. Jahrb.* **20**, 354.

Rothamsted Experimental Station, Harpenden, Herts. Lawes Agricultural Trust. Details of the Classical and Long-Term Experiments up to 1962. Numerical Results of the Field (Experiments), 1963, 1964, 1965 and 1966.

Smith, L. P. (1967). "The Effect of Weather on the Growth of Barley". *Field Crop Abstr.* **20**, 4, 273–278.

Sylvester, R. (1947). *Agriculture (London)* **54,** 422.

Thompson, H. L. (1966). "Malting Quality—Now and in the Future". *J. nat. Inst. agric. Bot.* X, Supp., 15–20.

Vavilov, N. I. (1926). "Studies on the Origin of Cultivated Plants". *Bull. Appl. Bot. Plant Breed., Leningrad* **16,** 2, 170.

Voss, A. (1885). *J. Landwirt., Berlin.*

Watson, D. J., Thorne, G. N. and French, S. A. W. (1958). *Ann. Bot. (London)* **22,** 321–52.

CHAPTER 3

Oats

I.	Introduction	122
II.	Botanical Classification	123
III.	Grain Production and Uses	127
IV.	Grain Quality	128
	A. Condition	128
	B. Milling Quality	129
	C. Chemical Composition and Nutritive Value	131
	D. Nutritive Value of Oat Flour	132
	E. By-products of Oatmeal Milling	132
	F. Oat Husks	133
V.	Straw Yield and Quality	134
	A. Yield	134
	B. Quality of Oat Straw	135
VI.	Soil	137
VII.	Climate	137
VIII.	Place in Cropping Sequence	138
	A. Winter Oats	138
	B. Spring Oats	139
IX.	Manuring	139
	A. Winter Oats	139
	B. Spring Oats	140
X.	Cultivations and Seed-bed Preparation	141
XI.	Drilling	141
XII.	Time of Sowing	143
	A. Winter Oats	143
	B. Spring Oats	143
XIII.	Seeding Rate	144
XIV.	Seed Dressing and Seed Treatment	145
	A. Seed Dressing	145
	B. Leaf Spot and Seedling Blight	146
	C. Seed Treatment	146
	D. Type of Seed	147
XV.	Varieties/Cultivars	147
	A. Winter Cultivars	147
	B. Spring Cultivars	148
	C. Potato Oat	149
XVI.	Spring Management of Winter Sown Crops	150
XVII.	Undersowing Oats with Grass and Clover Seeds	151

XVIII. Control of Insect Pest.   151
  XIX. Manganese Deficiency   153
   XX. Chemical Weed Control   153
  XXI. Harvest   155
        A. By Binder   155
        B. Windrow Harvesting   156
        C. Combine Harvesting   156
 XXII. Drying and Storage   157
        A. Polythene Sack Storage for Moist Grain   160
XXIII. Production Economics   162
XXIV. Yield of Grain and Straw   164
        A. Yield of Grain   164
        B. Yield of Straw   164
        Appendix   166
        References   174

I. INTRODUCTION

It is generally agreed that the cultivation of oats began much later than that of barley and wheat, since little of the history is known prior to the dawn of the Christian Era. Several authors have credited the Greek Dieuches (400 B.C.) with the first writing about oats (Körnicke and Werner, 1885; Zade, 1918 and Nicolaissen, 1940), whilst Haussknecht (1885), Schulz (1913) and Zade (1918) suggest that Theophrastus (371–286 B.C.) the Greek philosopher and naturalist was one of the earliest to mention this cereal. Those readers wishing to go further into the earliest writing on oats should consult the excellent summary of the classical literature by Coffman (1961) as a preliminary to their studies.

According to De Candolle (1882) neither the ancient Egyptians nor the Hebrews cultivated oats and there is no Sanskrit name or any name in modern Indian language for the genus. The ancient Greeks first recognised oats as a weed amongst the other cereals and they ascribed some medicinal properties to the grain and used the whole plant as green fodder or made hay for horses with it.

In reviewing the literature on the archaeology of oats, Malzew (1930) pointed out that *Avena strigosa* ssp. *strigosa* was found in the remains of lake dwellings (Bronze Age, 3500–1000 B.C.), *Avena fatua* ssp. *fatua* was associated with the Iron Age (from 1000 B.C.) and sub-species *Sativa* was recorded at several ancient Slav sites (900–500 B.C.). Thus in Europe oats have been linked with man's activities for about 4000 years and Coffman (1961) suggests that they developed concomitantly with barley.

In spite of being introduced to agriculture more recently than barley and wheat the time and place of origin of oats is relatively obscure. It was assumed that presently cultivated types (*Avena sativa*) originated in

Europe, until Vavilov (1926) came across them as a weed in Emmer wheat in Persia. Vavilov's assistants then studied grain from Southeast Europe, Asia Minor and North Africa and found amongst the wheat many types of oats as weeds, some of which were clearly transitional forms of this cereal.

From a study of the early classical literature one deduces that oats forced themselves upon the early cultivators of the soil. It would appear that many of the seed stocks of wheat and barley carried oats as an impurity and with greater adaptability to soil and climate they soon multiplied until notice had to be taken of them. Theophrastus records spelt wheats changing to oats and Pliny (first century A.D.) was convinced that barley degenerated into oats.

Wild oats were cultivated by the cave-dwellers in Switzerland before 1000 B.C. and in Britain carbonised grains of both cultivated and wild forms mixed with wheat and barley were found in Wiltshire, Somerset and Dorset, a relic of the Iron Age (about 400–250 B.C.). Similar finds were recorded a century or so later in the Forth and Clyde canal and at Camphill, Glasgow (Findlay, 1956). It is known that cultivated oats were being grown much earlier than this in several European countries and it has been concluded that the cultivated forms for human consumption arrived in Britain from our near neighbours on the Continent.

II. BOTANICAL CLASSIFICATION

After an extensive review of the literature including the initial work carried out by Schulz (1913), Carleton (1916) suggested the following outline or classification which links the present cultivated oat forms with their wild progenitors.

AVENA

fatua L.	*sativa* L.	Common or spreading oats.
	orientalis Schreb.	Side oats.
	nuda L.	Hull-less oats.
sterilis L.	*algeriensis* Trab. (also called byzantina)	Algerian or Red oats.
barbata Brot.	*strigosa* Schreb.	Rough or sand oats.
	brevis Roth.	Short oats.
wiestii Steudel	*abyssinica* Hochst	Abyssinian oats.

Although several authorities are in agreement with the above classification there are others who would link the present cultivated forms with *Avena sterilis* (Coffman, 1946). It seems clear that yesterday's weeds, "wild oats", have become today's crop "cultivated oats", and since transition will certainly still be going on somewhere in the world, precise classifications could soon become outmoded. It is therefore proposed to list the major species of oats, whether wild or cultivated, which are found in the world today, with a brief description indicating their relative importance and then to concentrate on those found or grown in Britain. Etherbridge (1916) published a classification of American oat varieties and the eight group or elemental types listed by him form the basis with one addition, namely *Avena ludoviciana*. (Readers interested in the phylogeny and taxonomy of *Avena* would be wise to consult the extensive investigations of Malzew, 1930.)

The basic chromosome number of the genus *Avena* is seven and amongst the various wild and cultivated species there appear diploid, tetraploid and hexaploid forms. Unlike many other plant groups, the chromosome numbers of members of the oat family are generally agreed by the various genetical authorities and these are given in brackets, with author reference, following the specific name. Other less important oat species can be found listed in Darlington and Wylie's book entitled "The Chromosome Atlas of Flowering Plants" (1955). A group of cultivated species of moderate importance in Europe, but with economic significance elsewhere, are grouped below.

	Species	Chromosome number	Reference	Where cultivated, purpose and common name
Diploids	*Avena nudibrevis*	14	Stanton (1936)	S. Europe for grain
	Avena ventricosa	14	Emme (1930)	Algeria for fodder
	Avena weistii	14	Emme (1930)	Egypt and Persia for fodder—Desert Oat
Tetraploid	*Avena barbata*	28	Huskins (1927)	Mediterranean area and Persia for fodder—Slender wild oat
	Avena vaviloviana	28	Emme (1930)	Russia for grain
Hexaploid	*Avena byzantina*	42	Huskins (1927)	Europe and N. Asia for grain—Red oat

From Darlington and Wylie (1955)

Avena abyssinica [28, Spier (1934)], known as the Abyssinian oat is both wild and cultivated in its surroundings. It is a particularly interesting species in the general phylogeny of oats since some of the evolutionary development of the palea is similar to that recognised on *Avena brevis* and *Avena strigosa* whilst the grain base resembles *Avena sativa*. In North Africa these oats are grown mainly for fodder but varieties with yellow, grey or brown-black grains have been recognised and Trabut (1911) considers them to be especially adapted to dry regions.

Avena brevis [14, Spier (1934)]. Short oats are characterised by two short awn points on the palea, and these are developments of two veins. The stems of this species are tall, thin and grass-like with compact unilateral panicles bearing plump grain which is brown-black in colour. It is grown in Southern Europe mostly for green fodder or hay since it produces bulky crops of stem and leaf.

Avena fatua [42, Philp (1933)]. The wild oat is cultivated for grain, green fodder and other purposes in South-west Asia but in Britain it is recognised as one of the most troublesome weeds of arable land. These spring germinating annuals send up long, fairly thin stems and the resulting panicles contain dark brown grains, each articulating separately in the spikelet and enveloped in brown hair. Colour variations from black, through dark brown to almost white exist but most of the grains carry a large, well-developed awn from half way up the back. The first part of this awn is heavily twisted and when variations in humidity occur or during rainfall the grain turns using the top part of the awn as a pivot and tends to bury itself when the soil is loose. With this built-in mechanism for survival plus the fact that the seeds can lay dormant in the soil for many years and still be viable once they have been brought up to the surface again, we must consider wild oats as one of our main arable-land weeds. (For photographic illustration see Fig. 9 in Chapter 1.)

Avena ludoviciana [42, Huskins (1927)], is called the Wild Red Oat around the Mediterranean and in South-west Asia where it is grown for grain, but in Britain it is simply known as the winter germinating wild oat. *Avena ludoviciana* is found in the same parts of the world as *A. sterilis* and botanically speaking, it resembles this species and also *A. fatua*. At present it is confined to Oxfordshire and several other counties of East and South-east England and for many practical purposes it is grouped with the spring germinating wild oat.

Avena chinensis (nuda) [42, Huskins (1927)], the Chinese Naked Oat is not cultivated in Europe but is grown extensively in the hilly districts of China for its grain. The kernel is only loosely held by the enclosing pales around it and is consequently very easily set free, thus providing the common name to this oat.

Avena sativa [42, Emme (1930)] is the name given to the common or cultivated oat in many parts of the world and according to Percival (1947) these present cultivated forms appear to have been derived from *A. fatua, A. sterilis* and *A. barbata.*

The spikelets of the common oat usually contain three flowers and the developing grain nearest the base is the largest one, often exhibiting a pronounced awn. Two and sometimes three grains are formed per spikelet but the distal one is usually small in comparison with the other two. The panicle in the case of *A. sativa* is equilateral, that is to say grain appears on both sides of the stalk and many cultivars exist which will be described at length later on.

Avena sativa, sub-species *orientalis* is simply a sub-species of Sativa and is mentioned since many, if not all, of the older books on oats have included it in their classification. The panicles are unilateral, with all the oat grains lying on one side of the stalk and in many instances they are rather contracted compared with the common equilateral oat. These one-sided panicle varieties were occasionally placed aside as a separate race and referred to as Tartarian oats. There are a few cultivars grown today which exhibit this form of inflorescence and since they are otherwise similar to the main group for the purpose of simplicity they will be dealt with under *A. sativa.*

Avena sterilis [42, Spier (1934)], the Animated or Fly Oat is found in the Mediterranean region and South-west Asia where it is grown principally for fodder. According to Findlay (1956), most of the oats grown in the Southern states of America, and in Argentina, South Africa, Australia and other countries with warm climates have been derived from this species. After a prostrate initial growth the plants develop short fine straws and the equilateral panicles contain few spikelets with large grains, giving the inflorescence a pendulous appearance. In respect of most other botanical characteristics they are similar to *A. fatua.*

Avena strigosa [28, Nishiyama (1936)], known as the Sand Oat is cultivated for fodder in the Mediterranean region. In Britain *A. strigosa* received the name Bristle-Pointed Oat and in former times it could be found cultivated on much of the poor, acid soil in Scotland and Wales. Findlay (1956) indicated that this species was also grown in Portugal and Uruguay and may be found today on the alkaline soils in Orkney, Shetland and the Western Isles. "Native Large Black" and "Small Black" oats in Orkney are blends of different proportions of *Avena strigosa* and the Murkle oat, according to the same author. Bristle pointed oats tiller freely, producing fairly short stems which support unilateral panicles of narrow grains which have two long bristle points at the tip. The grains can vary in colour but are usually brown or black, carry a twisted awn and like the wild oat fatua are partly enclosed by tufts of brown hair. The cultivated bristle-pointed oats (Gaelic–Core beag) have now largely disappeared from the mainland in Scotland but plants occurring as weeds in other cereals may often be found in the extreme North and West. A recent survey carried out by the West of Scotland Agricultural College showed that these oats occur quite frequently as a weed in barley crops in the West of Scotland along with other forms of "wild oats".

In the Western Isles, particularly the Uists, most of the oats grown are still *Avena strigosa* and on the machair land (coarse sandy soils close to the shore) they are usually mixed with rye.

III. GRAIN PRODUCTION AND USES

In recent years there has been a marked swing away from oat growing in Britain which has been taken up mainly by barley.

TABLE 1

Oat Acreage Statistics

	1939	1949	1959	1965
England and Wales	1,358,237	1,946,271	1,112,592	485,000
Scotland	777,057	931,603	720,352	432,872
N. Ireland	291,354	374,123	199,070	96,000
	1966	1967	1968	1969
England and Wales	426,000	522,000	526,000	565,000
Scotland	398,504	407,536	345,745	322,541
N. Ireland	81,800	82,400	72,800	57,700

Before the Second World War there were $1\frac{1}{3}$ million acres of oats in England and Wales, about $\frac{3}{4}$ million acres in Scotland and slightly over $\frac{1}{4}$ million acres in Northern Ireland. With the war-time agricultural policy of ploughing up as much grassland as possible the oat acreage naturally extended until about $3\frac{1}{2}$ million acres were being grown. Even in 1949 the effects of this ploughing up policy could be seen via the oat statistics but during the next 17–18 years the acreage of this cereal dropped dramatically in each of the counties of the U.K. The present figures indicate that about one-third of the pre-war acreage is being grown in England and Wales, approximately half the corresponding figure for Scotland with only about one-quarter of the oats remaining in Northern Ireland.

Recently, the *Farmer and Stockbreeder* (January, 1968) made a comparison of our national self sufficiency in respect of agricultural commodities with the pre-1939 figures and their findings in respect of oats have been reproduced in Table 2.

From the data in Table 2 it is clearly seen that little change has occurred in respect of the proportion of imported grains and that total consumption is now only slightly above half the pre-war figure.

The percentage used for human consumption in the past has amounted to approximately 10% (Kinsey, 1959), with 90% being fed to various classes of stock or retained for seed. A more recent estimate, quoted by Kent (1964), indicated that 7·5% of the total home crop in the United

TABLE 2

Home Production and Importation of Oats

| | Pre-1939 | | Provisional 1966 | |
	Tons	%	Tons	%
Home Production	1,940,000	94	1,116,000	97
Imports				
Canada	99,000	5	11,000	1
Other countries	18,000	1	18,000	2
Total	2,057,000	100	1,145,000	100

Kingdom in 1961–62 was used for human food, whilst the corresponding figures for other countries were 4% in the U.S.A., 2% in Canada and 0·2% in France. It was further estimated that 6–7·5% of the crop in all these countries was used for seed, the remainder for animal feed.

In Britain, oats are used in numerous ways for human food usually following milling. Oatmeal and rolled oats for porridge provide a large outlet and the meal is also used for baking oatcakes. Oat flour is used in the preparation of baby foods and in the production of ready-to-eat breakfast cereals, whilst cut or whole groats are a constituent of "black puddings".

IV. GRAIN QUALITY

A. Condition

This applies equally well to milling or feeding oat samples and defines in general terms the physical attributes which are important. Firstly, the moisture content should be low enough to ensure satisfactory storage (14–16%) and when figures in excess of these are obtained, bulk grain should be dried as soon as possible. Secondly, the sample should be free from significant quantities of foreign material, e.g. chaff, straw, rodent droppings, soil, stones, weed-seeds and other cereal grains. It must also be remembered that neither sprouted nor severely weathered grain are suitable for milling and these must be retained solely for livestock feed.

B. Milling Quality

(i) KERNEL CONTENT

This is probably the most important grain character determining the milling quality of a sample of oats. The yield of oatmeal is directly related to this and whenever possible millers will choose samples with the highest kernel content or looking at it from the opposite viewpoint, they will take the one with the smallest husk. The husk or outer coating of the grain is removed by "shelling" or "hulling" as it is sometimes termed, and during this process the grain is passed between two horizontal circular carborundum or emery stones. The bottom one is stationary, whilst the top one revolves at fairly high speed and the distance between these stones is closely adjusted so that the husk is cracked, but the grain remains undamaged. Quite large variations in the percentage husk of oat grains have been recorded, with variety playing a very large part and the likely range has been divided into four categories.

Very low	Low	Medium	High
20% or below	21–23%	24–26%	27–30%

(The figures in the above classification represent the weight of husk compared with kernel + husk, expressed as a percentage.)

(ii) GRAIN SIZE

The most sought after samples for milling are those which contain uniform, well-filled grains of average or medium size; very large or very small ones make for difficult shelling and are avoided. Before grain is offered for milling it should be properly cleaned and screening to remove very thin and very small grains will nearly always improve the appearance and the market value. The two measurements indicating grain size, used frequently to classify samples, are firstly the 1000 corn weight and secondly the 1000 kernel weight. Estimates of the former are more easily and quickly obtained and will range from 20 to 50 grams or even higher. Oats grown in the West of Scotland (Quality of Oats 1958) were classified according to their kernel size as follows:

Very heavy	Over 30 grams per 1000 kernels.
Heavy	28–30 grams per 1000 kernels.

Average	24–28 grams per 1000 kernels.
Light	22–24 grams per 1000 kernels.
Very light	Less than 22 grams per 1000 kernels.

(iii) BUSHEL WEIGHT

Millers prefer samples of oats with a bushel weight of 42 lb or more and this is usually indicative of plump, well-filled grains.

(iv) NEEDLE AND DOUBLE GRAINS

An otherwise acceptable sample of oats may be spoiled by the presence of quantities of needle oats or double grains. Empty-husked are referred to as needle oats and when present will reduce the yield of groats to a miller. "Doubles" or "double grains" are terms used to describe the condition when two grains are not properly separated and the husk of the outer one partly envelops the smaller bosom grain to produce what looks like a double sized grain. The actual husk percentage of a "double" is nearly twice that of an ordinary grain and thus with appreciable amounts present either the milling or feeding value is reduced. These can occur in excessive quantities when the oat crop is grown under poor conditions, is checked by an adverse climate and lacks moisture or nutrients during the grain swelling period. It is also known that some varieties are more prone to double grains than others and in this connection the recommended lists should be consulted.

(v) SHELLED GRAINS

Samples of oats offered for milling should not contain large quantities of shelled grains, i.e. groats, for several reasons. Naked kernels have a reduced keeping quality and are easily damaged when the bulk is being shelled. They present additional difficulties when the grain is cleaned, or kilned and in shelling and whether the grain is destined for milling or livestock feed the shelled grains should be kept to a minimum. The use of correct concave settings and drum speeds in threshing machinery or with combine harvesters will help to reduce this.

(vi) DECAYED GRAINS

Partial or totally decayed grain, even in very small quantities, can ruin an otherwise acceptable sample of oats and if possible the cause should be eliminated or the affected grain sent for stock feed. Damage to grain by the larvae of the frit fly, *Oscinella frit* (L.), can be extensive when the crop is late sown and this should be avoided if milling quality grain is sought.

(vii) HUSK COLOUR

This characteristic is often referred to as grain colour, but the kernels of most oats are fawn and show little variation. Considerable variation however occurs in respect of husk colour and most of the European cultivars fall into one of the following groups—white, yellow or black.

Millers prefer white husked oats, will tolerate ones with yellow husks but do not normally accept the black ones since even the smallest quantity of a black husk can be detected in oatmeal and therefore will spoil its appearance.

C. Chemical Composition and Nutritive Value

Oat grains are much higher in both fibre and oil compared with barley and because of the former have a much reduced starch equivalent value.

	% Crude protein	% Oil	% Crude fibre	% Carbohydrate	% Ash	Starch equivalent (per 100 lb)	Protein equivalent (per 100 lb)
Oats	10·4	4·8	10·3	58·4	3·1	59·6	7·6
Barley	9·0	1·5	4·5	67·4	2·6	71·0	6·5

After Evans (1960).

Estimates of gross energy were made during the period 1964–66 on grain from these two cereals grown in the West of Scotland and for oats and barley the average figures obtained were 4·77 and 4·35 kcal/g respectively. The higher figure for oats is ascribed to the significantly greater oil content.

Large varietal differences in respect of oil and protein content were recorded from oat crops grown in the West of Scotland (Quality of Oats, 1958), and cultivars were grouped on those two chemical characteristics as follows:

Chemical Analysis Based on Whole Grains

	Oil				Protein	
High	Medium	Low	Very low	High	Medium	Low
Over 5%	4–5%	3–4%	Below 3%	Over 10%	8–10%	Below 8%

Protein content in the West of Scotland appeared low by Evans standard, however it must be remembered that under high rainfall conditions applications of fertiliser nitrogen are usually minimal to avoid excessive lodging and although varietal differences in protein content obviously exist, these can be easily and markedly influenced by nitrogenous manuring. In this classification of varieties, the control Sun II contained 4·8% oil and 8·8% protein, based on a 15% moisture content grain. A high protein content is required in both animal and human nutrition and figures of 10–12% are desirable. This is particularly important in milling when the product is destined for baby food and although millers are prepared to elevate protein levels by additions from other sources, they may in future consider a quality payment based on the amount of protein which the grain contains.

High oil content means high energy value of the grain and for livestock feeding this is important. With oats for milling, however, medium to low values are desired because of the better keeping quality of the milled products.

At the time of writing, oatmeal, porridge oats and oat products from traditional Scottish companies were being produced only from oats grown in Scotland. Most of these crops could be located in the East and North-east and the average protein and oil content of the groats proved to be 11–12% and 7–8% respectively.

Several samples of oats, as used by one of the leading Scottish oat-milling companies, were specially analysed in February, 1968, and the average protein and oil content on a basis of whole grains was 8·5% and 4·8% respectively (Williams, 1968).

D. Nutritive Value of Oat Flour

The chemical composition and thus a guide to the nutritive value of oat flour may be obtained from the following table:

Moisture %	Protein % (N × 6·25)	Oil %	Ash %	Fibre %	Carbo-hydrate %	Niacin content (μ g/g)	Riboflavin content (μ g/g)
9·3	14·1	7·2	1·8	1·0	66·6	7·2	1·1

From Kent (1964) and Hellstrom and Andersson (1953)

There is an enzyme lipase adhering to the outer coat of the oat kernel and when oatmeal is stored for some time this enzyme breaks up part of the oil content into fatty acid which gives the characteristic "flavour" or "nip" which can be detected in some meals.

E. By-products of Oatmeal Milling

The output of oatmeal will depend on many factors. These include percentage husk, proportion of double grains, percentage empty husks and free grains, the amount of diseased or "fritted" grains and the proportion of trash and inert material in the sample. On average about 14 stones of oatmeal can be obtained from a quarter of oats (3 cwt) which means a yield equal to approximately 60% of the initial or gross weight. The main by-products of oatmeal milling are oat husk, oat dust and meal seeds, which constitute approximately 70, 20 and 10% respectively of the total by-product.

Oat dust and meal seeds have quite a useful feed value to stock, the latter containing approximately half its weight of broken kernels and half husk.

The "meal seeds" or "mealy sids" or simply "sids" in the past were used to make the Scottish farm beveridge known as sowans, sowens or flummery. They were steeped in water for two or three weeks until sour and any remaining particles (sids) then removed by straining through a sowans sieve. Supernatant water was poured off and the pulpy material left was boiled to form a jelly-like porridge. Sowans were eaten as porridge after addition of syrup and milk or could be taken diluted as a drink and in the latter form, the oatmeal being in such a finely divided state, was considered admirably suited for invalid feeding.

F. Oat Husks

These form the largest proportion of the by-products and following much research by the Quaker Oat Company of America, have found considerable use in the chemical industry on account of the furfural which they contain. According to Findlay (1956), furfural or its derivates are produced only in America and have found a wide use, for example:

(i) as a raw material for nylon manufacture;
(ii) as a selective solvent in the purification of butadiene (starting point for manufacture of synthetic rubber);
(iii) in refining lubricating oils and resins;
(iv) in the manufacture of fungicides, germicides, herbicides, disinfectants and preservatives.

Oat husks are ground and added to other feeding stuffs to produce oat feed meal in Britain and here and in America they have been put to a variety of other uses:

(i) As a filter-aid in breweries;
(ii) As a filler in linoleum manufacture;
(iii) Source of fibre in paper and board;
(iv) As a source of hemi-cellulose.

(v) For deep litter associated with battery hens.
(vi) For making activated carbon.
(vii) As an abrasive in air-blasting for the removal of oil or corrosion on metal components.

From: Kent (1964) and Coffman (1961).

V. STRAW YIELD AND QUALITY

A. Yield

In the past much has been written and said about the value of oat straw and even today in North and West England, Wales, and many parts of Scotland it forms quite a large part of ruminant diets.

Varieties of oats were classified as grain-producers or straw-producers and Findlay has listed the older cultivars under these two headings, and has indicated their relative straw : grain yield.

Grain-producers compared with Victory		Straw-producers compared with Castleton	
Castleton	185	Sandy	228
Marvellous	162	*Castleton*	185
Royal Scot	160	Bell	183
Onward and Ayr		Pure Line	182
Commando	159	Ayr Line	181
Early Miller	158	Quality	174
Yielder	157	Gordon and R.30	173
Golden Rain	153		
Victory	150		
Sun	149		
S.84	148		
Star	146		
Eagle	145		

(In each case grain yield is considered to be 100 and the corresponding straw yield is expressed as a %)

With much of the oat crop combine-harvested the present cultivars tend to be short straw grain types and apart from some notable exceptions, the yield of straw is now very much reduced. In the high rainfall areas of Britain the straw yield will always be higher on average than in the low rainfall areas and similarly for wet rather than dry years. From rotational experiments using the same cultivar Victory, Findlay (1956) showed grain and straw yield to be similar in a dry sunny year whereas under high rainfall and low sunshine straw production was twice that of grain.

Accurate assessments of straw and grain yield appear in Section XXIV and readers should note the differences in grain : straw ratios between countries.

B. Quality of Oat Straw

Voelcher (1861) was one of the first to realise that there were only very small differences in straw quality between varieties, even in extreme types. He concluded that major differences were likely as a result of harvesting the crop at different stages of maturity and his data well illustrates this.

Chemical Composition of Oat Straw at Varying Stages of Maturity

Stage of cutting	% Oil	% Albuminoids	% Soluble carbohydrates	% Digestible fibre	% Indigestible fibre
Green	1·9	6·6	19·1	31·4	29·6
Fairly ripe	1·3	3·1	12·6	35·9	37·8
Past full maturity	1·5	1·5	3·8	33·0	49·8

Examination of a large number of analyses of oat straws in Scotland between 1903 and 1931 (reported by Findlay, 1956) showed no significant differences in quality even when the straw of grain varieties was compared with so-called straw types, thus confirming the earlier theories of Voelcher.

More recently Dent (1957, 1959) and Dent and Boyd (1965) have looked into the chemical composition of oats under varying conditions when cut green or as grain crops. From the first series of experiments it was concluded that differences between varieties of oats in the chemical composition of the straw were negligible but differences between environments were substantial. Confirmation of this was obtained in the second series and it was demonstrated that the traditional Welsh hill variety Radnorshire Sprig was useful for fodder in this environment but more modern varieties showed a greater grain potential. The third set of experiments conducted in the West of Scotland between 1957 and 1961 under upland conditions again revealed little difference between varieties, the only consistent feature was a lower fibre content in the variety Blenda at both hay stage and at maturity.

Thus it is clear that environment and stage of cutting are the two factors affecting the quality of oat straw. Upland conditions will produce high quality soft straw as will early cutting when a high proportion of the plant nutrients remain in the vegetative parts.

TABLE 3

Analysis of oat straw—Percentage of Dry Matter

(Mean of same three varieties over several centres)

Year and centre		Crude protein	Oil	Crude fibre	Ash	Carbo- hydrate	Digestible protein	% Digestibility
1955	Upland	2·9	1·9	43·3	5·6	46·3	1·5	53
	Lowland	2·3	1·8	45·4	4·5	45·8	1·1	46
1956	Upland	5·2	1·3	42·1	5·2	46·3	3·0	57
	Lowland	3·4	1·7	46·7	3·4	44·8	1·8	54
1957	Upland	5·5	1·4	42·1	4·4	46·5	2·7	48
	Lowland	3·7	1·7	43·2	5·7	46·2	1·8	47

The results obtained by Dent (1959) have been summarised in Table 3 and show that eliminating the effect of different years and varieties, upland straw quality can be defined as higher crude and digestible protein and a higher digestible material.

TABLE 4

Average Chemical Composition of Oat Straw

	% Crude protein	% Oil	% Crude fibre	% Carbo- hydrate	% Ash	% Digestible crude protein	Starch equivalent	Protein equivalent (Per 100 lb)
From spring- sown crop	2·9	1·9	33·9	42·4	4·9	1·0	20	0·9
From winter- sown crop	1·9	1·5	34·6	43·1	4·9	0·6	21	0·5

Evans (1960) in "Rations for Livestock" gave the analyses shown in Table 4 for winter and spring oat straw and it may be assumed that these were taken from lowland crops. Apart from the figures for crude fibre these agree well with the data of Dent (Table 3).

VI. SOIL

The cultivation of oats may take place on a wide range of soil types with a reasonable degree of success. In the distant past they were grown under marginal conditions in the North and West of Scotland on the poorest alkaline soils and these oats belonged to the species *Avena strigosa*, what we would now term a wild oat. On the light sandy soils (machair) of the Western Isles where manganese deficiency is common, black oats are grown since they are relatively tolerant of this trace element deficiency. Oats can be found on all the light to medium soils in the higher rainfall areas and will give high yielding crops of good quality given a reasonable spell of weather at harvest. They can also produce good crops on some of the heavier soils, clay and silts, in the drier areas of Britain where there are significant moisture reserves in the soil which can be drawn upon during a particularly dry time. They may also be found to a limited extent on the fens and the re-claimed peaty soils but here the very short stiff-strawed varieties must be chosen otherwise there is a serious risk of extensive lodging, which results in a difficult harvest, not to mention potential grain losses. This cereal may also be found associated with hill peats and in this connection the pH level of the soil is important. It is true that oats will grow well where barley will fail completely and wheat produce only moderate crops but extreme acidity even with oats cannot be tolerated. Small (1946) has listed much of the experimental data linking soil pH and crop yield and this includes many references to the oat crop. Arrhenius (1926) placed the optimum pH at between 4·7 and 6·0 for oats, indicating poor growth below 4·5 and injury at 4·3 with all but the alkatolerant varieties. Trénel (1927) showed good oat yields at 5·6 down to 4·9 with injury occurring below 4·4.

VII. CLIMATE

The oat crop is particularly suited to the cooler, more humid climate of the western and northern regions of Britain where growth is relatively slow and as a result the grains have plenty of time to fill out to produce good plump samples. Oats produced in Scotland are better filled than those grown in England, a fact which explains the widespread use of Scottish oats for seed in the past. Findlay (1956) imported some Plate oats from the Argentine and grew them for two years in Scotland. The effect of change in environment is clearly seen in his data, which is reproduced in Table 5.

Oat crops which do not suffer through lack of moisture will produce high grain yields of good quality and on average the straw will weigh slightly more than the grain. They are the only cereal to be found in any quantity in the high rainfall districts where many crops are still handled by binder. Under these circumstances farmers are interested only in those varieties which show marked resistance to sprouting in the stook.

According to Lewis *et al.* (1951), most of the world's oats are produced at elevations below 2000 feet and probably half below 1000 feet.

TABLE 5

Environment and Grain Characteristics of Oats

	% Large grains	1000 corn weight (g)	% Husk
Original sample from Argentina	6	41·3	30·4
Once-grown in Scotland	56	47·5	26·8
Twice-grown in Scotland	69	51·3	26·5

In Scotland the crop will be encountered at all heights up to approximately 1000 feet but at this altitude the growing season is limited, harvest as such is quite often much delayed and fairly frequently the crop is cut green, since full maturity may never be reached. In England much of the crop is grown just above sea-level to about 500 feet but many useful crops may be found between 600 and 1000 feet on the Pennines. In Wales oats predominate on the upland stockrearing and dairy farms and they can be grown successfully up to 1000 feet and under exceptional circumstances at higher elevations.

VIII. PLACE IN CROPPING SEQUENCE

A. Winter Oats

These will frequently follow another cereal and under most circumstances this will be either barley or wheat. They grow well after a crop of beans and in order that autumn sowing is not delayed an early harvested winter bean crop is best. The harvesting of maincrop potatoes and sugar beet is often too late for them to be followed by winter oats, but after early potatoes, vining peas or turnip seed crops a very satisfactory establishment can be obtained. It will be noted from the range of suitable previous crops that the winter sowing of oats takes place largely in the arable areas of Britain mainly associated with South-east, South and South-west England and South Wales.

B. Spring Oats

By far the largest proportion of oats are spring-sown and although in the past they have been widely distributed in British farming they are now becoming concentrated in the areas most suited to their cultivation, that is to say the West and North. In these areas they are mainly associated with livestock farming where both grain and straw are valued feeds for beef and dairy cattle and where the grain is fed to sheep. Although working horses have largely disappeared from farms, their places have been taken by the various classes of riding horses, all or most of which receive oats as the main ingredient of their daily food. These areas of cultivation are dominated by grassland and therefore the oats are frequently taken as the first crop out of grass. Occasionally two consecutive oat crops may be taken after leys and swedes or turnips are often followed by this cereal. Oat crops are often described by means of the previous cropping and examples of these are given below, but no doubt many other local descriptions exist.

Previous crop	Nomenclature of oat crop which follows
Ley	Ley or lea-oats
Roots such as Swedes and turnips	Redland Oats
Oats	Yaval oats (Scottish)

One final point must be made and this is to warn against too frequent cropping with this cereal. Whereas barley and to a limited extent wheat may be grown continuously for considerable periods, oats should not appear frequently in cropping sequence due to their extreme susceptibility to cereal root eelworm.

IX. MANURING

A. Winter Oats

Where these are to be sown under high fertility conditions, or following crops which leave behind high residual levels of fertilisers, then winter oats should be drilled in the autumn without inorganic fertiliser and approximately 30–40 units of nitrogen will be sufficient as a spring top-dressing. The standard manurial recommendations suggested by

Cooke (1960) for average fertility conditions will apply to many winter oat crops and are given below in units/acre.

	N	P_2O_5	K_2O
Combine drilled in the autumn	—	30	30
Spring top dressing	50	—	—

When winter oats are brought into a mainly cereal rotation as a break-crop, then higher levels of fertiliser may be required, with a little autumn nitrogen to aid in establishing the crop. Average recommended levels, expressed in units/acre, are as follows:

	N	P_2O_5	K_2O
Combine drilled in the autumn	20	30–50	30–50
Spring top dressing	50-60	—	—

The above total level of fertiliser nitrogen would also be applied to a short stiff-strawed winter cultivar when grown under low fertility conditions.

B.   Spring Oats

Examples of fertiliser levels required by the spring sown crop under a wide range of climatic and edaphic conditions are given in Table 6 and these should be worked into the seed-bed prior to drilling.

TABLE 6

Fertiliser Levels for Spring Oats

Conditions	N	Units/Acres P_2O_5	K_2O
1. Following heavily grazed grass–clover ley in high rainfall area	0–20	0–20	0–20
2. For average conditions in high rainfall area	30	30	30
3. Oats in rotation in a low rainfall area	40	30	30
4. In a mainly cereal rotation in low rainfall area	50–60	40	40
5. Exceptionally short, stiff-strawed varieties, grown at low fertility	70	40	40
6. Crop to be undersown with grass and clover seeds in a high rainfall area	15–20	50	50

The average levels of fertiliser nitrogen suggested in Table 6 may be supplemented by a top dressing of 10–20 units should damage occur as a result of attacks by frit fly or leatherjackets.

Oats will grow well under moderately acid conditions but if the soil is strongly acid, damage can result due to calcium deficiency and very low yields will result. However, it is better to apply lime to preceding crops since a significant rise in soil pH can lead to manganese deficiency in oats, a condition often referred to as "grey speck". This deficiency is widely reported where oats are grown on the Fens and on Romney Marsh. It is recognised in many English counties and also in Wales and Scotland associated with lime application.

X.   CULTIVATIONS AND SEED-BED PREPARATION

Oats are said to be the best cereal to follow the ploughing up of grassland but this is only true if the grass is turned in timely and well, with the furrow slice properly inverted and no large air pockets left which tend to accentuate drying out.

Winter sown crops require well consolidated seed-beds which have been left rather rough on the surface to give some protection from frosts and winds and to safeguard against surface capping. It is necessary to provide the same degree of consolidation for spring sown crops to aid moisture uptake and to restrict the movement of pests such as wireworms and leatherjackets, thus reducing the damage they cause.

Ploughing depths should be in the region of 6 inches except where grass or surface trash need to be buried deeper and then 8–9 inches plough depths may be required. Cultivations may be necessary to break the furrow slice on the stiffer soils but where early ploughing has been carried out it is often only necessary to give the land a light–medium harrowing in order to obtain the desired tilth in the top 3 inches of the soil. On light land flat rollers can be used with safety to give the soil its correct degree of compactness and Cambridge rollers should be used for the same purpose on the medium to heavy textured areas.

Disc harrows may prove a useful implement in breaking up clods and at the same time giving the land a degree of firmness but this tool should not be used when large quantities of underground rhizomes of couch are present since the chopping up will more often increase the problem with this difficult weed grass.

XI.   DRILLING

Optimum drilling depth with oats lies in the region of 1·5–2 inches. It is a compromise between theory and practice, since the theory suggests a shallower depth of planting but practice refutes this on grounds of low soil moisture and increased severity of bird damage. Norwegian experiments quoted by Kinsey (1959) showed that spring oats drilled at 1 inch yielded 27·5 cwt and only 20 cwt at a depth of 4 inches. Without the limiting factor of soil moisture, Findlay (1956) showed that field germination was much reduced with deeper sowing and the number of days to brairding rose with increased planting depth and these facts in themselves could account for the yield reductions reported from Norway (see Table 7).

When seed-beds have an irregular surface tilth, drilling depth becomes uneven and in order to ensure that all the seed is covered it will often go in well below the optimum in many areas and this will be responsible for patchy brairds.

Drilling in the winter tends to be at slightly lower depths than in the spring on account of the rougher soil surface which is purposely left to

TABLE 7

Sowing Depth and Establishment of Oats

Depth Sown	% Brairded	Days taken to braird
0·5 inch	82	15
1 inch	80	16
2 inches	70	19
3 inches	65	21

reduce the dangers of surface capping. Early spring drilling may also go in deeper than normal in an attempt to reduce losses through bird activity. These are the crops which often receive the full attention of all the birds within a district and thus it could prove beneficial to cover the seeds with a little more soil. Mid and later spring sowings grow away much more quickly on account of the higher ambient temperatures and with so many more fields being drilled they seldom suffer the bird concentrations which befall the early seedings. Although not often possible, harrowing-in at right angles to the line of drilling can be effective in deceiving the birds, which tend to follow the final harrow marks and thus miss the majority of the seed.

The two main advantages with combine drilling are firstly, the seed and fertiliser go on in one operation and secondly, it is possible to obtain maximum benefit from the minimum amount of fertiliser, notably the phosphate and potash content. The main disadvantage lies in the fact that the rate of sowing is much reduced and since delays in drilling usually mean lower yields the faster method of seeding using a wide drill following a fertiliser spinner is more often employed for winter seeding and where large spring cereal acreages are being handled by a small labour force. Some farmers are even using the fertiliser spinners to sow the grain and are claiming high outputs, in the region of 35 acres/day (Lucas, 1967). This method is certainly very speedy and may be useful under some circumstances provided the grain is properly harrowed-in afterwards, and so long as the pattern produced by the spinner is reasonably uniform.

XII.   TIME OF SOWING

A.   Winter Oats

These should be sown early enough in the autumn for the plants to be well developed by the onset of winter conditions which bring further growth to a halt. This usually means that the optimum sowing time is during the first two weeks of October, but late October drilling will prove equally satisfactory particularly in a mild autumn. It is unlikely that sowings later than the end of November will result in a proper establishment and under normal circumstances are to be discouraged. In the past, many winter cultivars have not proved to be very winter hardy, in fact their vernalisation requirements have been minimal and it is for this reason that early autumn brairds are desirable. Many of these winter varieties (e.g. S.147) can be planted in the spring and although they will produce grain it is not a recommended practice. Harvest is later, straw heights or weights are lower and the same is usually recorded in respect of grain yield, together with the fact that many of the spring varieties have a higher genetic potential for yield than spring sown winter ones. Should the autumn prove to be exceptionally mild and growth very lush, i.e. winter proud, then the spring grazing of winter oats can be beneficial provided the soil is light and does not poach badly during the defoliation.

B.   Spring Oats

The months for sowing spring varieties are late February, March and April according to season and locality. Lowland oats are often best sown in February particularly where soil conditions are light and liable to later spring drought but on average most of the crop will go in once a reasonable tilth has been obtained in March. Under marginal hill conditions April sowings are usual sometimes extending into May especially when spring is late and the snow lies well into the year.

Experiments with spring oats have clearly demonstrated the value of early sowings in respect of yield and those carried out by Findlay (1956) in Scotland over a period of 28 years, where weekly seeding was practised, rank amongst the most important on this topic.

TABLE 8

Effect of Time of Sowing on Yield of Grain and Straw with Oats

| Date | Variety–Potato Oat | | Variety–Victory | |
	Grain cwt/acre	Straw cwt/acre	Grain cwt/acre	Straw cwt/acre
March 22	—	—	24·6	43·1
29	—	—	24·3	41·7
April 5	20·3	41·9	23·3	43·1
12	18·5	39·8	21·6	42·3
19	17·6	38·2	20·8	41·2
26	16·1	36·7	18·0	36·4
May 3	14·1	33·8	16·0	33·2

Besides the obvious benefits demonstrated in the above figures, early sowings lead to earlier harvests, quite often better standing crops and better quality grain.

Finally, and by no means least, early sown crops are much less susceptible to damage by frit fly, in fact only the late sown oats actually suffer economic damage from this pest. Under lowland conditions when seeding is of necessity very late, it is often worthwhile changing to barley providing the land is not too acid. However if the sowing of late oats is inevitable, the earliest maturing variety must be chosen and everything possible must be done to see that growth is never impeded.

XIII. SEEDING RATE

The average seeding rate for lowland oats is 4 bushels/acre and for those who prefer units of weight this is the same as 1·5 cwt, 12 stones or 168 lb/acre. There are, of course, a number of factors governing the optimum seeding rate, the most important being time of seeding and this is summarised in the table on p. 145.

Varieties with large grains usually require higher weights of seeds per acre since they are often associated with a lower tillering capacity, whilst the smaller grained varieties usually tiller freely and have in the past been classified as straw types and can be sown at lighter weights.

Under marginal hill conditions, when the soils are often cold and wet higher seeding rates are required in order to obtain satisfactory brairds and 5–6 bushels will usually be needed. Experiments carried out by the Welsh Plant Breeding Station using the same stock of spring oat seed in each case, showed that multiplication under upland conditions could be responsible for lower grain yields compared with the seed obtained on lowland farms, even though the former was harvested under good conditions and was graded in a similar manner. These results appear to indicate frequent, if not annual seed-change for the hill farmer and when one considers how unripe hill oats can be, together with a much reduced vigour of germination compared with their lowland counterpart, this would seem a reasonable practice.

Broadcasting rather than drilling cereal seed has resulted in poorer

Time of Seeding	Winter Drilling	Spring Drilling
Earlier	−	+
Optimum	=	=
Later	+	=

− less required; = average; + more seed required

field germinations and therefore an additional 10% is usually recommended.

In summarising the seed rate experiments in the northern half of Britain, Boyd (1952) concluded that the optimum seed rate for spring oats was 4–5 bushels and with the introduction of organo-mercury seed dressings, then the optimum could well be slightly lower than this. Experiments in the West Midlands of England, referred to by Kinsey (1959) put the optimum range between 1·5 and 2 cwt/acre and there was also evidence to support the view that higher seed rates would produce better yields under less fertile conditions. These conclusions follow closely those reached earlier by Findlay (1956) who suggests 1·5–1·75 cwt of seed for land in good heart at sea level. A further point made by Findlay which has since been verified was the increased risk of lodging with higher seed rates.

XIV. SEED DRESSING AND SEED TREATMENT

A. Seed Dressing

Dressing oat seed with organo-mercury compounds either in powder or liquid form has now become almost standard practice in the trade, for the control of many fungus diseases. Covered Smut and Loose Smut of oats are effectively eliminated and some control is obtained in respect of Leaf Stripe and Brown Foot Rot with these dressings. They are best administered by seed merchants who are properly equipped for this operation which incidentally carries a health hazard to those who are closely connected with it and the Ministry Safety Regulations must be observed at all times.

B. Leaf Spot and Seedling Blight

This is often referred to as Leaf Stripe on account of the symptoms produced and has recently developed a resistance to the normal organo-mercury dressings. As a result, the Department of Agriculture for Scotland is now insisting on the use of the fungicide Maneb (dithio-carbamate) for the control of this disease, on all oats which are entered for the Stock Seed Certificate.

It is possible to obtain organo-mercury seed dressings combined with the insecticide gamma-BHC for the additional protection against wireworms and a list of these single and combined seed dressings which have been officially approved can be obtained from the most recent Ministry Publication "Agricultural Chemicals Approval Scheme, List of Approved Products for Farmers and Growers".

C. Seed Treatment

When farmers are intending using part of their crops for next year's seed it is important that harvesting is done in the best possible conditions and a germination test must indicate a high percentage viability. Grain as harvested is variable in size, but grain used for seed should be uniformly large, as Findlay (1956) showed from twenty-two field experiments.

Seed size	Yield/acre	
	Grain (cwt)	Straw (cwt)
Large grain	23·4	41·1
Small grain	20·3	35·8

The interaction between oat plants which were produced from small and large grains was also studied by this author by mixing a known proportion of small white grains with large black ones and vice versa. He showed that plants derived from small seeds in competition with those from large ones yielded only half the grain and straw compared with similar plants taken from uniform population. Thus the plants produced from small seeds are adversely affected by those from the large grains indicating the necessity for graded seed. In the grading process the small grains, naked grains and much unwanted inert matter, and weed seeds are removed and up to 1 bushel of material consisting of these fractions has often been removed without reductions in subsequent grain yield. Small grains fair very badly, naked grains (groats) seldom germinate and ungraded seed often lodges whereas the large grains produce strong heavy yielding plants which stand well.

In the past some seed merchants have clipped their seed samples and this has meant the removal of part of the husk distal to the embryo. It gives the sample a plumper, more attractive appearance and allows the seed to run more freely but has little or no effect on yield.

D. Type of Seed

Numerous types of seed are available in respect of oats. They are similar to those already described for wheat and have been listed in Chapter 1. Since most of the oat crop is destined for livestock feed, farmers should choose the cheapest type of seed as far as possible and only purchase the expensive quality grades when a change of seed is deemed necessary or when the crop is being multiplied for seed under one of the Official Schemes. Scottish Certified or Scottish Field Approved seed is available in respect of a limited number of varieties. These include Astor, Blenda, Condor, Forward and Yielder whereas Field Approved Stocks of the varieties on the official lists for England and Wales can be obtained from these countries through normal commercial channels.

XV.   VARIETIES/CULTIVARS

The various agronomic characters relating to the presently recommended varieties for each of the countries of Britain can be obtained from the lists in the Appendix.

A.   Winter Cultivars

Winter varieties are only important in the southern half of Britain and the small list of N.I.A.B. recommended varieties for use in England and Wales is a reflection on the comparatively low acreage devoted to winter oats. Where farmers in Northern Ireland wish to grow this crop, they are referred to the N.I.A.B. list as the Ministry of Agriculture in Northern Ireland do not consider the special testing of winter cultivars worthwhile.

Pendrwm, Maris Quest and Peniarth are varieties for general use whilst Padarn is becoming outclassed. It is interesting to note that all these varieties are products of the Official Plant Breeding Stations and details concerning each can be obtained from the Recommended List which appears in full in the Appendix.

B. Spring Cultivars

A list of the most suitable varieties for the various regions and conditions in Britain have been put together in Table 9 in order that the reader can see at a glance those cultivars which have wide powers of adaptation to our soils and climate.

TABLE 9

Varieties of Spring Oats which are Recommended for General or Specific use in U.K. and their Current Relative Yields

England and Wales		Scotland		Northern Ireland
Mostyn G	103	Selma G(N)	106	Astor N
Selma G	103	Astor G	100	Condor G
Astor G	101	Condor G	100	Stormont Sceptre G
Condor G	99	Blenda S(O)		Stormont Iris O
Manod S	90	Yielder S		Castleton Potato O
		Maelor S(O)		Tyrone Tawny A
		Karin		Tyrone Tawny 8
		Tarpan S(O)		
		Forward S	*a*	
		Ayr Commando S		
		Shearer S		
		Bell S		

a Additional varieties for North and North-west Scotland.

Abbreviations used: G = Recommended for General use; S = Recommended for Special use; O = Becoming outclassed by other recommended varieties; N = New Recomendation for a variety on which further trials are still in progress.

The above table has been compiled from:

1. N.I.A.B. Farmers' Leaflet No 8. (1971). "Recommended Varieties of Cereals (England and Wales)".
2. "Varieties of Cereals for Scotland". Issued by the Scottish Agricultural Colleges (1970).
3. "Varieties of Cereals". Leaflet 96. Government of Northern Ireland, Ministry of Agriculture.

Condor is the mostly widely recommended variety followed by Astor and both are for general use under lowland conditions, particularly under high fertility and where the crops are to be combine-harvested.

Forward is an example of a unilateral panicle variety which is easily recognised in the field since the whole crop tends to lean over and often lodges. The grain has a high 1000 corn weight, giving it a bold appearance but the husk percentage is well above average. Yielder is a useful variety for upland conditions in Scotland where harvest is late and difficult and Maelor should be chosen when straw is an important consideration. In reporting spring oat variety trials carried out during the period 1954-60 Bell (1962) indicated the range of grain : straw ratios amongst some of our present day cultivars as follows:

Maelor	Sun II	Max	Angus	Vigor	Phoenix	Condor
1:1·22	1:1·15	1:1·13	1:1	1:1	1:0·87	1:0·82

Manod, another of the Welsh Plant Breeding Station products, is resistant to crown rust and stem eelworm, making it a useful variety for the West and South-west and Mostyn combines high yield, stiff straw and mildew resistance.

In Scotland and Northern Ireland a few locally bred varieties appear on the Recommended Lists and quite often these will be for some special condition such as upland farming in wet areas.

At the time of writing Condor and Blenda were the two most important varieties being used by Scottish Millers for the production of oatmeal and rolled oats.

C.   Potato Oat

The variety of oat known as Potato was discovered in 1788 (Hunter, 1951) and it was so called since it was found growing in a field of potatoes. No cereal variety has remained so long under cultivation as Potato oat and it therefore merits a special mention. It is a spring oat, with fairly long fine straw, broad leaves and with a high tillering capacity. The grain is small and white, usually well filled, with a low husk percentage. Because of the high kernel content it became popular with millers and with a high oil content and high straw yield of good quality it was preferred by most farmers in the Northern districts in the British Isles.

Both R.30 and Castleton are selections from the original Potato oat and the latter still appears on the recommended list of varieties in Northern Ireland. Ayr Commando and Bell are fairly similar to Potato

F

and they presently appear on the list of varieties for the North of Scotland. Ayr Commando is useful for upland conditions where earliness is important and Bell is recommended for extreme conditions of soil and climate in the Islands and Highlands.

XVI. SPRING MANAGEMENT OF WINTER SOWN CROPS

Early sown crops of winter oats can produce a large volume of vegetative material which can be utilised by stock at the beginning of the year provided the environmental factors are favourable for such a defoliation. As long as this grazing is carefully controlled, limited in its duration and carried out with due regard to weather and soil conditions then a valuable contribution to animal keep can be obtained with minimum effects on the subsequent grain yield. Holliday (1956) reviewed this subject and incorporated data from many parts of the world and Kinsey (1959), quoting data from North Wales associated with the variety S.147, showed how little grain yield was affected by sheep grazing (200 sheep days/acre) provided 30 units nitrogen were added in the form of a top dressing afterwards.

Mean Yield of Grain and Straw from Grazing Experiments 1950–52
(cwt/acre)

Fertiliser treatment	No grazing (control)		Spring grazing	
	Grain	Straw	Grain	Straw
Without nitrogen	21·1	37·8	19·3	29·2
With nitrogen	23·0	47·2	21·1	35·7

From the figures above a reduction of slightly more than 0·5 ton/acre of straw may be expected as a result of spring grazing and this would obviously reduce the risk of lodging. Grazing will normally take place in February and March and under normal circumstances should not be continued beyond mid April. In other countries bloat has been observed in animals grazing winter cereals in the spring and whilst stock are on this low fibre herbage it would be advisable to keep a close watch on them and to give them daily access to a small quantity of hay or straw.

Spring cultivations are normally restricted to rolling with flat or Cambridge rollers in March since harrowing is not generally recommended for this crop on account of its greater susceptibility to mechanical damage compared with the other cereals.

XVII.   UNDERSOWING OATS WITH GRASS AND CLOVER SEEDS

When winter oats have to be undersown with "seeds" then a reduction in the cereal seeding rate and the total amount of fertiliser nitrogen should be considered. Both of these will reduce the amount of flag leaf present and thus allow more light to penetrate the crop canopy to facilitate a better establishment of the grass and clover seedlings. If this is not done then there will be little chance of a satisfactory "take" and an expensive seed mixture could well be wasted. Undersowing takes place in the early spring before a full ground cover is obtained with the winter crops and in respect of spring sown oats, this operation should take place at cereal drilling time or soon after brairding. Slightly higher levels of phosphate and potash may be needed especially if a legume is included in the seed mixture and this fertiliser dressing is best applied broadcast. Combine drilling simply places the fertiliser close to the cereal roots and since the establishing herbage species are between the coulter rows, then a deficiency could quite easily occur from the point of view of the grasses and clovers. This is a major consideration when the soil reserves of these two major nutrients are low on analysis.

XVIII.   CONTROL OF INSECT PEST

With quite a large proportion of the oat crop following the ploughing up of grassland, wireworms and leatherjackets often present serious problems in the spring. Gamma-BHC applied to the soil or as a seed dressing is effective in controlling the former.

Leatherjackets. Crane flies or daddy long legs belong to the insect family Tipulidae and in the larval stage (leatherjackets) are injurious to many agricultural crops, particularly oats, in the wetter districts of Britain and Ireland. Several species may be involved in crop damage, the most commonly occurring ones of economic importance are underlined in the list below.

Tipula paludosa (Meig.) Marsh crane fly.
Tipula vernalis (Meig.)
Nephrotoma (*Pachyrhina*) *maculata* (Meig.)
Nephrotoma (*Pachyrhina*) *flavescens* (L.)
Tipula oleracea (L.)
Tipula lunata (L.) and *Tipula lateralis* (Meig.)

Leatherjackets are associated with marshland and grassland where spring populations may reach 3 million and 1 million/acre respectively. Populations are much smaller under arable crops and although the extent of damage is not so widespread, leatherjackets being gross feeders can quickly reduce plant numbers since they eat their way through the plants at ground level. Most of the damage caused by these pests is to spring cereals, usually oats, when grown immediately after ploughed up grassland. Leatherjacket numbers quickly rise under grassland since they are not disturbed and predators are working under great difficulties. Damage will be severe the first year out of grass when populations are high, but unlike the wireworm, leatherjackets complete their life cycle in one year and thus are only of economic importance in this first year.

TABLE 10

Chemical Control of Leatherjackets by Spraying or Baiting

A. M.A.F.F. APPROVED

Spraying (Materials to be applied in not less than 20 gallons water per acre)

(a) 10 fluid ounces of 80% gamma BHC Suspension

or

(b) 3 pints of 25% DDT

or

Only in the case of DDT susceptible barley varieties

(c) 2 pints of 30% Aldrin

Baiting (Based on 28 lb/acre of slightly moistened bran plus *one* of the following leatherjacket poisons)

(a) 1 lb of Paris Green (also controls slugs)*

or

(b) 5 fluid ounces of 80% gamma BHC

or

(c) 1 lb of 50% BHC wettable powder

or

(d) 1 lb of 50% DDT wettable powder.

* M.A.F.F. specification: Not less than 30% copper as cupric oxide; not less than 55% arsenic calculated as arsenious oxide, not more than 1·5% arsenic as arsenious oxide—soluble in water and not less than 10% acetate as acetic acid.

B. NEW MATERIALS RECENTLY APPROVED (R) OR AWAITING APPROVAL (A)

Sprays

(a) FOLITHION 2 lb active ingredient per acre—(A)

Granular Insecticides

(a) Phorate 1 lb/acre (R)

(b) Birlane 2 lb/acre (fairly expensive) (R)

(c) Parathion 1 lb/acre (A)

(d) Diazinon 2 lb/acre (expensive) (R)

For further information on Leatherjackets consult M.A.F.F. Advisory Leaflet No. 179.

Cultural control can be obtained by early ploughing in July or August (before the main egg-laying period of the crane flies) but this restricts the grassland output in the final year and is not often considered.

Chemical control can be obtained by either spraying or baiting and details concerning the insecticides appear in Table 10.

XIX. MANGANESE DEFICIENCY

Deficiency of manganese causes Grey Speck in oat crops. The leaves turn yellow and grey and in severe cases they die. The following should be considered to avoid and/or remedy this mineral deficiency.

(i) Apply lime to crops other than oats in rotations and make sure that excess quantities are not used.

(ii) Where Grey Speck has been recorded in the past, apply 20–30 lb/acre of manganese sulphate well before drilling

(iii) When young oat crops exhibit manganese deficiency, spray with a 1–1·5% solution of manganese sulphate, plus a wetter, at up to 100 gallon/acre.

XX. CHEMICAL WEED CONTROL

A wide range of herbicides exists today for the control of broad-leaved weeds in oats and these are listed in Table 11.

The chemical 2,4-D and mixtures containing it may be used on winter oats but they should not be applied to spring oats since the majority of varieties are susceptible to damage, which in some cases has amounted to almost total destruction of the crop. Bearing this in mind and the range of weeds to be removed, the appropriate weedkiller should be selected after due consultation with Table 11 and also Table 9 in Chapter 2. As far as possible the cheapest one should be chosen so long as it is recommended for the particular circumstances under review and it is important to read carefully the manufacturer's instruction and to comply with them. When oat crops have been under-sown with grass and clover seeds then the list of suitable weed-killers is much smaller since the establishing herbage seedlings are very susceptible to many of the stronger chemicals. Under most circumstances the young clover plants will need to have at least one or two trifoliate leaves formed before spraying can be safely carried out.

Finally a word of warning in connection with chemical weed control spraying on oats. Less than half of the materials can be recommended

TABLE 11

Weed-killers That May be Used on Oats in Spring[a,]

	Winter oats Crop fully tillered but before the "jointing" stage	Spring oats	
		3–4 leaves	5 leaves to "jointing"
I Weed-killers for the control of broad-leaved weeds			
(a) Cereals undersown with clover			
MCPA	√	√	√
2,4-D	√	NO	NO
mecoprop	√	√	√
mecoprop + fenoprop	√	NO	√
mecoprop + 2,4-D	√	NO	NO
dichlorprop	√	√	√
dichlorprop + 2,4-D	√	NO	NO
dichlorprop + MCPA	√	√	√
dicamba + mecoprop with or without MCPA	√	NO	√
2,3,6-TBA + MCPA	√	NO	√
2,3,6-TBA + dicamba + mecoprop + MCPA	√	NO	√
dicamba + benazolin + MCPA	√	NO	√
ioxynil + mecoprop	√	√	√
ioxynil + dichlorprop + MCPA	√	√	√
bromoxynil + ioxynil + dichlorprop	—	√	√
bromoxynil + MCPA	√	√	√
(b) Cereals undersown with clover			
MCPB + a little MCPA[c]	√	√	√
2,4-DB + a little MCPA[c]	√	√	√
2,4-DB + a little 2,4-D or MCPB and MCPA[c]	√	NO	NO
benazolin + 2,4-DB or MCPB and a little MCPA[c]	√	√	√
dinoseb[d]	√	√	√

[a] For trade names see the list of approved products.
[b] For the dose to use read the manufacturer's instructions.
[c] Clover should have its first trifoliate leaf before spraying.
[d] Clover should have two trifoliate leaves before spraying.
√ Indicates suitability.
NO = Not suitable.

This extract from M.A.F.F. Short Term Leaflet No. 19, 1970, is reproduced by kind permission of the Ministry of Agriculture, Fisheries and Food.

for spraying on young seedling oat plants in the 2–4 leaf stage and farmers must make sure that both chemical and growth stage are correct for the job in hand.

XXI. HARVEST

A. By Binder

It has already been pointed out that for straw to have additional feeding value the crop should be cut early and this will mean bindering one to two weeks before full ripeness. If harvested at an earlier stage than this, the grain will have a thin pinched appearance even though it has been shown that some food materials pass from the straw to the ears after cutting. Very tall crops may be cut earlier than usual to make the harvesting easier and to avoid risks of late lodging and this will also apply to crops which have been undersown. In order to get the best "take" of seeds, the highly competitive cereal nurse crop should be cut and removed from the field as soon as possible. Badly laid crops which go down early will smother the herbage seedlings which soon rot, leaving bare patches which can often be detected years later if the ley survives. If, on account of prolonged wet weather, the sheaves have to remain in the field for a considerable period then they should be moved from time to time in order to eliminate this decay of the young grass and clover plants. Quick maturing varieties and those unilateral panicle types like Onward and Forward will often be cut before the rest to avoid grain losses and lodging and this is particularly true in those areas which experience high winds around harvest time due to their situation or elevation. Findlay (1956) suggests that the best time to cut a good standing crop of oats is a few days before it is fully ripe, when there is still a tinge of green and he goes on to remind us to wait until the dew has gone in the morning before commencing and to discontinue immediately there are signs of it again in the evening. Since oats are much more prone to losing their grain by shattering than the other cereals then bindering must always be before full maturity.

Once cut the sheaves are stooked usually in groups of six or eight to keep their heads off the ground and to allow the sun and wind to dry them out. In order to obtain direct radiation from the sun on all sheaves in the stook the line should point in a north–south direction so that one side receives the rays in the morning, the other side in the afternoon. The sheaves from undersown crops will naturally contain a large amount of green herbage in the butts and when the weather is dry they may be left on the ground for a day or two for this material to dry out before stooking.

In the high rainfall areas of the country oat sheaves are sometimes stacked on tripods in the field to try and prevent the grain from sprouting and it may be advantageous to stook along both sides of a wire fence which may go round part or all of the field in question. This is of particular value when the oat crop has been undersown and to be able to remove all or part of the crop will give the young herbage plants a much better chance to establish. In any case the stooks would have to be moved weekly to avoid the rotting which has already been described.

The stooks will need to stand for about a fortnight in the field before they are ready to be brought into the stackyard or steading otherwise they will quickly overheat and go mouldy. Stacks are usually left to "settle" for several days before being thatched but this is a dying art with tarpaulins and plastic sheets taking the place of this picturesque but time-consuming farm craft.

In Central Scotland, specially constructed barns were erected for sheaf drying (already referred to under wheat) with a set of wires running from end to end. These wires left passageways in the stack through which air could circulate and dry out the sheaves.

B. Windrow Harvesting

Oats shatter more easily than barley or wheat and high winds when the crop is ripe can result in extensive loss of grain. In some parts of the United States and in parts of Canada a high proportion of the oat crop grown specifically for the grain is windrowed to avoid these losses due to shattering. This method of harvesting has not been widely adopted in Britain, but has been practised to a limited extent for many years. It is a two-stage harvesting technique designed to allow the crop to be cut earlier than with direct combining. The first operation involves cutting the crop with a windrower to lay it evenly on a high stubble, well off the ground. This windrow is gathered by a combine harvester fitted with a pick-up attachment several days later when the straw and grain have dried out.

C. Combine Harvesting

The proportion of the oat crop harvested by combine has risen markedly over the past few years and with the higher labour costs involved in binding, together with an ever-decreasing labour force in agriculture, this trend is almost certain to continue. The advantages of combining the crop lie firstly in the speed at which the operation can be carried out, secondly grain losses can be kept to a minimum and thirdly it is possible to salvage badly lodged crops which would be extremely difficult by any other means.

The grain should be left to dry out as far as possible in the field to reduce artificial drying costs, but not at the expense of grain losses. The straw should be baled as soon as possible following the combine and since there is always a ready market for this commodity it can easily be disposed of should it not be required. Good baled oat straw will always command a higher price than the equivalent quality barley or wheat straw especially in late winter or early spring.

Fig. 1. Simplex expanded metal radial airflow drying silo; 18 feet high × 12 feet 2 inches in diameter, with a net capacity of 45 tons. The central drying cylinder is illustrated in the empty silo in the foreground and the full ones to the rear demonstrate the centrally heaped grain. Photograph by courtesy of Simplex of Cambridge.

XXII. DRYING AND STORAGE

Drying of oats will be necessary for bulk grain storage when the moisture content is over 15–16% and the methods and details given under wheat and barley are also applicable to this cereal.

The storage of small quantities of oats on concrete or wooden granary floors can often be successful at moisture contents between 16 and 20%.

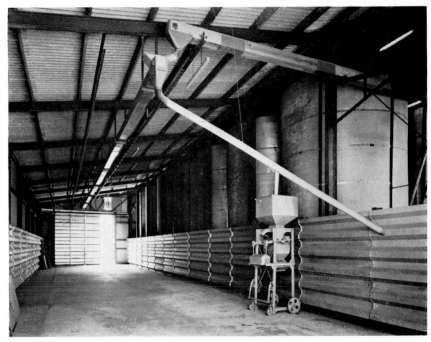

FIG. 2. Photograph of a double row of seven 24 feet × 14 feet 7 inches expanded metal radial airflow silos (each 85 tons capacity) and floor storage space for dried grain at the side. Tichwell Farms Ltd, Norfolk. Photograph by courtesy of Simplex of Cambridge.

FIG. 3. External view of drying and storage plant shown in Fig. 2. Total bin capacity 1200 tons plus 1400 tons floor storage space. Tichwell Farms Ltd, Norfolk. Photograph by courtesy of Simplex of Cambridge.

Fig. 4. Photograph of two rectangular self-emptying bins (18 feet high, capacity 44 tons) with ventilating floors and built-in sheet metal air duct running underneath. Central passage shows emptying points and grain conveyor. Photograph by courtesy of Simplex of Cambridge.

Farmers, however, must be prepared to turn the grain fairly frequently and storage depth should not greatly exceed 9–12 inches. This is the simplest and cheapest method of holding grain but it must be emphasised that it really only applies to relatively small amounts which are to be fed to stock soon after harvest.

Fig. 5. External views of drying plant illustrated in Fig. 4 showing it to be a self-contained unit. All equipment concerned with this drying set-up comes in a package deal from the manufacturer. Photograph by courtesy of Simplex of Cambridge.

A. Polythene Sack Storage for Moist Grain

This is probably the second cheapest method of grain storage and applies equally well to oats as it does to barley although most of the experiments and trials on a commercial scale have been done on barley. Polythene sacks which previously held fertiliser can be used once cleaned and tested for air leaks, but if this ultra-simple way of holding grain is to be a permanent feature then it is better to start with new ones which are specially manufactured for this purpose. The initial outlay will probably be in the region of £1–2 per ton of grain stored but these containers can be used for several years provided that due care is taken in the filling and handling of them. Various methods have been tried for tying up the bags. Heat sealing is probably the most effective but since they have to be broken open to remove the grain this usually means that the capacity in subsequent years is reduced. Metal wire clips are often favoured since they are easily and quickly put on and released. As much air as possible must be expelled from the polythene bags before they are sealed for the grain to keep its condition and rodents must be discouraged from gnawing holes in them otherwise the principle

FIG. 6. The Berwyn acid applicator coupled by two 45 gallon drums of propionic acid. This unit, operated by a 2 h.p. electric motor (mains 240 AC) shows the grain hopper and throughput auger (left). Maximum rating 20 tons/hour. Photograph by courtesy of Berwyn Engineering Ltd, Chippenham, Wiltshire.

FIG. 7. Easily transportable acid applicator—Condor Spraypak Unit. By courtesy of Condor (Agricultural) Ltd. Photograph by Jack Casselden, Ely, Cambridge.

of storage in a minimal oxygen environment will be lost. Even small quantities of air getting to the moist grain will induce mould growths which quickly develop and seriously impair the feeding value and the possibility of producing mycotoxins (referred to earlier under barley) is always present under these circumstances. Heavy-duty polythene of 500 and 1000 gauge is used but even this material does not altogether stop the passage of some of the volatile organic compounds associated with slight fermentation of the grain and these are probably instrumental in attracting mice and rats. This method of storage can be recommended for grain which is to be used fairly soon after harvest on the farms where it has been grown but it cannot be emphasised too strongly that the control of rodents is imperative for it to be successful once the properly sealed sacks are in granaries.

XXIII. PRODUCTION ECONOMICS

Much less attention has been paid to the production economics of oats than in the case of barley and wheat since the amount of land

TABLE 12

Average Production Costs for Bindered and Combined Oats

Item	Average cost/acre for combine cut crops	Average cost/acre for binder cut crops
	£	£
Fertiliser, seed and other materials	5·32	6·16
Farm tractor work and contract services	3·61	3·13
Farm Labour	3·43	6·68
Machinery depreciation, repairs, special equipment	2·74	3·17
Rent	2·39	2·17
Overheads[a]	5·75	7·95
F.Y.M. and application cost	4·20	—
Lime	0·05	0·89
Slag	0·25	—
	27·74	30·15
Adjustment for manurial residues	+0·83	+1·16
Total cost/acre	28·57	31·31

[a] Overheads have been calculated excluding F.Y.M.

devoted to the oat crop is much smaller. Also in recent years in all the counties of U.K. there has been a significant decline in the acreage and the majority of farmers grow this cereal for home consumption. Munro (1963), using data collected in Ayrshire and Wigtownshire, has given a guide to the cost of combined and bindered oats, details of which appear in Table 12.

Only a quarter of the combined crops actually received farmyard manure on this survey and in order to obtain a fair comparison in the field between oats handled by these methods, it was thought necessary to modify the figures by omitting the F.Y.M. charge and increasing the

TABLE 13

Oat Crops, 1962: Production Costs, Output and Margin/Acre

	Combined Crop £	Bindered Crop £
Gross Cost	26	32
Approximate straw value	5	8
Unsubsidised net cost	21	24
Subsidised net cost[a]	14·50	17·50
Straw yield cwt/acre	19	23
Grain yield cwt/acre	25	24
Value of grain[b]	22·50	21·50
Margin/acre (grain sales only)	8	4

[a] Deficiency payment of £6·50 acre assumed in both handling methods.
[b] Value of grain taken to be £18 per ton.
Based on original data by Munro (1963).

fertiliser costs slightly to compensate. In addition the value of the straw has been assumed to be approximately £5 and £7/ton for combine and bindered material respectively and the figures have been rounded-off in Table 13.

From the data in Table 13 it can be seen that oat crops in the West of Scotland would leave very little profit should the grain be sold and on average this margin in 1962 was of the same magnitude as the Government support through the deficiency payment. Present costs of production will be higher than those shown even when the acreage grown is large and for a higher financial output per acre where oats are grown as a cash crop, farmers must aim for higher yields and a better market price for the grain.

XXIV. YIELD OF GRAIN AND STRAW

A. Yield of Grain

Estimates of grain yield are made annually by the Ministry of Agriculture, Fisheries and Food for England and Wales, the Department of Agriculture for Scotland and for Northern Ireland their Ministry of Agriculture. These figures are published annually in the bulletin on Agricultural Statistics. Individual harvest years are quoted for each country and the 10-year averages have been calculated to facilitate decadal comparisons. In the early to mid 1930's oat yields in Scotland (16 cwt/acre) were marginally higher than in England and Wales (15·5 cwt/acre) but both were easily topped by those in Northern Ireland (19–19·5 cwt/acre). Improvement in national yield was slow in Scotland, moderate in England and Wales, and apparently non-existent in Northern Ireland from the mid 1930's to mid 1960's as can be seen from the figures below.

10-Year average	England and Wales (Average yield cwt/acre)	Scotland (Average yield cwt/acre)	Northern Ireland (Average yield cwt/acre)
1934–43	16·2	16·3	18·8
1944–53	17·9	16·7	18·0
1953–62	21·3	19·5	18·1

There have been significant increases in the national average yields during the 1960's, from 21 cwt to 28 cwt in England and Wales and from 21 to 25 cwt in Scotland, but again there has been little improvement in Northern Ireland.

B. Yield of Straw

It is interesting to note the differences in straw production over the 10-year period 1954–63: England and Wales, 17 cwt; Scotland, 22 cwt and Northern Ireland about 25 cwt/acre. The significantly greater amounts of straw produced in the North and West are largely due to the influences of climate namely the cool, humid condition of much of Ireland and Scotland. It is also due in part to the varieties which are grown and this is particularly applicable in Northern Ireland where the straw types have been recommended over a long period. The relatively small amount of straw obtained from oat crops, particularly in England, is due mainly to the lower rainfall and higher sunshine experienced there together with the fact that most farms under lowland conditions choose a combine oat which normally is a poor straw producer.

The corresponding grain yields are given above the average straw production figures for the period 1954–63 in order to examine the grain : straw ratios for each of the countries in turn. These indicated that oat crops in England and Wales produced 20% less straw than

10-Year period		England and Wales	Scotland	Northern Ireland
1954–63	Grain yield (cwt/acre)	21·6	19·8	20·6
	Straw yield (cwt/acre)	17·1	21·8	25·2
	Ratio grain : straw	1 : 0·79	1 : 1·10	1 : 1·22

grain whereas in Scotland and Northern Ireland the corresponding figures were 10% and 20% more straw respectively. The actual range of grain : straw ratios from 1 : 0·79 to 1 : 1·22 is similar to that reported by Bell (1962), referred to earlier, which represents the extreme grain and straw types respectively.

Appendix
Recommended Varieties of Winter Oats for England and Wales—1971†

The control for yield comparisons is the mean of Maris Quest and Peniarth. Comparisons other than with the control are not strictly valid, and any difference of 3% or less should be treated with reserve.

Varieties classified for General use G, Becoming Outclassed O	Recommended			Becoming outclassed
	Pendrwm	Peniarth	Maris Quest	Padarn
	G	G	G	O
Agricultural Characters:				
Yield as % of control	102	101	99	94
Standing power	5	6½	7½	5
Shortness of straw	4½	5	7½	4½
Earliness of ripening	6	7	5	7½
Winter hardiness	5	7	5	5
Resistance to mildew	6	7	4	6
Resistant to stem eelworm (R)		R	R	
Quality of Grain:				
Kernel content	7½	7	6	7
1000 corn weight	8	6	7	7
Year first listed	1968	1965	1966	1962

A high figure indicates that the variety shows the character to a high degree.

VARIETIES RECOMMENDED FOR GENERAL USE

PENDRWM

Yield high and straw moderately stiff.   V.998/4 × 4745/33/3. Welsh Plant Breeding Station, Aberystwyth.

PENIARTH

Combines high yield with stiff straw, resistance to stem eelworm and good resistance to mildew.   S.172 × (S.147 × 01747/10/7). Welsh Plant Breeding Station, Aberystwyth.

†Farmers' Leaflet No. 8. Issued by the National Institute of Agricultural Botany, Huntingdon Road, Cambridge.

MARIS QUEST
Combines very short, stiff straw with   Blenda × S.172.
high yield. Resistant to stem eelworm.   Plant Breeding Institute,
                                                          Cambridge.

VARIETY BECOMING OUTCLASSED

PADARN
Moderately high yield and early ripen-   Powys × (S.147 ×
ing.   01747/10/7).
                                           Welsh Plant Breeding
                                                Station, Aberystwyth.

Recommended Varieties of Spring Oats for England and Wales—1971†

The Control for yield comparisons is the mean of Condor and Astor. Comparisons other than that with the control are not strictly valid, and any differences of 3% or less should be treated with reserve.

	Recommended				
Varieties classified for General use G, Special use S	Mostyn G	Selma G	Astor G	Condor G	Manod S
Agricultural Characters:					
Yield as % of control	103	103	101	99	90
Standing power	7	7	8	6	6
Shortness of straw	7½	7	8	7	4
Earliness of ripening	5	5½	5½	5½	5
Resistance to mildew	6	2	3	4	6
Resistant to stem eelworm (R)					R
Quality of Grain:					
Kernel content	6	6½	5½	6	6½
1000 corn weight	7	7	7	7½	6½
Year first listed	1968	1970	1962	1960	1960

A high figure indicates that the variety shows the character to a high degree.

VARIETIES RECOMMENDED FOR GENERAL USE

MOSTYN
High yield and moderately good resistance to mildew. Stiff straw. 05443 × Condor. Welsh Plant Breeding Station, Aberystwyth.

SELMA
High yield, especially in the north, and stiff straw. Susceptible to mildew. Palu × Saxo. Weibull, Sweden.

ASTOR
Combines very stiff straw with high yield. Susceptible to mildew. Marne × Minor. C.I.V., Holland.

CONDOR
High yield with stiff straw. Susceptible to mildew. Minor × Expres. C.I.V., Holland.

†Farmers' Leaflet No. 8. Issued by the National Institute of Agricultural Botany Huntingdon Road, Cambridge.

VARIETY RECOMMENDED FOR SPECIAL USE

MANOD

The resistance of this variety to crown rust and oat stem eelworm makes it of special value in the south-west and west. Moderately good resistance to mildew.

01750/11 × Tama.
Welsh Plant Breeding Station, Aberystwyth.

1970 Varieties of Cereals for Scotland

Recommendations in this leaflet are based on trials and experience of the Scottish Agricultural Colleges.

In case of doubt, or for information about varieties not on the list, farmers should consult their Agricultural Adviser.

BRITISH CEREAL SEED SCHEME
(SCOTLAND)

Certified seed, Multiplication seed or Field Approved seed of the following recommended varieties will be available through normal commercial channels for sowing in 1970:

Oats—Astor, Condor, Forward, Blenda, Yielder, Ayr Commando, Karin, Tarpan.

SPRING OATS

G	(N)	Selma	106
G		Astor	100
G		Condor	100
S	(O)	Blenda	
S		Yielder	
S	(O)	Maelor	
S		Karin	
S	(O)	Tarpan	
S		Forward	
S		Ayr Commando	
S		Shearer	
S		Bell	

Additional varieties for North and North-west Scotland

SPRING OATS

SELMA

A few inches taller than Astor and ripens slightly later. Susceptible to mildew. Grain of high kernel content. In West College trials to date, yields no higher than Astor.

ASTOR

A very stiff, short strawed oat of average maturity. Mildew susceptible.

CONDOR

Stiff, short straw. Average maturity.

BLENDA

Suitable for binder-harvesting under a wide variety of conditions but rather late maturing for the uplands. A quality oat which can produce a good milling sample.

YIELDER

An adaptable variety suitable for binder harvesting in wet or late districts. Ripens early but gives only moderate yields of husky grain. Stands less well than more recent introductions.

MAELOR

Moderately good yield of large grain with high kernel content. Very long but stiff straw; fairly early. Recommended as a forage oat.

ADDITIONAL OAT VARIETIES FOR NORTH AND NORTH-WEST SCOTLAND

KARIN

Grain yield similar to or slightly higher than Astor. Straw breaks readily at maturity. Suitable for combining or if cut early, for bindering. Maturity 3–7 days earlier than Astor. Suited to high fertility conditions.

TARPAN

Grain yield similar to Astor. Straw which is weaker than that of Astor breaks readily at maturity. Ripens up to 4 days earlier than Astor. Suited to moderate fertility conditions.

FORWARD

May be binder-harvested before fully ripe or allowed to mature (straw breaks over) for combining. Fairly good yield of large but husky grain, not well suited for milling. Straw of medium length has only fair resistance to lodging.

AYR COMMANDO

A variety of moderate yield suitable for the uplands and high rainfall districts because of its early maturity. May be binder-harvested before fully ripe or allowed to mature for combining. Strong straw.

SHEARER/BELL

For extreme conditions of soil and climate in the Islands and Highlands.

VARIETIES OF CEREALS†

   This leaflet is issued as a guide for cereal growers. Since 1964, a further series of trials has been carried out, and present recommendations are based on results of these and previous trials. As in the previous

† Leaflet No. 96. Issued by the Northern Ireland Ministry of Agriculture.

leaflet, where no local trials have been carried out (that is, for winter cereals and spring wheat), data from the National Institute of Agricultural Botany (N.I.A.B.), Cambridge, are supplied as an indication of possible performance in Northern Ireland.

Each recommended variety is placed in one of four classes:

G—for general use
N—new recommendation
S —for special use
O—becoming outclassed by other recommended varieties.

SPRING OATS

The varieties Astor, Tarpan, 6290 Cn4/34, Aa737, Aa739 and Shearer were compared to Condor in trials in 1965, 1966 and 1967. Condor is retained as a recommended variety. Only Astor, of the others, is of sufficient merit for inclusion on the list. Tarpan yielded quite well but has relatively weak straw. Milford, previously on the Recommended List, has been removed as outdated.

Recommended List of Spring Oats, 1967

Agricultural characters	Astor	Condor	Stormont Sceptre	Stormont Iris	Castleton Potato	Tyrone Tawny A	Tyrone Tawny 8
Grain yield	7½	7½	7	5	4	3½	3½
Standing power	7½	7	7	4½	3½	3½	3½
Shortness of straw	7½	7	6½	4½	3	3	3
Earliness of ripening	7	7	6½	7	4	5	5
Quality of grain							
Kernel content	7½	7½	7	5½	7	7	7
Size	7½	7½	6	5½	6	5	5½

ASTOR

This variety has outyielded Condor slightly in trials during the last three years. The grain is large, rather long and looks similar to that of Condor. The straw is short, and stiffer than that of Condor. It is suitable for combining. The upper part of the straw tends to break down when dead ripe, thus protecting the grain from wind. For highest yields this variety requires high fertility conditions.

CONDOR (G)

Condor has given consistently high grain yields in trials in Northern Ireland over the last nine years. The grain is large, rather long, but of high kernel content. The straw is stiff. The only weakness of the variety is a tendency to sprout rather readily in a wet harvest. Condor requires high fertility conditions for highest grain yields.

STORMONT SCEPTRE (G)

This variety has a short, plump grain. Kernel content and yields of grain, while not quite as high as those of Condor, are better than those of Stormont Iris. Straw is short, stiff and resistant to lodging. This variety also needs high fertility conditions for highest grain yields.

STORMONT IRIS (O)

Stormont Iris is recommended for conditions of medium fertility. Grain is short and fairly plump. Kernel content is lower than that of Stormont Sceptre. Straw is medium length and less resistant to lodging than that of Stormont Sceptre. The grain of Stormont Iris usually ripens before the straw. Early harvesting is essential to avoid excessive shedding.

CASTLETON POTATO (O)

Grain is short, plump and of high kernel content. Straw is long and relatively weak. Potential grain yield is low. Recommended only if the fertility is relatively low.

TYRONE TAWNY A (SHORT GRAIN)/TYRONE TAWNY 8 (LONG GRAIN) (S)

Straw is fine and weak. Grain is usually brown but occasionally almost white. Recommended only for wet and low fertility conditions.

WINTER OATS (N.I.A.B. DATA)
CERTIFIED SEED

The importance of sowing good seed cannot be over-emphasised. Good seed makes better use of fertilisers and labour. It produces heavier crops of grain.

High quality seed is made available under the Ministry's Seed Certification Schemes. This seed is available commercially in sealed bags. It carries an assurance that it has been grown and processed according to the Ministry's rules for maintaining purity, freedom from weeds and germination. It is disinfected with a mercurial dust.

Merchants who make contracts with farmers for the production of certified cereal seed in Northern Ireland are:

Messrs J. Coburn and Son, Ltd—Newry Street, Banbridge.
Messrs J. Morton (Banbridge), Ltd—Commercial Road, Banbridge.
Messrs Samuel McCausland, Ltd—36 Victoria Street, Belfast, BTI 3GT.
Messrs Hurst, Gunson, Cooper-Taber, Ltd—High Street, Lurgan.

MINISTRY OF AGRICULTURE
DUNDONALD HOUSE
BELFAST BT4 3SB
April. 1968

References

Arrhenius, O. (1926). "Kalkfrage, Bodenreaktion und Pflanzenwachstum". Leipzig.
Bell, R. A. M. (1962). *J. nat. Inst. agric. Bot.* **9**(2), 126–138.
Boyd, J. (1952). *Emp. J. Exp. Agr.* **20**, 78, 115–122.
Carleton, M. A. (1916). "The Small Grains". Macmillan & Co., New York (reprinted 1920).
Coffman, F. A. (1946). "Origin of Cultivated Oats". *J. Am. Soc. Agron.* **38**, 983–1002.
Coffman, F. A. (1961). "Oats and Oat Improvement", pp. 15–40. American Society of Agronomy.
Cooke, G. W. (1960). "Fertilisers and Profitable Farming". Crosby Lockwood, London.
Darlington, C. D. and Wylie, A. P. (1955). "Chromosome Atlas of Flowering Plants", pp. 447–448. Allen and Unwin Ltd, London.
De Candolle, A. (1882). "Origin of Cultivated Plants". International Science Series. D. Appleton and Co., New York.
Dent, J. W. (1957). *J. Agric. Sci.* **48**, 336.
Dent, J. W. (1959). *J. nat. Inst. agric. Bot.* **8**, 614.
Dent, J. W. and Boyd, A. G. (1965). *J. nat. Inst. agric. Bot.* **10**(2), 180–187.
Emme, H. (1930). *Züchter* **2**, 65.
Etherbridge, W. C. (1916). "A Classification of the Varieties of Cultivated Oats". Cornell University Agric. Exp. Sta., Memoir 10.
Evans, R. E. (1960). "Rations For Livestock". M.A.F.F. Bulletin No. 48. H.M.S.O.
Farmer and Stockbreeder (1968). Farm Facts No. 14. January, 1968.
Findlay, W. M. (1956). "Oats. Their Cultivation and use from ancient times to the present day". Aberdeen University Studies 137. Oliver and Boyd, Edinburgh and London.

Haussknecht, C. (1885). "Uber die Abstammung des Saathabers". *Mitt. Georgr. Gesell.* **3**, 231–242.

Hellstrom, V. and Andersson, R. (1953). "Content of B vitamins in Oats and Oat Flour". *Var. Föda* **10**, 41.

Holliday, R. (1956). "Fodder Production From Winter-Sown Cereals And Its Effect Upon Grain Yield". *F.C.A.* **9**(3), 1–13.

Hunter, H. (1951). "Crop Varieties. Varieties of Cereals, Flax, Potatoes, Beans and Field Peas". Farmer and Stockbreeder Publications Ltd, London.

Huskins, C. L. (1927). *J. Genet.* **18**, 315.

Kent, N. L. (1964). "Technology of Cereals With Special Reference to Wheat", p. 211. Pergamon Press, London.

Kinsey, C. (1959). Ministry of Agriculture, Fisheries and Food Booklet. "Oats". H.M.S.O.

Körnicke, F. and Werner, H. (1885). "Handbuch des Getreidebaues" (2 volumes). Paul Parey, Berlin.

Lewis, C., Campbell, J. D., Bickmore, D. P. and Cook, K. F. (1951). "The American Oxford Atlas". Oxford University Press, New York.

Lucas, N. G. (1967). "Broadcasting Cereal Crops". *Farm Mech.* March, 47–48.

Malzew, A. I. (1930). "Wild and Cultivated Oats" (Sectio Euavena Griseb.). *Bul. Appl. Bot., Genet. and Plant Breed.*, Suppl. 38. Leningrad. (English translation pp. 473–506.)

Munro, R. F. (1963). "Oat Costings, 1962 Crop". West of Scotland Agricultural College, Glasgow. Economics Department Report No. 91.

Nicolaissen, Von W. (1940). "Hafer, *Avena sativa* L.". Paul Parey, Berlin; *Handbuch Pflanzen-Züchtung* **2**(13–18), 224–288.

Nishiyama, I. (1936). *Cytologia* **7**, 276.

Percival, J. (1947). "Agricultural Botany. Theoretical and Practical" (eighth edition). Duckworth, London.

Philp, J. (1933). *J. Genet.* **27**, 133.

Schulz, A. (1913). "Abstammung und Heimat des Saathafers". *Zeft. Gesam. Getreidewesen* **5**, 139–142; "Die geschichte des Saathafers". *Jahresber. Westfalischen Prov. Ver Wiss. und Kunst* (1912–1913) **41**, 204–217.

Small, J. (1946). "pH and Plants. An Introduction for Beginners", p. 144. Baillière, Tindall and Cox, London.

Spier, J. D. (1934). *Canad. J. Res.* **11**, 347.

Stanton, T. R. (1936). *Yearb. Agric.* U.S. Dept. Agric., p. 375.

The Quality of Oats (1958). "For Milling Into Products for Human Consumption". Issued jointly by West of Scotland Agricultural College and Research Association of British Flour Millers.

Trénel, M. (1927). "Die wissenschaftlichen Grundlagen der Bodensäure frage". Berlin.

Vavilov, N. I. (1926). "Studies on the Origin of Cultivated Plants". *Bul. Appl. Bot. Plant Breed.* **17**(2), 139–245.

Voelcher, A. (1861). "On the Composition and Nutritive Value of Straw". *J. Roy. Agric. Soc.*

Williams, H. T. (1968). Private communication. Production manager, A. & R. Scott, Ltd, West Mills, Colinton, Edinburgh.

Zade, A. 1918. "Der Hafer. Eine Monographic auf Wissenschaftlicher und Praktischer Grundlage", pp. 1–355. Jena, Vertag von Gustav Fischer.

CHAPTER 4

Rye

I. Introduction 178
   A. Rye Growing in the Rest of Europe 179
II. Botanical Classification 179
III. Utilisation of Rye 181
   A. Grain Uses 181
   B. Straw Uses 182
IV. Grain Quality 182
   A. Chemical Composition and Nutritive Value of Grain and By-
     products 182
   B. Rye Protein 184
   C. Grain Size 184
V. Soils and Soil pH 185
   A. Soils for Rye 185
   B. Optimum pH 185
VI. Climate and World Distribution 186
VII. Rye for Grain 187
   A. Position of Rye in Cropping Sequence 188
   B. Fertiliser Levels 188
   C. Cultivation and Seed-bed Preparation 189
   D. Drilling 190
   E. Seed Rate 190
   F. Seed Dressing 191
   G. Type of Seed 191
   H. Varieties of Rye 191
   I. Crop Management 194
   J. Chemical Control of Weeds 199
   K. Harvesting 200
   L. Grain Disposal 200
   M. Yield of Grain and Straw 202
   N. Production Economics 202
   O. Government Support for Rye 202
VIII. Rye for Forage 203
   A. Introduction 203
   B. Varieties of Rye for Forage Production 204
   C. Cultural Details for Forage Rye 205
   D. Grazing Management 206
   E. Yield and Quality of Forage 207
   F. Production in Terms of Cow-grazing Days and Economic Aspects .
     of Rye for Spring Grazing 208

G. Ergot in Rye 209
Appendix   211
References 213

I. INTRODUCTION

Wheat, barley, oats and rye were all grown for bread-making in early times. However, by the middle of the eighteenth century, Charles Smith (1758) (quoted by Percival, 1934) showed that by far the largest proportion of the population in England and Wales ate bread made from wheat. Bearing in mind the details of quality comparisons made in the chapter on wheat this is hardly surprising.

Population in England and Wales (1758)

6 million

	Wheat bread	Rye bread	Barley bread	Oat bread
Approximate %	3·75 million	880,000	739,000	623,000
total population	62%	15%	12%	11%

According to O'Brien (1925), rye bread formed the cereal food of at least one-seventh of the population of the British Isles in the time of Adam Smith (1723–90), and over 1 million acres were grown annually to supply this demand. Since then the acreage under rye has gradually decreased and by the end of the nineteenth century only 10% of this crop remained compared with the mid eighteenth century. From 104,000 acres in 1894 the acreage dropped to 53,000 in 1908 and to 51,500 by 1912, but it rose again sharply with the advent of the First World War when the national policy was geared to maximum home production and in 1920, 108,000 acres were grown. Between the First and Second World Wars there was a further decline in the acreage grown and this would have been even more marked, had there not been a considerable demand for seed rye to be used for catch cropping. Winter-sown rye put in during the late autumn and early winter was fed green to stock in the spring and there is still considerable interest in it today for this purpose.

Nowadays in comparison with wheat, barley or oats, rye is insignificant even when the acreage for grain and fodder are added together. This has been done in respect of the 1939 figures when a total of about 18,000 acres were sown. From Table 1 it will be seen that the majority of the rye acreage is listed under England and Wales although in practice rye for grain is nearly all located in England.

TABLE 1

Rye Statistics from the Annual June Returns (acres)

	1939	1949	1959	1965	1966	1967	1968	1969
England and Wales	16,269[a]	60,317	12,961	18,000	10,000	11,000	11,000	8000
Scotland	1130[a]	3632	808	NSR	NSR	NSR	NSR	NSR
Northern Ireland	249	538	300	350	210	220	140	131

[a] Includes rye for grazing.

NSR signifies that no separate return was made in these years.

During the Second World War, another impetus was given to the production of home-grown cereals and the acreage remaining in 1949, namely 65,000, is a reflection of this. Present production appears fairly stable at about 10,000 acres, the produce of which appears to satisfy a fair proportion of the country's requirements.

A. Rye Growing in the Rest of Europe

In the past rye has been cultivated much more widely in several European countries and although in general there has been a decline in the acreage devoted to it, particularly in France, it is still a very important source of bread flour. In the mid 1920's when rye bread constituted the cereal food of about one-third of the total population of Europe it was preferred to wheat in Russia, Germany, Poland and the Scandinavian countries. The "black bread" which rye produces is still often preferred by many in the Northern European countries.

II. BOTANICAL CLASSIFICATION

The cultivation of rye goes back at least 2000 years. Pliny mentions it as a new cereal to the Romans and records it as being grown by the conquered "barbarian" races. According to O'Brien (1925) this fixes the origin of rye about the beginning of the Christian era, thus making it more recent than wheat or barley.

In spite of it being a "relatively new" cereal there are several theories as to its origin. The first, put forward by Häeckel, suggested that the original form of rye was *Secale montanum* (Guss), a wild species growing in the mountainous countries bordering the Mediterranean and in parts of Central Asia. This species is a perennial, with a jointed rachis and these two botanical characters are thought to have disappeared through cultivation to give the present-day form.

This theory was disputed by De Candolle, who would not accept the present-day wild species as the origin of the cultivated forms. In explanation he suggested that the wild ones could have arisen via self-seeding of the ordinary forms, as they had developed the power of self-fertilisation.

Vavilov (1926) named Afghanistan, Turkistan and Eastern Persia as one of the possible areas of origin and Armenia, Georgia, Asia Minor and South-west Persia as another. Both wild and cultivated

TABLE 2

Species of Rye

Specific name	Common name	Chromosome number	Authority	Remarks
Secale ancestrale	—	14	Kostoff (1937)	Grown for grain in Anatolia
Secale vavilovii	—	14	Kostoff (1937)	Grown for grain in Armenia
Secale kuprijanovii	—	14	Nakajima (1954)	Grown for grain in the Caucasus
Secale africanum	African rye	14 / 14 + 1B	Gouws (1950) / Emme (1928)	Grown for fodder in North Africa
Secale fragile	Hungarian rye	14 + 1–2B	Emme (1928)	Found in South-east Europe and North Asia; grown for grain
Secale montanum	Wild rye	14 + 1–2B	Emme (1928)	Grown for fodder around the Mediterranean and Central Asia and also occurs wild
Secale cereale	Rye	14 + 0–8B / (7) 21 / 28	Müntzing (1943) / Müntzing (1937a, b) / Müntzing (1951)	Cultivated forms

forms of rye appear as weeds in the other cereals in these regions and large differences in form exist between the ryes of these regions thus suggesting more than one point of origin. From these primary areas of origin, rye cultivation moved West and North and since it is much less exacting as regards soil requirements and on account of its good resistance to low temperatures one would suppose the early expansion of rye growing to have been very rapid. The basic chromosome number of Secale is 7 and the different species as listed by Darlington and Wylie (1955) are all basically diploid, with the occasional addition of one or more chromosomes as indicated in Table 2.

Apart from the triploid and tetraploid forms of *Secale cereale*, indicated by Müntzing, the other forms of *Secale cereale* and the remaining species are diploids grown for grain or fodder over a wide area of Northern and Southern Europe, Asia and many other parts of the world. *S. montanum* (Guss.) is a wild species associated with the Mediterranean area and it is also cultivated for fodder. St John's Day or Midsummer Rye the spring variety of rye grown several decades ago, was thought to be related to this wild form on account of its rapid growth and marked tillering capacity.

III. UTILISATION OF RYE

Rye flour for the production of bread is second only to wheat and in Scandinavian countries and in Eastern Europe it is the main bread grain. In Britain and most other parts of the world wheat is preferred for its greater palatability and higher digestibility. Loaves from wheat are more attractive in their appearance compared with the "black bread" from rye. Nevertheless there is a demand for rye crispbread and rye starch-reduced crispbread in Britain and the production from the limited acreage has to be supplemented by grain from Europe.

In Western Germany rye was more important than wheat from 1939 to 1957; thereafter the production declined and larger proportions were fed to stock whereas in Eastern Germany rye production still exceeds that of wheat (Kent, 1964). In other parts of the world, rye grain is mostly associated with livestock feed, with a small amount used in distilling and in many areas the crop is simply cultivated for forage.

A. Grain Uses

1. Production of rye flour, coarse rye meal and rye flakes for the manufacture of soft bread, hard bread and breakfast cereal respectively. (Speciality bread is known as Pumpernickel.)
2. In the production of malt flour and malted grain for the distillation of rye whisky in Canada and U.S.A.
3. Rye flour is used as a filler in Britain and the U.S.A. in the manufacture of sauces, soups and custard powders.
4. Rye starch is one of the main ingredients of adhesives.
5. Production of gin in Holland.
6. In Britain rye grain is used at up to 15% in animal foodstuffs.

G

B.   Straw Uses

Crops of rye usually achieve a height of between 5 and 8 feet and in the past this long straw was valued highly for thatching purposes. It is almost solid in cross-section, tough and wiry and is probably the best cereal straw in its resistance to decay through climatic weathering. Average yields of grain and straw during the early 1940's were estimated at 12–16 cwt and 30–40 cwt respectively by Moore (1944). With about two and a half times more straw than grain and a good market for the former, it was easy to see why many rye crops were looked upon almost exclusively as straw producers. In fact under these circumstances, threshing was often dispensed with to obtain long, undamaged straw which would command the very highest price.

In the past rye straw was used for a variety of purposes in commerce. Until replaced by wood pulp, it was used in the production of coarse brown paper and nowadays, when available, it is used as a packing material for a variety of articles, e.g. furniture, bottles, nursery stock. It was also in considerable demand for the manufacture of such articles as hats, harnesses (padding), baskets, beehives and in brick-making (O'Brien, 1925), but little, if any, of this trade exists today.

It was also eagerly sought by stable owners for bedding purposes and today this is one of its major uses (associated with most classes of live-stock), since the feeding value of rye straw is considered to be even poorer than that of wheat. Another factor which has been responsible for this major change in straw use is due to increasing use of combine harvesters on the crop. In the threshing process the straw is broken up into small pieces and is often split, thus prohibiting its use for anything other than bedding or for some packing purposes.

IV.   GRAIN QUALITY

A.   Chemical Composition and Nutritive Value of Grain and By-products

The chemical composition of rye and its associated products are listed in Table 4 and author references are included for readers who may wish to seek further information on this subject.

"The Technology of Cereals" by Kent (1964) contains a section on the processing, nutritional aspects and technological uses of this cereal and should be consulted along with the above references when additional detail is needed in respect of the manufacturing or utilisation aspects.

Mineral and vitamin contents of rye flours have been estimated by McCance et al. (1945) and details are summarised by Kent (1964)

TABLE 3

Chemical Composition of Rye Grain, Flour and Offals

Material	Moisture (%)	Protein (%)	Fat (%)	Ash (%)	Fibre (%)	Carbohydrate (%)	Reference
Grain	15.0	9.9	1.6	1.7	1.7	70.2	Neumann et al. (1913)
Grain (English—weak)	16.5	6.5[a]	—	—	—	—	Kent-Jones and Amos (1947)
Grain (English—strong)	16.0	10.2[a]	—	—	—	—	Kent-Jones and Amos (1947)
Grain (Russian)	13.6	11.8[a]	—	—	—	—	Kent-Jones and Amos (1947)
Grain (English)	13.4	11.5	1.7	2.0	1.9	69.5	O'Brien (1925)
Grain	13.0	11.6	1.7	2.0	1.9	69.8	Evans (1960)
Grain (English)	—	8.0	2.0	—	—	75.9	Ryvita Company (1968)
Offals							
Fine Bran[b]	15.0	15.0	3.1	4.1	4.9	58.0	Neumann et al. (1913)
Coarse Bran[b]	15.0	17.8	5.1	3.7	9.2	49.2	Neumann et al. (1913)
Germ[b]	15.0	38.0	10.1	4.7	3.3	37.4	Neumann et al. (1913)
Bran	12.5	16.7	3.1	4.5	5.2	58.0	Evans (1960)
Flour							
75% Extraction	15.0	6.7	1.3	0.7	0.5	75	McCance et al. (1945)
100% Extraction	15.0	8.0	2.0	1.7	1.6	69	McCance et al. (1945)

[a] Protein arrived at by N% × 5.7.

[b] Offals after 73% flour extraction—fine bran 23.4%, coarse bran 2.4% and germ 0.2%.

and this author suggested that the nicotinic acid (niacin) content of a 60% rye flour was of the order 9μg/g.

Evans (1960) has indicated a starch equivalent value of 72 for rye, which is similar to that quoted for both wheat and barley and the protein equivalent suggested was 9·1 as compared with 9·6 for wheat and 6·5 for barley, although one would expect a higher figure for barley particularly when grown specifically for livestock feed.

B. Rye Protein

Rye protein differs from wheat protein in that it does not contain glutenin. Wheat gluten is the protein complex which is arrived at by combining glutenin and gliadin and is responsible for the elastic properties of wheat dough. Without glutenin and thus a lack of gluten, the dough from rye flour is inelastic and tends to produce a heavy, dense bread which is not preferred by the majority as standards of living rise. Quality grain is that which contains a high protein content and according to Hunter (1951) this is usually a reflection of the fact that the rye has been grown under good fertility conditions. In the older literature, weak and strong are terms used to signify low and high levels of protein in grain and usually refer to the dough which is produced from them.

C. Grain Size

The grain of rye has probably the greatest size range of all the cereals and according to Kent (1964) it can vary between 15 and 40 g/1000 grains. From the data of Kent-Jones and Amos (1947) in their book "Modern Cereal Chemistry", 1000 corn weights have been calculated and these largely fall at the lower end of the range suggested by Kent.

Country or place of origin	1000 grain weight (g)
Plate	14·3
Australia	15·4
Canada	16·9
Russia	17·8
Rumania	22·2
Lithuania	26·3
Holland	27·4
Poland	28·2
England	30·8

It is interesting to note that the grain produced under English conditions was the largest in the above data and will to some extent reflect better soil and growth conditions generally. The older varieties (presumably all diploids) were notoriously poor in the uniformity of the grain which they produced and no doubt part of this trouble could be ascribed to the fact that rye is cross-pollinated, rather than self-fertilised like wheat, barley and oats.

With the introduction of newer varieties, there appeared an improvement in the uniformity of grain, and in respect of the selections from older cultivars and of recent tetraploids there was a very significant change in grain size. So much so, that these were and are still referred to as "large-grained ryes".

V. SOILS AND SOIL pH

A. Soils for Rye

In Britain the cultivation of rye has been almost exclusively associated with light, sandy soils which could seldom support economic production of the other three cereals. These soils are normally very short of lime, a feature which does not appear to unduly affect the growth of this cereal, but it must be recorded that rye would grow very much better and produce higher yields on somewhat better land. Under extreme sandy conditions, the establishment of winter rye has often been successful in avoiding the blowing or drifting of the sand particles so often a hazard to spring sown crops. Practical experiments have shown that rye will grow very satisfactorily on all mineral soils except the very heavy clays and it is not at all suited to the reclaimed high organic fen or moss land. This is due to the large volume of straw which this cereal normally produces. Crop heights of 5–7 feet are easily reached under poor fertility conditions and thus on the organic soils with significant releases of mineral nitrogen severe lodging would nearly always result.

B. Optimum pH

Most literature dealing with rye has stressed the fact that it will grow under conditions of extreme acidity and one is apt to think of it

Authority	Optimum pH range or	Maxima
Trénel (1927)	4–7	
Hiltner (quoted by Small, 1946)	5–7	
Arrhenius (1926)	5–6	
Arrhenius (quoted by Trénel, 1927)		4·5 and 7·8
Olsen (1923)	6–6·5 and 6·5–8	

as growing well only on acid soils. Small (1946), however, indicated that it was an amphi-tolerant species, with most authorities quoting a pH range below 7 for good yields.

Trénel's extensive field-plot data demonstrated more luxuriant rye growth between 3·8 and 5 than from 5–7 but a better quality rye grain between 5 and 7 compared with pH 4–5. The relative performance of this crop, grown on sandy soils in France, with varying pH values has been given in La Potasse (1962) and is reproduced below.

pH value of soil	4	4·5	5	5·5	6	6·5
Relative yield of rye	20	80	97	100	98	96

VI. CLIMATE AND WORLD DISTRIBUTION

Due to an extensive root system, rye can be grown on soils with very limited water resources and in areas of very low rainfall. It is also very tolerant of low temperatures and on account of this and its drought resistance this cereal is widespread in both northern and southern

TABLE 4

World Production of Rye—1930

Rank	Country	Million acres	Million bushels produced	% Total production
1	U.S.S.R.	65·29	881·29	47·60
2	Germany	14·61	400·76	21·65
3	Poland	14·20	254·38	13·74
4	France	1·75	32·02	1·73
5	United States	2·92	31·27	1·69
6	Hungary	1·58	28·48	1·54
7	Lithuania	1·23	22·62	1·22
8	Spain	1·49	22·16	1·20
9	Belgium	0·52	20·07	1·08
10	Sweden	0·56	16·79	0·91
11	Netherlands	0·44	15·66	0·85
12	Finland	0·55	13·77	0·74
13	Rumania	0·94	13·73	0·74
14	Latvia	0·63	12·40	0·67
15	Argentina	0·94	9·87	0·53
16	Canada	0·86	8·94	0·48
	All others	3·78	67·08	3·63
	World total	112·29	1851·29	100·00%

From Klages' data it will be noted that over 80% of the world's production in the 1930's came from Russia, Germany and Poland.

countries in Europe where most of the world's production is located. Rye statistics associated with the early 1930's which were compiled by Klages (1942), showed the total area of land under rye to be 112·29 million acres, with a production of 1,851·29 million bushels, made up as shown in Table 4.

Table 5, which has been compiled by the F.A.O., indicates the 1965 position with regard to area, production and yield.

TABLE 5

World Statistics on Rye—1965[a]

		1948–52	1961	1962	1963	1964	1965
Europe	a	12,154	9858	9438	8915	8942	8989
	b	17,794	16,752	15,733	15,895	16,508	16,914[b]
	c	14·6	17·0	16·7	17·8	18·5	18·8
U.S.S.R.	a	23,592	16,700	16,938	15,025	16,807	16,000
	b	17,961	16,700	17,024	11,878	13,619	16,100
	c	7·6	10·0	10·1	7·9	8·1	10·1
Canada and	a	1258	851	1055	907	961	896
U.S.A.	b	993	860	1340	1067	1135	1268
	c	7·9	10·1	12·7	11·8	11·8	14·2
Latin America	a	760	736	330	694	811	369
	b	553	540	196	569	686	279
	c	7·3	7·3	5·9	8·2	8·5	7·6
World Totals	a	38,027	28,894	28,528	26,339	28,332	27,084
	b	37,702	35,491	35,051	30,359	32,746	35,390
	c	9·9	12·3	12·3	11·5	11·6	13·1
U.K.	a	25	8	7	8	8	7
	b	52	18	17	22	25	21
	c	20·8	23·8	23·6	26·3	29·9	29·3

[a] F.A.O. Production Yearbook (1966), Vol. 20.
[b] Of the European production in 1965 (16,914 metric tons), 77% was grown in Poland and Germany.
a = area in 1000 Hectares; b = production in 1000 metric tons; c = yield 100 kg/ 1 Hectare. Figures given in 100 kg/Hectare roughly correspond to cwt/acre.

VII. RYE FOR GRAIN

In the United Kingdom most of the very much reduced rye acreage is located in South-east England and in general it is found in the low rainfall, high sunshine arable cereal growing districts in association with light soils.

A. Position of Rye in Cropping Sequence

In the past rye appeared in a crop rotation where winter wheat and barley would normally be grown and it took their place particularly on the very light soils. In the neighbourhood of Escrick (West Riding of Yorkshire) where the morainic drift had deposited sand and stones to form a very sharp infertile soil it was the only cereal to feature in a cropping sequence. In this area it was grown after roots, potatoes and seeds, producing large grains and good grain yields, but the chances of lodging were fairly high.

Since most of the rye grown today is in the arable cereal growing areas, crops will normally follow other cereals. On the light to medium soils after a cereal sequence, with low fertility, rye could be taken with advantage before potatoes and roots are employed to build up fertility again. Rye is now considered as a "break crop" by many farmers on the light marginal soils. Although a cereal, it is not susceptible to many of the pests and diseases which attack wheat and barley.

B. Fertiliser Levels

With a crop height of 5–8 feet, phosphate and potash have nearly always been advocated at 30-60 units/acre but most authorities recommend little or no nitrogen, with the exception of Garner *et al.* (1945).

Units/acre			Comments	Reference
N	P_2O_5	K_2O		
8–12	35–50	28	Nitrogen applied in the form of nitrate of lime as a spring top-dressing	O'Brien (1925)
21	35	60	Nitrogen as sulphate of ammonia applied as a top dressing in the spring	Moore (1944)
16	—	—	Nitrogen as nitrate of soda as a spring top dressing, and crop follows roots or seeds	Watson and More (1949)
Up to 40	Up to 50	—	With an additional 20 units N in the seed-bed following a cereal crop	Garner *et al.* (1945)

Farmyard manure at 10–12 tons/acre has been recommended in the past for the very light soils where fertility needs improving and under these circumstances no additional fertilisers were suggested.

From the results of National Trials (Bell, 1954) conducted on light loams during the late 1940's and early 1950's, with varieties of varying straw lengths, compound fertilisers which supplied between 15 and 30 units of nitrogen/acre together with adverse conditions of climate were responsible for lodging in many of the years. Thus it would appear that even using cultivars with the shortest straw, lodging will occur in a number of seasons even though fertiliser nitrogen levels are low by the standards set by the other cereals.

The phosphate and potash should be applied before or at the time of drilling in the autumn or early winter and with few exceptions all the nitrogen should be in the form of an early spring top dressing.

TABLE 6

Fertiliser Recommendations for Short-strawed Rye Grown for Grain

(Units of N, P_2O_5 and K_2O/acre)

	N	P_2O_5	K_2O
A. Very coarse, sandy soils in low fertility	30	40–60	40–60
B. Similar soil type to A. but with medium fertility	10	20–30	20–30
C. Following cereals in a mainly cereal cropping sequence	20–30	30	30
D. Following leys, legumes, roots or potatoes *or* For rye on land of very high fertility.	0–10	Nil	Nil
E. Details as in C. above but with spring grazing	30 10–20[b]	30 —	30[a] —

[a] Autumn application with the seed.

[b] Spring top dressing.

In the near future, spraying with the chemical cycocel (2-chloro-ethyltrimethylammonium chloride), also known as C.C.C., could well revolutionise the cultivation of rye for grain. Using plants of the variety Petkus, Linser and Kühn (1962) showed that an application of C.C.C., produced shorter stems with increased resistance to lodging and Geering (1965) found that C.C.C.-treated plants of the spring varieties Berna, Beka and Karlshulder also exhibited shorter stems and at the same time produced slightly larger ears. With significant reductions in straw height, it may be possible to apply a little more fertiliser nitrogen in an attempt to raise the relatively low grain yield obtained with this cereal.

C. Cultivation and Seed-bed Preparation

The roots of the rye crop must be capable of penetrating deeply in the light soils which are associated with its cultivation and it is therefore important to see that plough-pans are not present. It is seldom necessary to plough more deeply than 5–6 inches for this crop and this operation should take place at the end of the summer or in early autumn. Should the crop follow the ploughing-up of a grass ley, then consolidation through the use of a furrow press will help to obtain the necessary firmness of seed-bed, a factor so widely and strongly recommended that it must be heeded. In the past, when rye followed potatoes, it was suggested that the cereal be broadcast over the soil and then ploughed in with a 2 or 3 inch furrow slice. A better establishment however will usually be obtained by cultivating and harrowing the potato land into a fairly rough but firm seed-bed, followed by drilling of the rye. Within reason, the shallower one is able to drill cereal seed the better will be the field germination and plant establishment.

D. Drilling

Being connected with the light soils, inferring free drainage conditions, rye drilling can usually be carried out at the optimum time in the autumn. Although broadcasting is suggested in most of the older books and literature, modern crop husbandry almost demands the use of seed-drills if only for the economy of seed. Uniform depth of drilling, followed by a quick, even emergence can often result in obtaining a useful spring grazing from this cereal before it is taken as a grain crop. August and September, in the past, have been suggested as the time to sow rye, but since it is the earliest of all the cereals to mature and because many of the harvests of the previous crops are not fully complete until perhaps the end of September, nowadays rye goes in at the same time as the other winter cereals in October.

In the past, a coulter spacing of 7 inches has been suggested for rye, but an examination of the experimental data leads one to the conclusion that higher yields would almost certainly be obtained at narrower intervals. In summarising only the European trials and using data from nine experiments, Holliday (1963) has shown via weighted means that a change from coulters spaced 7–8 inches apart to those of 4–5 inches, could result in an additional 9% in grain yield.

E. Seed Rate

Sanders (1947) states that the normal seeding rate for rye is 3 bushels/ acre with a range from 2–4 bushels. This was also suggested earlier by Porter (1929). Watson and More (1949) and O'Brien (1925) suggest 2–3 bushels as being optimal, the lower figure to be used for early sowings whilst the higher one is necessary for either poor soil conditions or late drilling. For those who prefer seed rates in units of weight,

2–3 bushels is approximately equal to 1–1·5 cwt. Most of the National Variety Trials, which will be referred to later when cultivars are considered, were sown in mid October at seeding rates around 1·5 cwt/acre and since plant establishments and subsequent yields were satisfactory, then except for some special circumstances, higher seeding rates should be avoided.

F. Seed Dressing

Rye seed should be dressed with one of the approved organo-mercury seed dressings. It is susceptible to attack by the two fungi which cause bunt and loose smut in wheat and also to stripe smut. These fungi can be prevented by treating the seed with organo-mercury compounds.

G. Type of Seed

With such a limited demand for seed in Britain many merchants do not stock rye and when requested for it they often have to approach others within the trade. The types of seed available are few in number which again is a reflection on the current acreage. Also, a large proportion of our requirements naturally come from those other European countries which have supplied the cultivars which we in the United Kingdom find to be the highest grain yielders. This is inevitable since it would not be profitable for any of our official or private plant breeders to embark on large-scale breeding and research programmes with rye, bearing in mind the limited home demand even for a cultivar with ideal agronomic properties.

H. Varieties of Rye

Winter and spring forms of rye exist as with the other three cereals, but in general it is only the winter cultivars which are of any importance in the United Kingdom. They are more capable of resisting lower temperatures than the most hardy winter wheats under cultivation, are capable of being grazed lightly in the spring and will produce higher grain yields than the spring forms and thus are much preferred. Surprisingly little attention has been paid to the breeding or selection of better varieties of this cereal and whilst this is understandable on economic grounds in Britain, it has not been so on the Continent where millions of acres are sown annually. It is proposed to list and briefly describe varieties of the past and present and in the case of the latter, the reader's attention is drawn to the Appendix in which will be found a full description of the presently recommended varieties of rye for England and Wales (published by the National Institute of Agricultural Botany).

The older varieties grown in England could well be called landrace types. They seldom carried a definite name but simply a description, e.g. Common Rye or Giant Rye and on occasions may have been

O'Brien (1925)	Hunter (1951)	N.I.A.B. Recommended List (1962) (Grain Ryes)[a]
Winter or Common Rye	Petkus	Petkus Normal-straw
Giant Rye	Star	Petkus Short-straw
Petkus	Steel	Petkus Tetraploid
Star Rye (Svälof)	King	King II
White Russian	King II	Borris Pearl
Danish Brattingsborg (Brittany)	Pearl	Petkus Spring rye—S[l]
Rosen Rye		Karlshulder Spring rye—S[b]
Mammoth White Rye		
St John's Day (Midsummer Rye)[b]		

[a] Most of these varieties appear on the N.I.A.B. Farmers' Leaflet No. 12.— Varieties of Rye (Revised 1965) which is reproduced in full in the Appendix.
[b] S denotes spring form. Remainder are winter cultivars.

referred to by the name of the locality in which they were grown originally or in which they were popular. These were characterised by early, bulky spring growth, exceedingly long straw with a long lax ear and were generally early maturing.

(i) VARIETIES LISTED AND DESCRIBED (O'BRIEN, 1925) AS BEING THE CHIEF VARIETIES GROWING IN THE 1920's

Winter or common rye: very winter hardy, slow growing but early maturing. Sown for forage or soiling alone or with vetches, successful even on very poor sandy soils.

Giant rye: heavier yields of grain and straw, cf. Common, requires better soil and climatic conditions.

Petkus: one of the highest yielding grain varieties from Svälof, the Swedish Plant Breeding Station. Good field characters and widely cultivated in Europe and America.

Star rye: selection from older Petkus variety, smaller plumper grain. Suitable for the heaviest soils.

White Russian: extremely winter hardy, grown within the Arctic Circle —very long straw.

Danish brattingsborg: one of the heaviest yielding ryes grown on Continent

of Europe and at the time of writing was extensively grown in Denmark.
Rosen rye: mostly cultivated in America.
Mammoth white rye: no distinctive characters—one would assume that
it was long-strawed.

(ii) CULTIVARS DESCRIBED BY HUNTER (1951)

In the 1950's, Hunter recalls that apart from the variety Petkus,
most of the rye stocks in the British Isles were the indigenous long-
strawed, lax-eared types. His list of varieties contained those from the
Continent which were showing up well in the N.I.A.B. Primary Trials
and the N.A.A.S.–N.I.A.B. Secondary Trials at that time, details of
which were published later by Bell (1954) and Bell and Price (1955).
Petkus. Selected by von Lochow from a Probsteier stock in Sweden
in 1881 and although it has formed the basis of several new varieties,
it is in its own right a most useful high yielding type.
Star. Selection from Petkus which was marketed in 1914 as being
superior in agronomic characters.
Steel. Selection from Star, released in 1921, higher grain yield, shorter
straw of better standing ability.
King: Selection from Star, released in 1933, superior grain yield,
shorter straw of better standing ability.
King II. A selection from Steel made at Svälof. At the time this was
considered to be the most promising variety of rye for grain. Compared
with all the other varieties it had a very short straw and exceptionally
high grain yield and from the trials it emerged as suitable for the more
fertile soils.
Pearl. Produced at the Borris Plant Breeding Station in Denmark and
considered similar to King II. In its country of origin this variety has
given good yields on loams, sands and peats, producing attractive
grain with a fine skin.

Many of the cultivars listed and described above appeared in the
National Trials, along with some others and at this stage yield comparisons show some interesting results (Bell, 1954):

Varieties Compared with King II (100%) in Respect of Grain Yield

				Commercial	RN. 12 (A selection	Gartons
Petkus	Pearl	Steel	King II	rye	made by Dr Hunter)	L.G.
119	*118*	103	100	*92*	90	*89*

The above data represent the average yields of each variety in the
trials carried out between 1947 and 1950 in the counties of Cambridge,
Hampshire, East Suffolk and East Yorkshire and the figures in italics

indicate that they differ significantly from the control—King II. Confirmation of these comparisons emerged with the results of the secondary trials reported by Bell and Price (1955) where again the grain yields are compared with the control variety King II.

Varieties Compared with King II in Respect of Grain Yield (average of 14 trials)

Petkus	Pearl	King II	Commercial Rye
111·2	112·3	100	96·1

Petkus and Pearl were signficantly higher yielding than King II which itself was superior to the local commercial ryes and from the period of these trials these three varieties have featured on the recommended list. Garton's L.G. (large grained) was extremely variable in grain yield but with a robust growth habit, it is capable of producing reasonable yields following spring grazing and although it has not received official recommendation for either grain or forage, is nevertheless a useful "dual-purpose" cultivar.

The recommended varieties of rye for grain production for the late 1960's have already been listed and it is interesting to note that only one new name appears amongst the winter types and that is Dominant. Petkus actually dominates the scene with three types—normal straw, short straw and a tetraploid. Further details of all winter varieties and a note on spring types appear in the Appendix.

The tetraploid version of Petkus, with double the normal chromosome number, has a larger grain, and under poor fertility conditions has given lower grain yield than Petkus normal straw although the straw is shorter and stiffer. Discussion of varieties particularly suited to spring grazing and grain production and those specifically bred for forage production appear later under "Rye for Forage".

I. Crop Management

(i) CULTURAL OPERATIONS IN THE SPRING

Virtually all the rye grown for grain is winter sown, in seed-beds resembling those for winter wheat and as a result the spring management is also very similar. Harrowing will give some measure of weed-control and the heavy rolling which must follow gives a flat surface to a well consolidated soil. Rolling is vital on all the light soils to conserve moisture and to facilitate capillary rise from the lower soil horizons.

(ii) SPRING GRAZING

Formerly, rye was drilled at the end of August and into September which meant that by November and December of the same year there

Fig. 1. Early spring bite from winter sown rye. Late March, 1963, Hartland, North Devon. Photograph by courtesy of *Farmer and Stockbreeder*.

was an appreciable amount of vegetative growth. This was due to the fact that rye produces an abundance of tillers and is capable of growing at lower temperatures than the other three cereals. December grazing with sheep was often practised and provided that this defoliation was controlled and limited in its extent, little reduction in subsequent grain yield resulted. Nowadays with normal sowing dates around mid October, crops of rye are hardly forward enough for winter grazing unless the autumn is unusually warm and therefore the defoliation, if practised, will take place in the spring.

Rye is perhaps the most suitable winter cereal for spring grazing. It produces the best ground cover in the autumn and under some circumstances the highest weight of forage in the spring. Holliday (1956) has reviewed the fodder production from winter sown cereals and its effect on grain yield and Kimber (1965), in assessing forage production from rye, has also shown how much grain yields would be affected, should this material be removed.

TABLE 7

Yield of Grain from Winter Rye (cwt/acre at 15% moisture)

Harvest year	South-east England (Cambridge)				South-west England (Seale-Hayne)			
	Sowing dates	A^a	B^b	Cutting dates	Sowing dates	A^a	B^b	Cutting dates
1957	12/10/56	12·5	12·6	29/3/57	26/10/56	14·9	39·1	4/4/57
1958	20/9/57	19·3	22·0	14/3/58	30/9/57	17·8	20·8	11/3/58
1959	26/9/58	18·6	20·0	8/4/59	23/10/58	22·5	41·5	14/4/59
Mean	—	16·8	18·2	—	—	18·4	33·8	—

[a] A = cut in the spring.
[b] B = not cut in the spring.

Kimber's data (Table 7) has been re-arranged on a regional basis, with mean values inserted to facilitate an easy appraisal of the effect of removing vegetative growth in the spring. Sowing dates and cutting dates are included since these will greatly influence the amount and period for growth. Removal of forage had little effect on the yield of rye in the Cambridge trials, whereas in the South-west it reduced the yield considerably as can be seen from the above figures. In 1957 and 1959, when the amount of green material removed from the trials in the South-east was small, subsequent grain yield was not significantly affected whereas in all three years in the South-west, with large forage cuts, grain yields were significantly lower. Early removal of forage from rye also had a smaller effect on grain yield than did later cutting, a feature which has already been stressed in relation to wheat.

(iii) VARIETAL REACTION TO SPRING DEFOLIATION

The reaction of different cultivars to spring defoliation was also

Fig. 2. In spite of a cold, late spring, cattle were strip-grazing winter sown rye on 11th March, 1962, on Mr A. P. Pick's Spring Farm, Hanwell, near Banbury. Photograph by courtesy of *Farmer and Stockbreeder*.

recorded by Kimber (1965) and the relative yields of each variety with or without forage removal, appears in Table 8 below.

TABLE 8

Comparative Yield of Grain (15% moisture) from Rye Cultivars—1957–59

Treatment	Petkus Normal	Petkus Short	Petkus Spring	Gartons Large Grained	Bernburg Fodder	C.R.D.
Cut in the spring	78	93	63	83	78	64
Not cut (vegetative growth intact)	110	108	88	98	106	88

Removing the vegetative growth in the spring adversely affected the yield of all varieties but Petkus Short and Garton's Large Grained being least affected by this technique were still able to produce economic quantities of grain and should be selected where spring grazing is envisaged and early establishment is possible.

With the spring grazing of autumn sown wheat, barley and oats most of the reduction in grain yield could be recovered by additional nitrogen in the form of a spring top dressing. In the case of winter rye which is being grown for grazing and later for grain, the autumn fertilisers should include a reasonable quantity of nitrogen to stimulate the early vegetative growth and a small quantity of nitrogen may be used as a spring top dressing. Farmers must exercise care and discretion particularly in respect of this spring nitrogen. Quantities in excess of the crop's requirement coupled with wet summers can play havoc with rye crops due to widespread lodging.

(iv) UNDERSOWING WITH GRASS AND CLOVER SEEDS

Undersowing winter rye crops with herbage seeds for the establishment of leys is problematical. On the one hand this cereal will produce a large volume of vegetative growth and good ground cover after sowing and in the spring, which is contrary to conditions required for a good "take" of small seeds. On the other hand, those grasses and clover which do manage to establish themselves will have a long period of growth after the cereal harvest since rye is the earliest to mature. On balance, undersowing the rye crop is not a recommended practice, but should farmers find that it has to be done then there are a few modifications which would assist in obtaining a reasonable field germination. A broadcast dressing of 40–60 units each of phosphate and potash should accompany the cereal in the autumn and in the spring, grazing the crop well down before broadcasting the grass and clover seeds will help to cut down the competition for light. No fertiliser nitrogen should be applied in order to give the cereal its best possible chance of remaining upright until harvest. Should it be possible to dispense with the rye as a grain crop, it will make a useful nurse crop for the small seeds and at the same time contribute much to the first spring bite or cut for silage.

J. Chemical Control of Weeds

By comparison with the other three cereals the cultivation of rye is very small and as a result a lot less attention has been paid to the chemicals which can be used for the control of dicotyledonous weeds. However, it is known that a number of herbicides which can be used

TABLE 9

Recommended Weed-killers for Spring Use on Winter Sown Rye
(Growth Stage of the Cereal—Fully Tillered but Before the Jointing Stage)

(a) *Crop not undersown with grass and clover*

        MCPA
          2,4-D
        Dichlorprop + MCPA
        2,3,6-TBA + MCPA

(b) *Crop undersown with grass and clover*

        MCPB + a little MCPA[a]
        2,4-DB + a little MCPA[a]
        2,4-DB + a little 2,4-D or MCPB and MCPA[a]

               Dinoseb[b]

[a] Clover should have its first trifoliate leaf before spraying.
[b] Clover should have two trifoliate leaves before spraying.

Based on M.A.F.F. Short Term Leaflet No. 19 (January 1970)
For up-to-date information on herbicides consult the most recent "M.A.F.F. Short Term Leaflet No. 19" or the latest edition of "The Weed Control Handbook".

on the other cereals are not suitable and as a result the list of recommended weed-killers is quite a short one. Details appearing in Table 9 have been based on the Ministry's Short Term Advisory Leaflet No. 19 (January, 1968) and since new, improved herbicides appear almost annually, it is important that the most recent Ministry Leaflet be consulted. For crops of rye which are not undersown the choice lies between the old established ones MCPA and 2,4-D, or MCPA with dichlorprop or 2,3,6-TBA. The selection should be made firstly with the weed flora in mind and secondly related to the cost of material and where the crop has been undersown Table 9(b) should be consulted.

K. Harvesting

In former days rye crops were taken for straw, for grain or both yield components. When quality straw was being produced, cutting was done with a reaper or binder often well before the ears were properly filled and on no account was grain considered. The straw thus produced was valued highly for a number of purposes, usually associated with thatching since it was not broken up by mechanical processes. If grain and straw were both required, cutting by binder took place after the ears were filled and since the straw and flag leaves of rye mature well in advance of the ears, the crop could be carted in within a few days of cutting, provided the weather was good. A modified threshing technique which did least damage to the straw was used to secure the grain. Sheaves were fed into a threshing mill, head first. The operator did not allow them to go through, but took them out once the grain had been "buffed off". The straw could then be re-tied into bundles and this also produced a quality product.

Nowadays, with fewer and fewer binders working, rye crops for grain are usually combine-harvested and this will mean waiting until the grain is mature and as dry as possible before the combine operation. The straw obtained is short and broken and can only be used for bedding.

Grain coming from the combine with moisture contents above 15% will need artificial drying if it is to be stored in bulk and the methods described under the other cereals will do equally well for rye.

L. Grain Disposal

Most of the rye grown in Britain is used in the production of flour for crispbread, biscuits and for sauces and has a limited industrial outlet. Grain which is unsuitable for manufacturing for human food will usually find its way into animal feeds where it can form up to 15% of the compounds. Crops which have lodged early resulting in sprouting will produce grain of high diastatic power and this is unsuited for industrial use (Drew, 1968).

A number of farmers contract to grow rye with some of the leading grain merchants, who in turn supply the manufacturing companies with grain of suitable quality.

FIG. 3. Six foot rye (in ear) and ryegrass cut and windrowed by forage-harvester at the end of June (1962) on Mr Peter Oldfield's Dinton Pastures Farm, Berkshire. Photograph by courtesy of *Farmer and Stockbreeder*.

M.   Yield of Grain and Straw

O'Brien (1925) estimated the average yield of rye to be 25–35 bushels/acre (56–62 lb/bushel) and 30–40 cwt of straw, and suggested that 60–70% of the weight of a crop was straw and 30–40% grain. Writing a few years later Porter (1929) indicated that wheat and rye would give similar grain yields and for average conditions 28 bushels/ acre (15 cwt/acre) could be expected. Moore (1944) and Watson and More (1949) state that rye grain yields vary between 12 and 16 cwt/ acre, producing 1·5–2 tons of straw. Recent Ministry statistics relating to England and Wales indicate that rye yields have increased over the past few years but this has not been anything like the improvement shown in wheat and barley.

Average Yield of Rye—England and Wales Statistics (cwt grain/acre)

10-Year average	5-Year average			
1954–63	1959–63	1963	1964	1965
20·3	21·0	21·5	23·6	23·9

N.   Production Economics

Estimates of production costs and returns are presented below and serve to indicate the very small profit margin with rye.

Cost/acre		Returns/acre	
	£		£ p
Labour costs	5	24 cwt grain @ £1.20	26.40
Tractor costs	2	36 cwt straw @ £2 per	
Machinery depreciation and		ton	3.60
repairs	3		
Rent	3		
Seed	5		
Fertilisers	2		
Sprays	1		
Overheads	3		
Adjustment for manurial			
residues	1		
Total	£25.00		£30.00

O.   Government Support for Rye

In accordance with the 1947 Agriculture Act, a guaranteed price is fixed annually for rye and when the average market price falls below this figure a deficiency payment, related to the difference, is paid to growers.

During the 1968–69 cereal year the average market price of rye was 23s. 11·2d./cwt and since the guaranteed price stood at 21s. 7d., no deficiency payment was due.

VIII. RYE FOR FORAGE

A. Introduction

The spring grazing of autumn sown rye crops for grain has already been considered and it was concluded that with reasonably early sowings it was possible to obtain a useful amount of herbage and if this was utilised early in the spring it had the least effect on subsequent grain yield.

Turpin and Marais in South Africa (1933), Washko in Tennessee, U.S.A. (1947), Stansel et al. in Lubbock, Texas, U.S.A. (1937) and workers at Rothamsted have all demonstrated that rye will produce larger quantities of forage than wheat, barley or oats, although at two other sites in Texas, rye did not show up so well in this respect (Stansel et al., 1937).

The actual figures obtained by the Rothamsted workers are interesting and being the only available British data are reproduced below.

Comparative Yields of Forage from the Four Main Cereals (Cwt dry matter/acre)

Cereal	Autumn	Spring	Total	Production from rye, wheat and barley as % of oat forage yield
Rye	3·7	18·0	21·7	161
Barley	3·1	13·1	16·2	120
Wheat	1·9	12·6	14·5	107
Oats	3·9	9·5	13·4	100

Each of the above cereals were sown with beans and vetches but the forage cut in autumn or spring was separated into its various components and the data above relates only to cereal forage production. An early sowing towards the end of July was responsible for 2–4 cwt of dry matter by mid November and approximately 10–18 cwt in the spring with rye featuring at the highest end of the scale on each occasion. The undoubted success of rye could be attributed to an early sowing date, its mode of growth and also to the fact that seed was probably year old stock which had overcome the normal dormancy.

With former sowing dates for rye earlier than those of today, the seed had to overcome its dormancy early to produce good growth during the autumn and winter of the year of sowing and for the particularly early sowing then year-old stocks were certainly used. Nowadays with rye sown in October, the grain is required almost immediately it has been harvested and there is bound to be considerable dormancy before it will germinate. The National Institute of Agricultural Botany have referred to this in their Farmers' Leaflet No. 12 and they also indicate that the fall in germination of rye stocks during the spring and summer after harvest could well be greater than the other cereals. Germination tests are therefore suggested as a routine measure for yearling seed.

One of the earlier writers on the value of winter rye for spring grazing by cattle and sheep suggested a September sowing of 1·5–2 cwt of seed to produce approximately 10–12 tons of green material by April and May (Porter, 1929). For sheep grazing in the spring Moore (1944) advocated sowing in August at 1 cwt Giant rye/acre and where green soiling material is required, 1 cwt rye plus 0·5 cwt vetches put in during September will give a heavy crop for cutting in May. The latter mixture can also provide useful forage for arable silage making on account of the early spring growth, but it must not be allowed to get too far advanced otherwise there will be a significant drop in digestibility on account of the rapid development of fibre in the cereal stems. Oats and vetches have often been preferred to rye and vetches for arable silage for later cutting without serious reduction in digestibility.

B.   Varieties of Rye for Forage Production

In the past many of the varieties or stocks of rye were used for forage purposes although the Giant or Common ryes were probably the most suitable. Reference was made to the polyploid varieties (N.I.A.B. Farmers' Leaflet No. 12, Rye' 1956) in the mid 1950's as being more suited for grazing purposes than for grain production on account of their more vigorous spring growth and greater bulk.

A number of varieties were tested by the N.I.A.B. during the period 1957-63 (Kimber, 1965) and those which were included for both forage and grain assessments have already been referred to. Comparisons with Westerwolds ryegrass indicated a marked superiority of all varieties of rye in the amount of spring growth. The following results have been extracted from Table III of Kimber's data where specific comparisons with this early grass were made.

Treatment A measured the amount of early spring growth and subsequent recovery whereas Treatment B gave an indication of late spring productivity. In each case the varieties of rye gave yields which were much superior to the grass and the range of additional forage yields from them are given on p. 205.

For early forage production Bernburg Fodder and Lovaszpatonai

*Additional Productivity Range from Rye Varieties Compared
with Westerwolds Ryegrass*

Treatment A		Treatment B
1st Cut	2nd Cut	(Late cut only)
68–142%	44–81%	102–177%

were the highest yielders and the former showed additional merit in its noted recovery after the first cut in the spring (Kimber, 1965). Limited data from trials in the West of Scotland (1965) help to confirm the superiority of Lovaszpatonai grown alone or in conjunction with 20-30 lb of an Italian/Perennial ryegrass mixture.

In the event of neither of these cultivars being available, Ovari, another Hungarian variety and New Zealand C.R.D. have been suggested as alternatives (M.A.F.F. Advisory Leaflet No. 501, 1961) and from other limited evidence, Garton's Large Grained, King II, Petkus Normal and Petkus Spring are all capable of worthwhile fodder yields. Bernburg, Lovaszpatonai, Milns Grazing Rye and Rheidol appear on the N.I.A.B. List of Rye Varieties (revised 1965) which is reproduced in full in the Appendix.

C.   Cultural Details for Forage Rye

Winter sown rye for fodder will do well on most types of soil provided they are properly drained. It is not restricted to the rather lighter soils, as for grain production and much of it used for cattle and sheep grazing may be associated with hill or marginal conditions at altitudes up to 1000 feet. Being more tolerant of low temperatures, it becomes the most suitable species for "early bite" under many circumstances.

Time of sowing is one of the most critical factors in securing good yields. If the rye is sown between mid August and early September, it enables a wide range of previous crops to be grown and allows plenty of time in the autumn and winter for growth but it can be difficult to put into practice due to the unavailability of seed. Growing one's own especially for this purpose has been suggested, but for those farmers constantly relying on rye for their first grazing, it would be wise to arrange forward contracts with merchants to ensure that seed is available for early sowing.

When sown alone 1·5–1·75 cwt/acre is the recommended range, the lower seed rate being suggested for the prostrate varieties such as King II and Petkus and 1·75 cwt for the mid-European cultivars

(M.A.F.F., 1961). In spite of the seed being quite costly, it would be foolish to reduce the rates suggested above; in fact, under nearly all circumstances 200 lb/acre (approximately 1·75 cwt) will be required. Dressed seed should be drilled in August or September, into a good seed-bed at a depth of approximately 1·5–2 inches. If the dormancy has been overcome, rye will germinate and braird rapidly on account of the grain being naked.

To assist in the early establishment, a compound fertiliser should be applied before or at drilling to supply 30–40 units of each of the major nutrients. In the spring an early top dressing of up to 60 units of nitrogen will further stimulate growth. In some cases this has doubled the yield of forage and at the same time it has greatly increased the protein content of the forage. Higher levels of fertiliser nitrogen would probably give greater yields but may lead to high levels of nitrate nitrogen in the plant which is undesirable from the view of animal health. The period of grazing can be extended to May and June by including 10–20 lb of Italian ryegrass along with the rye and although this increases the cost of production, additional feed over a longer period can be obtained. After an early grazing, the increasing quantities of Italian ryegrass in the forage will help to reduce the drop in digestibility which occurs particularly with rye during May. Average digestible dry matter figures have been calculated from Kimber's data (1965) and are shown below.

% Digestible Dry Matter of Rye

Early March	April/May regrowth	April/May first growth
80·4	77·6	73·2

Rye and ryegrass should not be broadcast or drilled as a mixture. Broadcasting will mean a poor rye coverage and a bad cereal "take" and the drilled mixture gives far too much competitive ability to the cereal which results in a poor grass "take". The rye should first be drilled into a good seed-bed and the ryegrass broadcast immediately afterwards.

D.　Grazing Management

A mild autumn following an early rye sowing will result in sufficient growth for a light grazing in November. Besides providing cattle or sheep with fodder during that period, this defoliation can stimulate spring growth of slightly superior quality and will enable better management control to be exercised over a fast-growing crop which, if not grazed, rapidly loses its quality.

It has been suggested that maximum green feed is obtained by a single grazing when the rye crop has produced 8–10 inches of growth. Defoliation will normally extend over a period of 7–14 days, depending on growth rate, stocking rate and availability of second grazings. In order to be grazing at 8–10 inches of growth about the middle of the defoliation period, an earlier start must be made when the rye is about 6 inches high. Strip grazing by dairy cows is the technique normally applied to once-grazed rye for early bite and as will be seen later when the economic aspects are dealt with, high stocking rates tend to produce the best "profit margins" with this enterprise.

When two grazings are required, the first, preferably using sheep, should take place in March when the rye is 4–5 inches high and should be completed quickly using a high stocking rate for a very short period. The amount for aftermath grazing in April or May will depend mainly on the spring weather, but can be stimulated by a top dressing of about 30 units of nitrogen immediately following the first defoliation.

With such a rapid drop in digestibility once the stems become visible, it is important to see that high stocking densities secure maximum utilisation before this maturity stage with rye.

E. Yield and Quality of Forage

The three main factors affecting the amount of forage available in the spring from winter rye are sowing date, weather and nutrient levels and the two variables under the control of farmers have already been discussed at length. Productivity in variety trials and field experiments can be measured in tons of green material or cwts of dry matter or protein and the most recent estimates of productivity are listed below.

Green Material Tons/acre	Dry Matter Yield Cwt/acre	Reference or Authority
3–6	10–20	M.A.F.F. Advisory Leaflet No. 501 (1961)
0·5–6	1·5–17	West of Scotland Agricultural College. Results of Experiments (1964). Crop Husbandry Department
	9·5	March ⎫ Kimber
	12·5	April/May regrowth ⎬ (1965)
	28·75	April and May first growth ⎭

In respect of quality, the spring growth of winter sown rye is reckoned to be similar to young grass, i.e. 14–18% dry matter and 20–28% protein in the dry matter (M.A.F.F., 1961). Once again the results from the National Variety Trials, reported by Kimber (1965) have been used to assess the average conditions in the field.

The results in Table 10 are unweighted means for eleven varieties which were grown in Cambridgeshire, Cardiganshire, Devon and Hampshire during the period 1957–63, and show how much the quality of the forage drops by the late spring, even with re-growth on plants which were defoliated early.

TABLE 10

Percentage Protein and Percentage Fibre in Early and Late Spring Growth of Rye and in April/May Re-growth

	March growth	April/May re-growth	April/May first growth
Crude protein	21·2	16·6	14·9
Crude fibre	16·5	22·4	24·4

F. Production in Terms of Cow-grazing Days and Economic Aspects of Rye for Spring Grazing

One acre of rye will provide approximately two weeks grazing for ten cows with a range of 120–150 cow-grazing days/acre on average. In the survey carried out in Wales, where the performance of winter rye for spring grazing by dairy cows was considered most satisfactory, between 96 and 178 cow-grazing days/acre were obtained from twenty-six farms. Exceptionally high productivity was realised on the other

TABLE 11

Financial Evaluation of Early Spring Bite from Winter Sown Rye

Season	Savings in foodstuffs per acre of rye grown	Surplus of receipts over cost/acre of rye grown
1955–56	£18 3s. 0d.	£16 7s. 10d.
1956–57	£19 9s. 1d.	£14 8s. 2d.
1957–58	£15 8s. 3d.	£12 0s. 0d.

After Prytherch (1960)

three farms in the survey, where 348 cow-grazing days/acre were recorded from rye sown during the first week of September (Prytherch, 1960). These data were collected over three grazing seasons 1956–58 and the average economic advantages are reproduced in Table 11.

The data also revealed a correlation between financial return and sowing date. When the rye was drilled during the first week of Sep-

tember, the surplus ranged from £14–47/acre but with sowing about the middle of the third week of that month returns varied from a loss of £8·50 to a surplus of £8/acre.

When other cereal crops show extensive "winter kill", rye is normally unaffected by even the lowest winter temperatures experienced in Britain and it will produce vegetative growth for early spring grazing two to three weeks before the earliest Italian ryegrass. It can therefore be recommended for early bite at a time of year when nothing else is available and provided that keep is assured once the rye grazing is finished, this autumn cereal can leave behind substantial profits particularly when grazed by dairy stock. During the period of the survey (1956–58 grazing seasons) the extra cost in establishing the rye crops amounted to just over £8/acre, which by 1968 standards would have risen to £9–10.

G.   Ergot in Rye

Ergot (*Claviceps purpurea* (*Fr.*) *Jul.*) was one of the major diseases of rye in the past. Today it is present only in small quantities in some of the rye crops and whilst it could seldom be considered a yield-reducing disease, it is important on account of the blackish-purple horn shaped bodies which are produced on the ears. These contain toxic alkaloids derived from ergotine, which are harmful to humans and animals. Ergot is mainly associated with rye; however, wheat and barley are sometimes attacked but oats are rarely troubled by this disease. It is widespread amongst the cultivated and wild grasses, being found along the hedgerows in large quantities in a number of areas, particularly Yorkshire. This disease is readily recognised by the blackish-purple bodies which appear in place of some of the grain in the ear. These foreign bodies contain a hard compact mass of white fungus mycelium, are about 0·5–0·75 inches long and mature about the same time as the grain. They either fall to the ground before or during harvest or contaminate the harvested grain. These fungal masses which are close to the soil surface germinate around cereal anthesis, producing spores which are carried by the wind to infect the ear and the blackish-purple ergots result.

Ergotism is the name given to the disease which is contracted by humans following ingestion of ergots. Convulsive and gangrenous forms of this disease occur in Germany where large quantities of rye are consumed annually and although control measures are not usually necessary in Britain, rye which contains ergots must be cleaned before being used by man or animals. Abortion can also be caused by ergot in humans and although it is widely believed that the same occurs in cattle, there is no conclusive proof. They are, however, poisonous to livestock and may cause other acute illnesses.

When growing rye, only clean, uncontaminated seed should be sown to prevent infection and consequent spread of this disease. Affected grain samples can be freed from ergot by floating in a saturated salt

Ergots

Rye

FIG. 4. Ergots removed from rye using the Electronic Sorting Machine illustrated in Chapter 7, Fig. 4.

solution or by passing them through an electronic Sorter (Fig. 4), but manufacturers using rye for human consumption will normally reject any samples containing ergot. Deep ploughing of land which has carried a heavily infected rye crop will bury the ergots and crop rotation to exclude rye for several years will help to control the disease should it become serious in Britain.

Appendix
VARIETIES OF RYE†

Rye is grown for grain and for cutting green or grazing. Sown in the early autumn, rye provides an early forage crop in the following spring and is therefore particularly valuable where extension of the normal grazing season is desired. The purpose of this leaflet is to help farmers to choose between the varieties of rye which are available.

The National Institute of Agricultural Botany tests the more important rye varieties available in Britain and the Continent. In some cases use has been made of information from trials on the Continent where rye is a crop of more importance.

WINTER AND SPRING RYE

Rye is usually autumn sown, and the majority of varieties are of the winter type. Like true winter wheats, winter ryes must be sown early enough to experience a period of growth at low temperature if grain production is to be successful. Some varieties cannot be sown after Christmas without some risk of failure. Although seldom grown in this country spring varieties do exist.

GRAIN PRODUCTION

Rye has a special value for grain production on poor acid soils in the drier districts of England where other cereals may not produce economic returns. It will grow well on more fertile soils but the straw tends to be excessively long and the crop may become difficult to handle. It may be treated as a dual purpose crop which is grazed in the early spring and then allowed to grow on and produce grain.

Genuine stocks of the older varieties of rye grown in Britain, described variously as Giant Rye, Winter Rye or English Rye, are now difficult to obtain. The Continental varieties which are now grown for grain production in Britain have somewhat shorter straw and a shorter, denser ear. They stand better and have given higher yields of grain.

Winter Varieties

Seed of the following varieties is imported from time to time:
BORRIS PEARL (Borris Plant Breeding Station, Denmark)
Similar to Petkus Short-straw in yield and maturity. Straw slightly longer.

† Farmers' Leaflet No. 12. Reproduced by kind permission of the National Institute of Agricultural Botany, Huntingdon Road, Cambridge.

DOMINANT (Cebeco, Holland)

A limited number of trials in England confirm Dutch results indicating slightly higher yield than Petkus Short-straw. Maturity and grain size similar. Straw slightly longer.

KING II (Svalöf, Sweden)

Lower grain yield and slightly later maturing than Petkus Short-Straw. Smaller grain is favoured by some processors. Shorter straw.

PETKUS NORMAL-STRAW† (von Lochow-Petkus, Germany)

The original "Petkus", now named Petkus Normal-straw to distinguish it from the more recently introduced Petkus Short-straw. Similar grain yield and longer straw than Petkus Short-straw.

PETKUS SHORT-STRAW† (von Lochow-Petkus, Germany)

Known also as Petkus II when first introduced. Straw short, but not as short as King II.

Spring Varieties

On the Continent the grain yield of spring rye is normally lower than that of winter rye. Seed of Petkus Spring and Karlshulder (Charles) Spring rye may be available in this country, but as there are winter varieties with similar names it is important to specify which is required when ordering seed.

GRAIN QUALITY

Rye grown in this country is used mainly for the manufacture of biscuits or crispbread. The most important characters for this appear to be even grain size, freedom from sprouting and high protein content of the grain. Protein content in cereals is influenced more by soil and cultural conditions than by the variety. Nitrogeneous manures are likely to increase both yield and protein content, but may cause too much growth of the straw. The risk of lodging may be lessened by applying the fertiliser when tillering has ceased.

VARIETIES FOR FORAGE PRODUCTION

For production of early spring forage it is essential to sow early in the autumn, at least by the middle of September, to get the best results.

---

† These varieties are often offered as "Petkus" without further distinction between them.

The main requirement of a forage rye variety is the ability to make growth while temperatures are still low. Ability to recover from early cutting or grazing may also be important if aftermath production is required or the crop is to be grown on for grain. In general, the more advanced the crop when cut green in the spring, the greater will be the reduction in subsequent growth. This is particularly true once the growing points have elongated sufficiently to be damaged by cutting or grazing. Nevertheless an early light spring topping may be beneficial in preventing undue length of straw even where the main purpose of the crop is grain production. Varieties for forage production are liable to develop rather longer straw than grain varieties under the same management.

Trials, in which green forage production was measured by cutting early in the spring, have shown the following four winter varieties to be especially suitable. Differences between them in earliness, yield of green forage and recovery after cutting are small.

BERNBERG	Bernburg, Germany
LOVASZPATONAI	Keszthely Breeding Station, Hungary
MILNS GRAZING RYE	Miln, Chester
RHEIDOL	Welsh Plant Breeding Station, Aberystwyth

SEED

Seed is often required for sowing almost immediately after it has been harvested, and germination may be low because of temporary dormancy. This condition, however, largely disappears by the time the seed is sown. A fall in germination during the spring and summer after harvest is more likely with rye than other cereals. When yearling seed is sown it is particularly important that it should first be tested for germination.

Rye is open-pollinated, unlike wheat, oats and barley. Isolation of at least 250 yards is therefore required between a seed crop and any other crop of rye.

References

Arrhenius, O. (1926). "Kalkfrage, Bodenreaktion und Pflanzenwachstum", Leipzig.

Bell, R. A. M. (1954). "Winter Rye Variety Trials". *J. nat. Inst. agric. Bot.* **7**, 80.

Bell, R. A. M. and Price, C. D. (1955). "N.A.A.S./N.I.A.B. Trials of Winter Rye Varieties". *J. nat. Inst. agric. Bot.* **7**, 303.

H

Darlington, C. D. and Wylie, A. P. (1955). "Chromosome Atlas of Flowering Plants" (second edition). Allen and Unwin Ltd, London.

Drew, N. (1968). Personal Communication from Managing Director, Ryvita Company Ltd.

Emme, H. (1928). *Z.i.A.V.* **47**, 99.

Evans, R. E. (1960). "Rations for Livestock". M.A.F.F. Bulletin No. 48. H.M.S.O.

F.A.O. Production Yearbook (1966). Vol. 20, pp. 40–42. Director of Statistics Division, P. V. Sukhatne. Editor, Nafiz Erus.

Garner, H. V., Hoare, A. H., Long, H. C., Stapledon, R. G., Rayns, F. and Wallace T. (1945). "Profit From Fertilizers". Crosby Lockwood & Son Ltd, London.

Geering, J. (1965). "Ein Standfestigkeits-Hilfsmittel für Weizen". *M. Schweiz. Landwirts.* **1**(13), 1–19.

Gouws, J. B. (1950). *Natuurwet Tijdschr.* **32**, 139.

Holliday, R. (1956). "Fodder Production from Winter-sown Cereals and its Effect upon Grain Yield". *Field Crop Abstracts* **9**(3), 1–13.

Holliday, R. (1963). "Row Width and Cereal Yield", F.C.A. **16**, 2, 1–11.

Hunter, H. (1951). "Crop Varieties". Farmer and Stockbreeder Publications Ltd, E. & F. N. Spon Ltd, London.

Kent, N. L. (1964). "Technology of Cereals, with Special Reference to Wheat". Pergamon Press, London.

Kent-Jones, D. W. and Amos, A. J. (1947). "Modern Cereal Chemistry". Northern Publishing Co. Ltd, Liverpool.

Kimber, D. S. (1965). "Trials of Rye Varieties for Forage—1957–63". *J. nat. Inst. agric. Bot.* **10**, 2, 195–203.

Klages, K. W. H. (1942). "Ecological Crop Geography". The Macmillan Co., New York.

Kostoff, D. (1937). *C. R. Acad. Sci. U.R.S.S.* **14**, 213.

La Potasse (September 1962), no. 297, 36th year, p. 146.

Linser, H. and Kühn, H. (1962). "Fertilizers for the Control of Lodging on the Basis of Gibberellic Acid Antagonists of the C.C.C. Group". *Bodenkunde.* **96**(141), 231–47.

M.A.F.F. Short Term Leaflet No. 19. (1968, 1970). "Choosing Selective Weedkillers for Use on Cereals in the Spring". M.A.F.F.

M.A.F.F. Advisory Leaflet No. 501. (1961). "Winter Rye for Spring Grazing". H.M.S.O.

McCance, R. A., Widdowson, E. M., Moran, T., Pringle, W. J. S. and Macrae, T. F. 1945. "The Chemical Composition of Wheat and Rye and of Flours Derived Therefrom". *Biochem. J.* **39**, 213.

Moore, H. I. (1944). "Crops and Cropping". Allen and Unwin Ltd, London.

Müntzing, A. (1937a). *Hereditas* **23**, 401.

Müntzing, A. (1937b). *Cytologia, F.J.N.*, 211.

Müntzing, A. (1943). *Hereditas* **29**, 91.

Müntzing, A. (1951). *Hereditas* **37**, 17.

Nakajima, G. (1954). *Bot. Mag. Tokyo* **67**, 69.

National Institute of Agricultural Botany. (1956). Farmers' Leaflet No. 12— Varieties of Rye. Issued by the N.I.A.B., Cambridge.

National Institute of Agricultural Botany.(1962). Farmers' Leaflet No. 12— Varieties of Rye. Issued by the N.I.A.B., Cambridge.

Neumann, M. P., Kalning, H., Schleimer, A. and Weinmann, W. (1913).

O'Brien, D. G. (1925)."The Rye Crop", p. 253. *Z. ges. Getreidew.* **5**, 41. The Gresham Publishing Co. Ltd, London.

Olsen, C. (1923). Studies on the Hydrogen Ion Concentration of the Soil and its Significance. *Compt. Rend. Lab. Carlsberg.* XV-1.

Percival, J. (1934). "Wheat in Great Britain" (re-issued 1943). Published by the Author. Leighton, Shinfield, Berks.

Porter, J. (1929). "The Crop Grower's Companion". Gurney and Jackson, London.

Prytherch, E. I. (1960). "Autumn Sown Rye for Early Spring Grazing—a Survey of Costs and Returns". *Experimental Husbandry* no. 5, pp. 27–31. H.M.S.O.

Rothamsted Experimental Station (1932). *Report*, p. 150.

Ryvita Company Ltd (1968). Personal Communication.

Sanders, H. G. (1947). "An Outline of British Crop Husbandry". Cambridge University Press.

Small, J. (1946). "pH and Plants. An Introduction for Beginners". Baillière, Tindall & Castle, London.

Stansel, R. H., Dunkle, P. B. and Jones, D. L. (1937). "Small Grains and Ryegrass for Winter Pasture". Texas agric. Exp. Sta. Bulletin 539.

Trénel, M. (1927). "Die wissenschaftlichen Grundlagen der Bodensäure Frage". Berlin.

Turpin, H. W. and Marais, J. C. (1933). "Grazing Winter Cereals Under Irrigation in the Karoo". *Fmg. S. Afr.* **8**, 297.

Vavilov, N. I. (1926). "Studies on the Origin of Cultivated Plants". Leningrad.

Washko, J. B. (1947). "Effect of Grazing Winter Small Grains". *J. Amer. Soc. Agron.* **39**, 659.

Watson, J. A. S. and More, J. A. (1949). "Agriculture. The Science and Practice of British Farming" (ninth edition). Oliver and Boyd, Edinburgh and London.

West of Scotland Agricultural College. (1965). Results of Experiments, 1964. Crop Husbandry Department.

CHAPTER 5

Maize

I.	Introduction	.	.	.	.	.	.	.	.	. 217
II.	Evolution and Botanical Classification	.	.	.	.	.	.	. 218
III.	Groups or Types of Maize .	.	.	.	.	.	.	. 220
	A. Dent Maize .	.	.	.	.	.	.	.	.	. 220
	B. Flint Maize .	.	.	.	.	.	.	.	.	. 220
	C. Sweet Corn .	.	.	.	.	.	.	.	.	. 220
	D. Soft or Flour Maize	.	.	.	.	.	.	. 222
	E. Popcorn	.	.	.	.	.	.	.	.	. 222
	F. Waxy Maize	.	.	.	.	.	.	.	. 222
	G. Pod Maize .	.	.	.	.	.	.	.	. 222
IV.	Maize in Britain	.	.	.	.	.	.	.	. 222
V.	Maize Development in Other Parts of the World	.	.	.	. 224
VI.	General Requirements of Climate and Soil	.	.	.	.	. 225
	A. Climate	.	.	.	.	.	.	.	.	. 225
	B. Soil Requirements	.	.	.	.	.	.	. 230
VII.	Fertiliser Requirements	.	.	.	.	.	.	. 232
	A. Crop for Grain	.	.	.	.	.	.	.	. 232
	B. Crops for Green Soiling or Silage	.	.	.	.	. 232
VIII.	Time of Application, Placement and Type of Fertilisers .	.	. 234
IX.	Cultivations and Seeding	.	.	.	.	.	.	. 235
	A. Cultivations	.	.	.	.	.	.	.	. 235
	B. Seeding	.	.	.	.	.	.	.	.	. 235
X.	Seed Dressings .	.	.	.	.	.	.	.	. 236
XI.	Variety Classification	.	.	.	.	.	.	. 237
	A. Cultivars for Grain	.	.	.	.	.	.	. 237
	B. Cultivars for Sweet Corn	.	.	.	.	.	. 237
	C. Cultivars for Silage	.	.	.	.	.	.	. 238
	D. Cultivars for Feeding Green .	.	.	.	.	. 239
XII.	Seed Rate, Plant Populations and Spacing	.	.	.	. 239
XIII.	Weed Control .	.	.	.	.	.	.	.	. 242
XIV.	Pest Control	.	.	.	.	.	.	.	.	. 244
	A. Frit Fly	.	.	.	.	.	.	.	.	. 245
XV.	Diseases of Maize	.	.	.	.	.	.	.	. 245
	A. Maize Rust	.	.	.	.	.	.	.	. 246
	B. Maize Smut	.	.	.	.	.	.	.	. 246
	C. Downy Mildew	.	.	.	.	.	.	.	. 246
	D. Damping Off	.	.	.	.	.	.	.	. 247
	E. Root Rot and Foot Rot	.	.	.	.	.	. 247
XVI.	Growth and Mid-season Management .	.	.	.	. 247

XVII. Harvest and Storage 248
    A. Sweet Corn. 248
    B. Silage Maize 248
    C. Grain Maize 250
XVIII. Yield and Crop Quality 252
    A. Grain Maize 252
    B. Silage and Green Fodder Maize 253
    C. Sweet Corn 253
XIX. Production Economics, Gross Output and Gross Margins . . 254
    A. Grain Maize 254
    B. Silage Maize 255
    C. Sweet Corn 256
    Appendix. 257
    References 260

I. INTRODUCTION

It is thought that maize has been cultivated for several thousand years and this view is supported by the archaeological excavations in New Mexico which unearthed maize grains and parts of maize ears from caves and rock shelters in use 4500 years ago (Berger, 1962). According to Weatherwax and Randolph (1955), pollen grains of Zea, Tripsacum and Euchlaena of even earlier origin were said to be found at a depth of more than fifty metres below Mexico City.

Several of the common names for maize in the European languages (Turkish, Asian and Indian corn) suggest the Eastern hemisphere as the area of origin but De Candolle (1884), after a critical examination of all the evidence, is didactic in his statement that maize originated in the New World. A few years later Harshberger (1893), in an exhaustive survey of the literature on the origin of maize through philology, concluded that it was introduced to the United States of America from the tribes of Northern Mexico and from the Caribs of the West Indies. It may well be that the name "Indian corn" was derived simply from the fact that some of the American maize had its origin in the West Indian Islands or that some of the first cultivation of maize seen by Europeans was that practised by the North American Indians.

With the discovery of the New World, maize cultivation was learnt from these Indian tribes. Christopher Columbus was probably the first to see and describe this new cereal, found by members of his crew in Cuba and he named it maize. It is most likely that Columbus or some of his contemporaries were the first to introduce maize to Europe. Once established in Spain and Portugal its passage to the African continent was simply a matter of time and with the Portugese exploring in India and the East, its transfer to Asia can easily be envisaged.

Maize is a relatively new crop to British agriculture. With its large bulk of cob and stover (leaves and stem), cutting for silage and green soiling in summer to cattle have been the main uses from a relatively small acreage. However, in recent years a number of farmers, notably residing in South-east England, have pioneered its cultivation for grain.

II. EVOLUTION AND BOTANICAL CLASSIFICATION

Using Vavilov's concept of centre of origin, the highlands of Peru, Bolivia and Ecuador would appear to have the best claim as the starting point of maize on account of the great diversity of the native forms in that region (Jugenheimer, 1958). On the other hand, maize may have

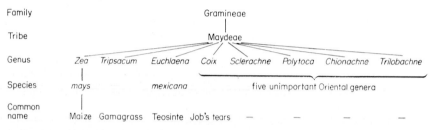

Fig. 1. Botanical family and relatives of maize (*Zea mays*; Linneaus, 1737). Maize is strictly speaking a grass, belonging to the family Gramineae with the following associated seven genera.

originated in Mexico or Central America because this area is considered to be the home of teosinte grass (*Euchlaena mexicana*, Schrad.), a near relative of maize, which shows a wide range of types in this area. Wild maize, to date, has not been found and so there has been much speculation on its origin as can be judged by the following list of authors responsible for summaries of evidence on this subject or original contribution.

De Candolle (1884)
Harshberger (1893, 1896)
Collins (1912)
Burtt-Davy (1914)
Vavilov (1931)
Kempton (1931, 1938)
Kuleshov (1933)
Weatherwax (1935, 1950, 1954)
Jenkins (1937)
Mangelsdorf and Reeves (1938, 1939)
Anderson and Cutler (1942)

Cutler (1946)
Anderson (1947)
Brown and Anderson (1947)
Mangelsdorf (1947, 1948, 1950, 1952)
Mangelsdorf and Smith (1949a, b)
Stoner and Anderson (1949)
Mangelsdorf and Oliver (1951)
Wellhausen (1952)
Anderson and Brown (1952)
Weatherwax and Randolph (1955)

Mangelsdorf (1965) in his essay on the evolution of maize, uses the similarity of fossil and present-day pollen to draw two important conclusions. Firstly, the ancestor of maize was maize and not one of its two American near relatives, teosinte and gamagrass. Secondly, maize is of American origin and not from the Old World or Asia as suggested by some botanists. A number of authorities, however, suggest that maize originated in Central America or Mexico from the chance hybridisation of *Euchlaena* and *Tripsacum* but Weatherwax and Randolph (1955) state that there is no support from cytogenetical or cytotaxonomic studies for this assumption. The tribe *Maydeae* consists of 8 genera of grasses, although Cobley (1956) restricts his list to 7 by omitting one of the oriental genera. They show the greatest degree of specialisation of all the tribes of the Gramineae in that the sexes are separated into different spikelets. This fact is probably responsible for the theories that maize developed from the related genera *Euchlaena* and *Tripsacum*.

Maize cultivation has extended to most of the warmer regions from the New World during the last three hundred years. It is grown for grain, as a fodder crop, for the production of oil and as a vegetable. As a grain crop it is quantitatively the third most important on a world basis behind rice and wheat.

Zea mays has a basic chromosome number of 10 (Jugenheimer, 1958) although Darlington and Wylie's list (1945) contains several references to maize chromosomes above this figure.

Common name	Chromosome number	Reference
Maize	20 + 1 − 7B	Randolph (1928)
Indian Corn	20 + 1 − 7B	Darlington and Upcott (1941)
Maize	10, 40, 80	Randolph (1932)
—	30	Beadle (1930)

Readers interested in the inheritance in maize would do well to consult the exhaustive summary compiled by Emmerson *et al.* (1935) or the one published later by Hayes *et al.* (1955). Rhoades (1950), (1955) outlined the cytogenetics of maize and Weijer (1952) catalogued the genetical types of maize and at the same time produced an extensive bibliography on all aspects of this crop.

III. GROUPS OR TYPES OF MAIZE

Several thousand varieties of maize are now grown throughout the world and most of these can be allocated to one of the seven most important groups, which are listed and briefly described below.

Zea mays indentata Sturt.—Dent maize
Zea mays indurata Sturt.—Flint maize
Zea mays saccharata Sturt.—Sweet corn
Zea mays amylacea Sturt.—Soft maize or Flour maize
Zea mays everta Sturt.—Popcorn
Zea mays ceratina Kulesh.—Waxy maize
Zea mays tunicata Sturt.—Pod maize.

The above classification is based mainly on the characteristics of the seed and this is self evident in the common name.

A. Dent Maize

This is the most widely grown type in America and Mexico. The seeds contain soft starch grains which shrink when ripe and are dried to give the characteristic "dent" in the crown.

B. Flint Maize

Predominates in Western and parts of Central Europe. The seeds are small, smooth and hard, containing little soft starch and plants of this type usually exhibit better agronomic characters than the Dent types. Higher germination, earlier maturity and better tillering capacity are the main features of Flint maize which finds its main outlet for poultry feeding.

C. Sweet Corn

Grown as a vegetable to be eaten fresh or for canning, and in some parts of the world it is used to produce syrup. It differs from dent maize by a single recessive gene which prevents some of the sugars being converted to starch, hence its common name. Sweet corn is mainly associated with the United States, but several other countries cultivate small acreages for direct human consumption.

FIG. 2. Maize for fodder and silage on Mr B. L. Koppenhagen's Church Farm, Wisley, Surrey. Photograph by courtesy of *Farmer and Stockbreeder*.

D. Soft or Flour Maize

This has been associated with the drier areas of the United States and parts of South America and South Africa. It is similar to the flint types in plant characteristics but the grain is very soft, like that of dent maize. According to Berger (1962), soft maize is one of the oldest cultivated forms, being frequently found in the graves of the ancient Aztecs and Incas.

E. Popcorn

Popcorn does not really constitute a separate group, distinct from the other forms of maize, but simply refers to those varieties which characteristically "pop" when heated in cooking fat or oil. The grains are small and round, containing a very high proportion of hard starch and are used in confectionery mainly in the United States.

F. Waxy Maize

This type of maize obtains its name due to the nature of the grains which exhibit a waxy appearance. The starch granules which they contain differ in structure from the other forms of maize in that they are sticky and resemble those of glycogen. Waxy maize is grown to produce starch which is said to have the same characteristics as tapioca. Cultivation is mainly restricted to countries of Eastern Asia.

G. Pod Maize

Pod maize, *Zea mays tunicata*, does not represent a group of presently grown varieties but is typical of some of the earliest cultivated forms of this cereal. It is therefore grown on a small scale in experimental studies concerned with breeding and the origin of maize. Each kernel is surrounded by a pod or husk and the cob is enclosed within husks like the other forms of maize.

Cultivars presently used in British agriculture include American Dent types and many of European origin which are newly introduced hybrids between the Dent and Flint varieties. (For current recommended list see Appendix.) Special varieties are used both in agriculture and horticulture for the production of sweet corn as a vegetable.

IV. MAIZE IN BRITAIN

Maize has been grown in Britain for several hundred years but the acreage devoted to it has never been large since it requires fairly high temperatures for optimum growth and for full maturity. In the past it has been grown mainly for green fodder to be fed to dairy cows at grass in late summer in the drier areas of Eastern England. At this time of year before the autumn flush of grass, quality and quantity of herbage are both low and require to be supplemented. Green soiling maize is valuable for this purpose and is particularly beneficial when rainfall is below average.

During the past twenty years maize has developed considerably as an arable silage crop in Britain on account of its great bulk, produced

Fig. 3. Twenty-two acres of maize on Mr Waid Helver's Little Langford Farm, near Salisbury, Wiltshire, on 5th September, 1960. Photograph by courtesy of *Farmer and Stockbreeder*.

in a relatively short growing season. Its use for silage has been stimulated by a range of new hybrid maize varieties, mainly from the Continent and still further improvement in cultivars is expected when the new cold-tolerant types have been properly evaluated. Over the period in question, quite a large number of the sealed silos, developed in America for maize silage, were imported into this country. A number of these were and are used for the storage of other crop products but with maize it has been found that these containers, together with the crop handling machinery both into and out of the silos, can produce a near automatic feeding system which is admirably suited to beef fattening. Much of the top-unloading machinery for removal of silage in Britain is of American origin or type and although they do not work so well with the grass silage, they are admirably suited to chopped maize—having been designed with this crop in mind.

In 1952 the National Farmers' Union sponsored a four-year trial period during which the various aspects of maize as a grain crop in Britain were critically examined. A Maize Development Association (M.D.A.) formed by commercial organisations and research establishments with special interests in maize for grain or silage, suggested that since 1956 the case for continuing and expanding the investigations was stronger than in 1952 (M.D.A., 1967). With significant developments and changes in the fields of plant breeding and herbicides and with the introduction into Britain of maize harvesting machinery, further work was thought necessary to evaluate these before making firm recommendations for agricultural practice.

A small quantity of maize is grown both in horticulture and agriculture, the cobs of which are used as a vegetable. Varieties chosen for this purpose are usually of the sweet corn type and have been specifically bred for this purpose. It is, however, possible to use the cobs of grain or fodder maize varieties for human consumption but the quality is much inferior to those of sweet corn and fewer cobs are produced per acre.

V. MAIZE DEVELOPMENT IN OTHER PARTS OF THE WORLD

Following the breeding and introduction of new hybrid varieties which are tolerant to much colder conditions in the spring, there has been a considerable increase in the amount of maize grown in Canada and North-west Europe and it is now found at more northerly latitudes in many countries. According to the Maize Development Association (1967), there has been a significant increase or spread of maize 40 miles north of Paris and it is being grown now in Brittany and Normandy, provinces of France which do not differ greatly in climatic conditions compared with much of Southern England. Maize for grain has crept northwards in Germany and Holland and in respect of the latter country the acreage has already reached 30,000. It is also moving northwards in Canada above the 49th Parallel mainly on account of these cold-tolerant varieties. (See Fig. 5 for position of 49°N latitude line in relation to Britain and France.)

VI. GENERAL REQUIREMENTS OF CLIMATE AND SOIL

A. Climate

In recent years there has been a significant spread of maize growing in the New World and Europe north of the 49th Parallel (Vancouver–Paris). Under Canadian conditions of approximately 110 frost-free days, the early maturing hybrids will ripen satisfactorily, but at the same latitude in Europe, although the growing season is longer (140 days), spring and early summer growth is much slower due to lower temperatures accompanied by fairly wet conditions. According to Hanna

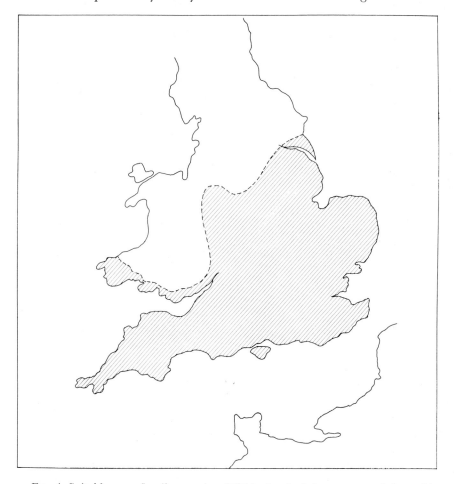

Fig. 4. Suitable areas for silage maize. Within the shaded area, most of the arable land at sea level, which is not susceptible to late spring frosts, would be suitable for silage maize.

(1924), once maize plants have appeared above ground they are unable to withstand temperatures below freezing point, at $-1.6°C$ (29°F) they are damaged and at $-4.4°C$ (24°F) they are killed. However the cold-tolerant cultivars will survive the cooler conditions (under some circumstances even a slight frost) and can be employed to produce bulky silage crops in the Southern half of Britain or grain in the extreme South. The temperatures for growth of maize compared with the other cereals, given by Hall (1945) and reproduced below show that on average maize needs temperatures which are 8–10°F (4°C) higher than those required by the two main cereals. This fact alone dictates the northerly limits for maize cultivation in Britain and Figs 4 and 5 will give a guide as to the suitable areas for grain and silage.

Growth Temperatures for Maize, Wheat and Barley

	Minimum		Optimum		Maximum	
	°F	°C	°F	°C	°F	°C
Wheat	41	5	84	29	109	43
Barley	41	5	84	29	100	38
Maize	49	9	93	34	115	46

To illustrate the importance of temperature and its effect on growth rate Hall (1945) demonstrated the extent of maize root development on both sides of the optimum and this data shows how little this cereal is likely to grow at temperatures below 50°F (10°C)

Temperature		Root growth of Maize in
°F	°C	24 hours (millimetres)
63	17	1·3
79	26	24·5
92	33	39·0
93	34	55·0
101	38	25·2
108·5	42·5	5·9

The majority of the North American maize is cultivated south of latitude 45°N and the cultivars enjoy a frost-free climate of nearly eight months. The greatest intensity of cultivation of maize in the Northern Hemisphere occurs in the region between the 21·1 and 26·7°C July isotherms (69 and 79°F) and the cultivars produced in U.S.A. for this type of climate are only of moderate use for grain in Britain although the early maturing ones may be useful for silage, where full maturity is not wanted.

Fig. 5. Suitable areas for grain maize. Soils which are early to warm in the spring and not susceptible to late spring frosts (hatched area) could be considered for grain maize since the accumulated daily mean temperatures are above 10°C (50°F). An important maize-growing area (cross-hatched) in Northern France is also shown.

(i) RAINFALL

In the United States, rainfall is the second limiting climatic factor in maize growing and according to Berger (1962), this point is reached when the level of precipitation falls below 200 millimetres (8 inches) and for optimum growth 460–600 millimetres (18–24 inches) are required. With lower temperatures in Southern England compared with America, less rain will be needed by the crop and only in exceptionally dry years will May to September rainfall be as low as 8 inches. However, in the future it may be demonstrated that additional water in the

TABLE 1

Irrigation Guide for Maize and Sweet Corn (M.A.F.F., 1967)

Crop	Period of irrigation at seed or early seedling stage	Main irrigation to be applied	Usual months of application	Rate of application		
				A	B	C
Maize	—	At the tasselling stage	July and August	1·5 inches at 1·5 inches or more S.M.D.	1·5 inches at 1·5 inches or more S.M.D.	None
Sweet corn	June	Throughout the whole growth period	June–August	1 inch at 1 inch S.M.D.	1 inch at 1 inch S.M.D.	2 inches at 2 inches S.M.D.

S.M.D. = "soil moisture deficit".

A = soils with low available water, i.e. not more than 1·5 inches/foot depth.

B = soils with medium available water, i.e. more than 1·5 inches and less than 2·5 inches/foot depth.

C = soils with high available water, i.e. more than 2·5 inches/foot depth.

Soils which fit into the groups A, B and C have been listed by Pizer (1963) and the calculation of available water capacity (AWC) was done on the lines suggested by Salter and Williams (1967).

form of overhead irrigation, if and when available, could be beneficial for maize in the driest areas of Britain. Soil moisture to a depth of several feet can be used by this crop with its extensive root system and this coupled with average summer rainfall figures of 10–12 inches (M.A.F.F. Bulletin 138, 1958), so long as it is reasonably evenly distributed, will usually be adequate for growth. The average summer potential transpiration between April and September is approximately 17–18 inches for the area suggested in Fig. 5, M.A.F.F. Bulletin 138 (1958), but since maize will not produce significant growth until May, it may be possible to consider total potential transpiration some 2 inches lower. With rainfall figures of 8–10 inches during this period, irrigation would probably give economic responses with maize provided that a cheap source of water was available.

Recommendations for the irrigation of both maize and sweet corn have recently been given by the Ministry of Agriculture in their new Short Term Leaflet No. 71. These however, must be treated as tentative recommendations only until the results of experiments or observations confirm or suggest modifications to them, and are thus reproduced in Table 1 only as a guide.

(ii) SUNSHINE AND SOLAR ENERGY

Little is said and written about sunshine or solar energy requirements with this cereal. It is, however, assumed that for satisfactory growth and ripening of the crop high levels of bright sunshine are required; in fact in Britain the maximum sunshine hours experienced in the South-east of England are still short of this plant's requirements, if full genetical grain potential is to be exploited. Not only do we experience lower temperatures in our northerly latitudes, but hours of bright sunshine and to a greater extent radiant solar energy are much reduced by comparison with the more southerly regions of Europe or America. Maize is quite unique in its mode of growth and extent and duration of its leaves. They grow in a manner which facilitates efficient utilisation of radiant energy by trapping most of the sun's rays and since the duration of full leaf extends almost to grain maturity, the sun's energy can be transferred to grain yield throughout virtually the whole life of the plant. A point close to optimum leaf area index (L.A.I.) is obtained early and maintained almost to grain maturity thus making maize one of the most efficient utilisers and converters of solar energy into plant energy particularly when the whole plant is considered as the economic yield.

B. Soil Requirements

(i) TEXTURE AND FERTILITY

Successful maize cultivation is more frequently and more easily achieved on soils which are of medium texture. As the soils become lighter the greater is the chance of them "drying out" in mid-summer and although there is really nothing else against them, the very light sandy soils should be avoided except possibly when overhead irrigation may be possible. Soils which are inherently heavy by nature will carry successful crops of maize provided good seed-beds are obtained and this will usually follow correctly timed ploughing and cultivation. It would be wise to avoid the really stiff, heavy clays which present such great difficulties following heavy rains during the growing season and at harvest. Having suggested light to medium textured soils for maize it must also be stressed that organic status and fertility should be high.

It was feared that chalk soils would not be suitable for maize but judging from the experiences gained in France there appears no justification for these apprehensions in respect of soil type alone (Milbourn, 1968). However, since maize is a gross feeder and crops of this nature require a high level of nutrition and a good depth of soil for self expression, the chalks in Britain, with their shallow profile and somewhat exposed situations, sometimes at quite considerable heights, are generally unsuited to maize.

One final point in respect of soil condition and that is to remind would-be growers of maize that the land should be free draining in order that as much of the heat as possible is employed in raising soil temperatures and not removing excesses of soil moisture. The soil should be naturally free draining or contain adequate tile drains to pin down the water table to enable a full rooting system to develop in a plentiful supply of oxygen.

(ii) SOIL pH

Early field trials with maize showed it to be an amphi-tolerant cereal and according to Arrhenius, 93–100% of its maximum yield could be obtained between pH 4 and 9. Pot experiments carried out by Kappen (1929) indicate that maize is fairly acid tolerant and his data, which is reproduced below, suggests an optimum pH of around 6·5.

	1st Estimate		2nd Estimate		
pH	4·69	6·24	4·85	5·74	6·70
Comparative yield	84%	100%	91%	89%	100%

More recently Becker (1952), in assessing the right soil conditions for hybrid maize in Holland, states that a pH of 5·2 or above is generally required. Berger (1962) indicates that maize can be successfully culti-vated on the moderately acid soils and gives the range pH 6–7 as optimal. From the results of Arrhenius's field experiments, together with the practical experiences in France referred to earlier, it may be assumed that maize growing can be successful under alkaline conditions provided there are no serious deficiencies of the micro-nutrients. With considerable quantities of free calcium in the soil there is a tendency towards lime-induced chlorosis amongst the cereals and maize is certainly no exception.

Fig. 6. A well developed maize cob just past the correct maturity stage for silage making, but not ready for grain harvest. Photograph by courtesy of *Farmer and Stockbreeder*.

VII.　FERTILISER REQUIREMENTS

A.　Crop for Grain

With a comparatively small acreage devoted to maize, little critical work has been done to assess the precise requirements in respect of the three major nutrients N, P and K. In connection with grain maize, the recommendations for French conditions, made in La Culture de Mais (1966), are for 100 units/acre of each of the major nutrients. Responses in cereal yield to phosphate and potash applications in Britain have usually been small and this has been verified in connection with maize by Bunting (1962). Milbourn (1968) suggests the following fertiliser levels for British conditions for grain maize.

	Units/acre		
	N	P_2O_5	K_2O
Soils with adequate reserves of phosphate and potash	100	80	80
Soils low in both of these major nutrients	100	100 or even higher	100 or even higher

It has been suggested that the phosphate and potash could be applied during the late winter or early spring and the nitrogen incorporated into the seed-bed at the time of drilling. In the past, fertilisers were normally broadcast but for the future, placement of fertilisers for maize would seem logical in view of the fact that for all purposes it is normally cultivated on wide rows.

B.　Crops for Green Soiling or Silage

When cutting green for direct feeding or silage the whole plant (i.e. stover plus cobs) is used and the criterion of yield is the total amount of dry matter produced per acre. During the past forty years, dry matter yields have risen considerably, to such a point that maize for arable silage must be considered by those farmers in suitable areas (Table 2).

When attempting to produce 4–5 tons of dry matter per acre from maize it must be thought of as a gross feeder, requiring considerable quantities of the major nutrients and fertiliser levels in excess of the figures recommended for the grain crop may be necessary. From observations in the field, rather than from experiment, Eddowes (1962) maintains that on average 6–10 cwt of a high nitrogen compound fertiliser will be needed for maximum green yield with maize except in

FIG. 7. Maize ready for grain harvest. Photograph by courtesy of the Maize Development Association (taken by Charles Howard, Chichester).

TABLE 2

Estimates of Green Yield and Dry matter Production from Maize

Green yield (tons/acre)	Dry matter yields (tons/acre)	Reference
14·6	2·41	Woodman and Amos (1924)
17·0	—	Woodman and Amos (1928)
10·5	2·50	Castle *et al.* (1951)
9·2–16·0	2·88	Bunting and Blackman (1951)
13·8	3·52	Bland (1953)
25·0	4·26	Eddowes (1962)
—	Slightly less than 4 to slightly over 7	Bunting (1962)
—	Over 5	Milbourn (1968)

the wetter west and north. In these areas the mineralisation of soil organic matter will release significant quantities of the plant nutrients and 4–6 cwt/acre may then be sufficient. Fertiliser recommendations for green maize, in absence of reliable experimental data, are tentatively suggested at 150 units N, 100–120 units P_2O and 100–120 units K_2O when the land does not receive farmyard manure with appropriate reductions when this is available.

VIII. TIME OF APPLICATION, PLACEMENT AND TYPE OF FERTILISERS

It has been suggested that phosphate and potash should be applied to the land well in advance of drilling (winter or early spring) and the nitrogen incorporated into the seed-bed just prior to drilling otherwise much of it would be lost by leaching. In practice high N compound fertilisers are normally broadcast and worked into the seed-bed before drilling as spring work is normally unhurried since maize does not usually go in until May. Being a gross feeder and grown in wide rows would suggest combine drilling but there are two main reasons why this is not practised. Firstly the amount of fertiliser required by maize is much in excess of the maximum which can be combine drilled satisfactorily and secondly germination is much retarded and reduced by fertilisers in contact with the seed. Starter dressings containing two or three of the major nutrients are applied close to the seed either as a band application below and to the side in solid or liquid form in America and France, to be followed by early top dressings of nitrogen or sideband placement of N, P and K. Results from a limited number of trials with both liquid and solid fertilisers indicated that with a high nitrogen compound, 1 cwt broadcast could be equalled by 0·66 cwt placed (Eddowes, 1962). The same author was unable to show different responses in silage maize crops to solid and liquid forms of fertiliser and concluded that the choice was largely governed by unit cost and convenience and it will also depend on whether or not contract services for liquid fertiliser applications are available. Contrary to expectations, liquid fertilisers have often been slower to apply and more costly to use in the past, and the resistance shown by farmers to adopt them appears well justified.

IX. CULTIVATIONS AND SEEDING

A. Cultivations

With a more extensive and deeper rooting system than the other cereals, maize will require deeper ploughing, cultivations and seed-beds to obtain maximum growth. Stubble cleaning followed by autumn ploughing to 9–10 inches on the stronger soils is advantageous but ploughing may be left until the early spring when textures are light. Cultivations which follow should be kept to a minimum in the preparation of a good seed-bed which should be to a depth of 4–5 inches and with restrictions in the number of operations. This should avoid excessive loss of soil moisture, a point of major consideration in the drier areas of Britain during April and May. Stale seed-bed techniques may be employed with this late planted crop with the intention of killing by cultural means the first batch of weeds after germination and although inter-row cultivation can follow crop emergence to obtain further weed control, chemical means are often preferred, again with moisture conservation in view. Seed-beds should be uniform and fine to obtain a quick germination of as many seeds as possible and to assist the action of herbicides in their control of weeds. Over-consolidation should be avoided on the heavier soils otherwise crop emergence will be delayed.

B. Seeding

Minimum temperatures for growth with maize are around 50°F (10°C) and thus early spring sowings are of little value except when the soils are warmer than usual. Under cool conditions seeds rot and final emergence figures are usually much lower for early April sowings compared with late April and May as the data from many field trials showed (Bunting, 1962), p. 236.

Sowing date	6th April	13th April	20th April	27th April	4th May
Final emergence %	59	71	80	85	96

The experiments from which the above data were taken also indicated that seeds sown at any time during April tended to germinate and emerge at much the same time in May, between 55 and 75% having come through by the 12th of the month. When the average temperatures are over 50°F (10°C) the emergence of maize will take approximately two weeks and during this period and in the early seedling stage this cereal is particularly susceptible to bird damage. Late spring frosts can also be damaging to seedling maize although with the cold tolerant varieties being introduced there is every chance that this crop may now survive the first few degrees of frost.

X. SEED DRESSINGS

Maize is very susceptible to attacks by soil inhabiting fungi and by wireworms and to obtain a satisfactory braird, seed dressings should be used to give protection against these two. Seed dressings are usually available combining fungicides and insecticides, but unfortunately they have not been designed for use on maize. According to Milbourn (1968) there is not sufficient fungicide in them to protect the young seedling maize and when the rate of application is increased the insecticidal part will probably be phytotoxic. In order to obtain satisfactory control of fungi and wire-worms, the fungicide and insecticide should be purchased separately and applied to the seed just prior to sowing at the following rates:

Chemical	Application rate	Commercial preparation or trade name	
Thiram (dithiocarbamate)	4 oz/cwt of seed (50% a.i.)	Fernasan A	(Plant Protection)
Gamma–B.H.C.	2 oz/cwt of seed (40% a.i.)	Abol Gamma Seed Dressing	(Plant Protection)

From Milbourn (1968)

One of the major problems with maize is the control of birds particularly corvids and pigeons and when effective bird repellent seed dressings are available they should also be incorporated, to minimise plant population losses in the early stages of the crop. Seed dressings based on anthraquinone are now available which are said to render cereal grains unpalatable to some birds and these may be used in conjunction with practical methods of bird scaring to avoid economic damage.

XI. VARIETY CLASSIFICATION

Maize cultivars have been tested by the National Institute of Agricultural Botany (Cambridge) in conjunction with the Agricultural Research Council's Unit of Experimental Agronomy (Oxford) for many years and the results appear in N.I.A.B. Farmers' Leaflet No. 2 "Varieties of Green Fodder Crops" (see Appendix). They have been classified according to maturity rate into early and medium (late types included for completeness) and this characteristic alone will usually determine the use.

Early maturity	Medium maturity	Late maturity
Grain or silage	Silage or green feed	Green feed
Caldera 535	Asgrow 77	Pioneer 383
Dekalb 202	Inra 321	Kingscroft KS2
Anjou 196	Austria 290	White Horse Tooth

A. Cultivars for Grain

The early American hybrid maize varieties, mostly of Wisconsin origin, although produced for a shorter growing season than that in Southern Britain, were not able to withstand the cool, wet springs so often experienced here. As a result they were unsatisfactory for grain, but found a limited use for silage in the 1950's. Recently plant breeders in North West Europe have produced cold tolerant varieties much more suited to our climatic conditions and INRA 200 from the official plant breeding station in France and Kelvedon 59A produced by Dr Pap, Inra 200, Anjou 210 and Pioneer 131, are suitable for grain. Earliness, one of the most important agronomic characteristics, is exhibited by both these varieties together with the ability to produce and support sufficient cobs per plant to give an economic yield. It is interesting to note that the French hybrids are crosses between local flint types and the high yielding American Dent types (Milbourn, 1968).

B. Cultivars for Sweet Corn

In the past a limited number of sweet corn varieties have been available for small scale production usually in horticulture, but with increased interest in maize from an agricultural point of view, both in Britain and on the Continent there has been a renewed interest in the growing of corn on the cob. Early maturing varieties must be chosen and although it is possible to use the cultivars listed for grain or silage, much better quality cobs are obtained from the sweet corn varieties and these should be selected in order to obtain a high financial output per acre.

John Innes Hybrid sweet corn has been available for many years and although initiated for the amateur gardener or small scale producer in horticulture it is a useful early variety producing a marketable cob from each plant. North Star, another early maturing type, is very popular with growers and appears to produce more cobs than the previous variety (J. I. Hybrid) and Northern Belle, although slightly later than the previous two varieties, is likely to outyield them by supporting two cobs per plant. Kelvedon Glory is a relatively new variety from the Hurst Gunson Cooper Taber Seeds Organisation, with a medium maturity rating and capable of producing about 1·33 cobs per plant whilst October Gold, as the name suggests, is late compared with the others but is capable of high yields.

Sweet corn variety trials are now being carried out by the National Institute of Agricultural Botany (Cambridge) and with a total of sixteen cultivars on test in 1967 and twelve in 1968, there is plenty of interest in this crop and probably sufficient information for definite recommendation to be made early in the future. Those varieties which are already being grown commercially have been commented upon above and of these Northern Belle and Kelvedon Glory appear to give good results.

C. Cultivars for Silage

Good quality maize silage is produced from crops with a high proportion of cob in the total green weight. The cobs themselves are the most nutritious part of the maize plant and it is from these that most of the soluble carbohydrates exude, following laceration and these are needed by the bacteria to produce the right condition for fermentation. It is therefore important on two accounts to choose varieties from the early or medium groups which will produce on average 1·5–2 cobs/plant which have reached the cheesy stage of maturity by the time silage making comes along. The N.I.A.B. Farmers' leaflet No. 2 "Varieties of Green Fodder Crops" list the following as highest yielding:

Early maturity	Medium maturity
Caldera 535	Asgrow 77
Dekalb 202	Inra 321
Anjou 196	Austria 290

D. Cultivars for Feeding Green

Varieties for summer or autumn feeding can be chosen from any of the groups, but those which are considered late maturing have been recommended only for green feed, or green soiling as it is sometimes termed. Cob production from them is normally very poor, but they will give a high output of stover (leaf and stem) per acre which can be fed to stock at grass when the growth of herbage has ceased. Pioneer 383 is the highest yielder within this late group.

XII. SEED RATE, PLANT POPULATIONS AND SPACING

In assessing plant population and grain yield with maize Bunting and Blackman (1951) showed that 6 plants/square yard were optimal for flint types and the early hybrid varieties and that individually spaced plants performed better than those set "on the hill". (This was a common practice in America, placing seeds in groups in order to facilitate cultivation in two directions at right angles.) More recent work in France (Chesneau, 1966) suggests a final population of 7·5/square

TABLE 3

The Effect of Plant Population on Dry Matter Yield with Maize

Plant density (number/sq. yd)	Comparative yield (expressed as % of the yield at 6 plants/sq. yd)	
3	75	
6	*100*	Control
9	116	
12	121	
15	123	

After Bunting (1962).

yard which is equivalent to 36,300 per acre for maize crops grown for grain.

When yield is taken as the whole plant and the highest yield of dry matter produced per acre is the goal, then higher plant numbers are required as indicated Table 3.

Dry matter production rose with each successive increase in plant numbers and was highest at 27 plants/square yard, the highest level studied by Bunting (1962). Excessive vegetative growth at this very high density severely restricts the sexually reproductive phase and cob production is poor. This in turn depresses the feeding value of the crop and at the same time it increases the risks of serious lodging. Thus for silage making a small proportion of maximum possible dry matter yield should be sacrificed to maintain crop quality and final plant numbers of 9–10/square yard will normally do this and at the same time give economies in seed. Translating those data to seed rate should take into account the number of seeds per pound and estimates of field germination. Assessments of these have been made to arrive at the weight of seed required per acre at planting for grain, silage and green feed maize in Table 4.

TABLE 4

Seed rate for Maize (lb/acre) Related to Sowing Date and Seed Size

Crop Use	Sowing Date					
	Mid April			Late April/early May		
	Large seed	Medium seed	Small seed	Large seed	Medium seed[a]	Small seed
Grain (8 plants/sq. yd)	55·3	39·5	30·7	43·0	30·7	23·9
Silage (10 plants/sq. yd)	69·1	49·4	38·4	53·7	38·4	29·8
Green Feed (12 plants/ sq. yd)	82·9	59·2	46·1	64·5	46·0	35·8

In calculating the seed rates given above, large, medium and small seeds are assumed to contain 1000, 1400 and 1800 seeds/pound respectively. Field germination following a mid April sowing is assumed to be 70% whilst the corresponding figure for late April/early May seeding is taken as 90%.

[a] Thus it will be seen that with medium seed planted around the optimum date 30, 38 and 46 pounds of seed are needed/acre for grain, silage and green-feed crops respectively.

Row widths for maize are governed almost entirely by the dimension of harvesting machinery whether grown for grain or silage. In other parts of the world row widths of 36–42 inches are commonly employed, but under British conditions 20–28 inches is the most suitable range for silage maize and 28–32 inches when grown for grain. These narrower row widths are considered suitable for the newer hybrid varieties. They are also satisfactory for the maize attachments for forage harvesters or for the cob picking attachments for combine harvesters (Figs 8 and 9).

Dressed seed should be drilled or precision seeded at a depth ranging from 1 to 2 inches. Increasing the depth of sowing below 2 inches, whilst affording a little extra protection from corvids, is likely to result

FIG. 8. Harvesting grain maize at Wye College, Kent, in 1968 by a combine harvester fitted with special maize attachments. Photograph by courtesy of *Farmer and Stockbreeder*.

FIG. 9. Close-up of maize attachments which are fitted to the front of a combine to guide the crop in and to facilitate easy harvesting. Three or four rows will normally be harvested at a time depending on crop row width and combine size. Photograph by courtesy of *Farmer and Stockbreeder*.

in much poorer field germination and thus is not recommended by most authorities.

XIII. WEED CONTROL

Sowing dates for maize associated with late April or early May enable some growers to practice weed control entirely by cultural means using a false seed-bed technique. This involves the preparation of a rough seed-bed several weeks before the actual date of sowing and as many weeds as possible are encouraged to germinate. These can be eliminated by ploughing or cultivation and the maize sown in a clean seed-bed. Inter-row cultivation following crop emergence will complete the cultural approach to weed control but unfortunately will lead to significant soil moisture loss in drier areas of England. As a result farmers are keen to use chemicals for weed control and in recent years some of the new materials have given excellent control of broad-leaved

TABLE 5

Recommended Herbicides for Weed-control in Maize

Chemical	Trade name	Rate of application (commercial product)	Time of application	Comments
SWEET CORN (M.A.F.F. STL. 52, 1968–69)				
Simazine	Gesatop	2–3 lb/acre	Immediately after drilling	Soil acting and needs moist conditions
Atrazine	Gesaprin	2 lb/acre	Before or after crop emergence	Soil and foliage acting. Post-emergence application should be before the weeds are 2 inches high

GRAIN MAIZE (Milbourn, 1968)
Chemical: Atrazine.
Dosage rate: 1–1·5 lb active ingredient/acre — Highest rate for heavy soils, organic
Commercial formulation rate 2–3 lb/acre — soils or lighter soils with heavy infestation of grass weeds.
Suggestion made that an additional 0·5 lb active ingredient applied post-emergence will give good weed control if weed growth is less than 1·5 inches in height.
MAIZE FOR GRAIN OR FODDER ("Weed Control Handbook". Issued by the British Weed Control Council. (Fryer and Evans, eds), fifth edition, 1968).
The following part of this table incorporates the recommended herbicides for weed control in maize, together with dosage rates and comments where necessary.

TABLE 5 (*Continued*)

Type of treatment	Herbicide	Rate of application (active ingredient /acre)	Comments
Contact pre-emergence	Cresylic acids	As per manufacturer's instructions	⎫
	Diquat plus paraquat	7–14 oz	⎬ For seedling weeds
	Paraquat	8–16 oz	
	Pentachloro-phenol	3–4 lb	
	Sulphuric acid	10–12·5% v/v at 100 gal/acre	⎭
Residual pre-emergence	Atrazine	1–1·5 lb	1 lb for light soils 1·5 lb for heavy soils
	Simazine	1–1·5 lb	As for Atrazine
Post-emergence	Atrazine	1–1·5 lb	As above
	2,4-D Amine	1 lb	Applied in medium to high volume when crop is 3–6 inches high. Direct spraying between the rows reduces the check to maize plants.
	2,4-D Ester	1·5–2 lb	Applied in medium to high volume. Some residual action can be expected.

For information only and not as a recommendation the previous edition of the "Weed Control Handbook" stated that linuron at 1–2 lb/acre sprayed between the rows, was giving good results on the Continent.

and grass weeds and it is often the latter which have emerged as our number one crop hazard amongst maize and the other cereals.

Chemicals for weed control fall naturally into two groups, pre- and post-emergence and these are listed below.

Pre-emergence	Post-emergence
Simazine	MCPA
Atrazine	2,4–D Amine
	2,4–D Ester

From trials in the West Midlands, reported by Eddowes (1962), it was established that most of the above chemicals gave fairly satisfactory weed control. The post-emergence herbicides were applied when the maize had four leaves and whilst the immediate control of weeds was good, there remained a few resistant ones, and those germinating later on tended to become a nuisance unless the cereal itself produced sufficiently dense and rapid growth which would suppress them. The pre-emergence herbicides Simazine and Atrazine, applied before seedling emergence and usually within a week of drilling, gave better weed control and in general resulted in higher maize yields. Atrazine gave better weed control in these trials than Simazine due partly to its higher solubility and there is some doubt as to the value of Simazine in very dry areas and in exceptionally dry years. Bunting (1962) and Milbourn (1968) confirm the superiority of Atrazine on the grounds of higher solubility and the latter suggests only Atrazine for weed control in grain maize. A summary of the recommended herbicides for weed control in the various types of maize is given in Table 5.

XIV. PEST CONTROL

Early insect predators such as wireworms and leatherjackets are likely to be much more troublesome when maize follows ploughed up grass and in addition to the combined fungicide/insecticide seed dressing suggested earlier, a soil application of gamma-B.H.C. in dust or liquid form may also be required.

Emerging maize plants are particularly prone to attack by rooks and anything which deters the activity of these birds during this very vulnerable stage in the crop should be employed. These will include the use of chemical bird repellents (i.e. anthraquinone-based formulations) and any bird scaring devices which give some degree of protection. Automatic acetylene operated bangers have proved effective over short periods and where necessary time clocks can be coupled to them so that they begin to operate at dawn, thus alleviating noise at night which can be a nuisance to nearby inhabitants. Periodic excursions round the crop with a shot gun will often enhance the value of these bangers and it has been suggested that hanging up rooks that have been shot may deter some of the others from attacking the crop. Long-playing records, the tracks of which contain bird warning cries at intervals, are being marketed on the Continent for farmers with bird problems and favourable reports on these have been received. However, it is a little premature at this stage to say whether they would prove effective over a long period and critical experiments under British conditions will be necessary to see whether they can make a significant contribution to crop protection.

One form of deterrent which nearly always seems to work is the erection of black thread over the crop. Terylene thread is strong and durable and according to Milbourn (1968), when placed at waist height in 50 yard squares, complete protection over three years was obtained at Wye and the erection needed only one-third of a man-hour/acre.

A. Frit Fly

Maize, like oats, is susceptible to damage by the frit fly (*Oscinella frit*). Infestation levels are not usually high and unless maize has been grown on the land for several years, control measures may not normally be considered. Where frequently grown for grain and especially where sweet corn is marketed as a vegetable then DDT sprayings or systemic insecticidal applications will form part of routine practice in the cultivation of maize.

Ministry of Agriculture recommendations through their annual publication "Approved Products for Farmers and Growers" (1968) for frit control are as follows:

Maize	Sweet corn
Chlorfenvinphos	Chlorfenvinphos
	phorate

(Chlorfenvinphos marketed in granular form:
   Birlane granules, Shellstar; Sapecron granules, Ciba;
Chlorfenvinphos marketed in liquid form:
   Birlane 24, Shellstar;
Phorate marketed in granular form:
   Thimet phorate 10% granules, Cyanamid)

Recent developments in agricultural engineering have produced precision applicators for use with these granular insecticides. Microband application can be obtained using hoppers which incorporate metering devices to give accurate dispensation and at least one machine uses microband meters coupled to an airstream (which are set up from hydraulically driven fan units) for delivering the insecticide to the soil. Small, hand-operated gravity feed barrow models are also available when small acreages are involved.

When maize was grown for silage in the West Midlands, approximately 10% of the plants were affected by frit fly and although there was a temporary setback, the crop had recovered by late June. There was no significant loss in yield and control measures were considered unnecessary (Williams, 1960).

XV. DISEASES OF MAIZE

At the present time diseases which attack maize under British conditions are few in number and those which occur seldom produce

I

damage of economic significance. It is also fortunate that this crop is not affected by the major root and foliar diseases of wheat and barley which have become troublesome in recent years with cereal mono-culture or close cereal cropping. As a result maize can be looked upon as a break-crop, but if grown frequently in rotation and should the acreage rise very drastically, then the disease position is almost certain to alter fairly rapidly.

During the period 1943–46, Moore (1948) reported rust (*Puccinia sorghi* Schw.), smut (*Ustilago zeae* (Bechn.) Unger), grey mould (*Botrytis cinerea* Fr.) and root rot (*Fusarium* spp.) on varieties of sweet corn. In 1961 Moore and Moore list rust, smut and downy mildew (*Sclerospora macrospora* Sacc.) as diseases of maize and Milbourn (1968) suggests that soil moulds (*Pythium* spp.), smut and foot rot (*Gibberella zeae* (Schw.) Petch) may be troublesome if maize becomes more widely grown.

A.   Maize Rust (*Puccinia sorghi* Schw).

An important disease when this crop is grown frequently and in large quantities, e.g. in U.S.A. It has been recorded on several occasions in England, usually quite late in the growing season and like most of the other diseases which do not occur early in plant life, it has little economic effect. This rust fungus may spend part of its life on species of Oxalis (Wood-Sorrel) but up to now it has not been thought of as a serious disease of maize.

B.   Maize Smut (*Ustilago maydis* (DC) Corda)

Previously known as *U. zeae*, this occurs frequently wherever maize is grown on a considerable scale. In North America the level of infection varies between a trace up to 6%. Smut has been recorded on maize in England but is of little consequence as a yield reducing disease. Unlike the other cereal smuts, it is not a seed-borne infection and therefore seed dressings are of no value. Maize smut is controlled by restricting the frequency of maize cultivation and it is suggested that a crop once in three or four years should keep this disease in check. However, at the moment with such a limited acreage maize smut should only be of academic interest.

C.   Downy Mildew

Downy mildew is caused by the fungus *Sclerospora macrospora* (Sacc.) and can attack all the cereals, including maize and several grasses but although it is frequently found in many other parts of the world, according to Moore and Moore (1961) it has yet to be seen in Britain.

D. Damping Off

A condition which is often responsible for poor braids with maize is caused by soil borne fungi known as *Pythium* spp. Control measures simply involve the use of the correct amount and type of seed dressing which according to Martin (1965) are as follows:

Captan at 2 oz active ingredient/cwt of seed (3 oz of 75% strength) or Thiram at 2 oz active ingredient/cwt of seed (4 oz of 50% material).

E. Root Rot and Foot Rot

Root rot caused by *Fusarium* spp. and Foot rot, *Gibberella zeae*, are diseases which could cause some damage to maize and whilst adequate drainage would help to reduce the amount of the former Milbourn (1968) suggests that resistant varieties may be needed to control the latter.

XVI. GROWTH AND MID-SEASON MANAGEMENT

Eddows (1962) in tracing the growth and development of maize for silage in the West Midlands of England, has indicated a height increase of only 2–3 inches/week in the first 6–7 weeks following emergence, but with higher July temperatures the rate increased to 1 inch/day until the tassels were fully developed. In Southern England silage maize plants show a daily growth rate of 1·25 inches in July and August (Bunting, 1962). At plant population between 30,000 and 60,000 per acre leaf area index (L.A.I.) ranged between 3 and 5 in July rising to between 7 and 10 in August. This large leaf cover without early senescence is responsible for the high rate of dry matter production per acre and it is important that individual plants remain standing to intercept the solar energy. In order to avoid lodging those varieties with the best field characters should be chosen and excessive levels of fertiliser nitrogen should be avoided.

On germination, maize produces primary and secondary roots below ground and once the plant itself is well established, groups of adventitious roots emerge from the stem nodes above ground. These penetrate the soil in a circle at the base of the stem and are often referred to as prop-roots because they help to keep the plant vertical. In many parts of the world inter-row cultivation is practised firstly for weed control and secondly to build up the soil around the base of the plant. The latter is thought to aid growth and general development and at the same time reduce the risks of lodging. Maize in Britain, whether for silage or grain is found in the drier areas and thus inter-row cultivation does not help in moisture conservation. Once the residual chemicals are adopted as an integral part of maize growing, subsequent cultivation become superfluous. However, where cultivation is relied upon for weed control, it is important that the hoes and tines are set at least 3–4 inches away from the stem in order to protect the prop roots from excision.

In respect of sweet corn, grey squirrels may be troublesome once the cobs approach ripeness and occasionally these pests may require to be controlled using shotguns.

XVII. HARVEST AND STORAGE
A. Sweet Corn

Sweet corn is produced on a relatively small scale by growers in Britain and as a result it is usually picked by hand.

Well presented, evenly graded cobs which are harvested early can result in good profit margins with maize, but samples which are not properly graded, contain poorly filled cobs or are late will seldom show satisfactory financial returns. Competition from Europe, particularly Holland and France, is severe and dictates that to be financially successful, British growers must be technically proficient in growing, harvesting and marketing as early a crop as possible, since they start out with several factors (notably climatic) against them. Sweet corn is ready for harvesting when the majority of the cobs are fairly close to maturity. Early varieties in exceptionally hot summers would be ready in August but the bulk of the crop would not normally be ready until September and when exceptionally late cultivars are chosen, this could even be as late as October.

B. Silage Maize

Bunting (1962) demonstrated increasing dry matter yield with silage maize up to the end of September or the beginning of October and this should be considered as the average harvest period. At this time of year the combine harvesting of the other cereals will normally have been completed and maize for silage can be taken before the potato or sugar harvest begins. It has often been said that the optimum stage of growth for cutting silage maize is once the cobs have reached the "cheesy" stage. This implies that cobs are well advanced towards maturity but the grain still contains a considerable amount of moisture and thus soluble carbohydrates will be made available for the fermentation process. In all events the crop should be ensiled before the onset of serious autumn frosts which tend to lower the feeding value of the leaves. Harvesting will normally be done with forage harvesters which usually give something approaching the right degree of laceration of cob and stover. This fine chopping is required firstly to release part of the soluble carbohydrates and secondly to arrive at the correct length of material to facilitate easy handling, adequate consolidation and easy removal from the silo by mechanical unloader or animals, should self-feeding be the method of utilisation. Ordinary forage harvesters of all kinds will require maize attachments to guide the plants into the cutting mechanisms in order to minimise field losses. Single or double row attachments are available and in the case of the latter, it usually necessitates the crop to be grown on wider than average row-widths and the power requirements of the tractor may be greater.

Flail type of forage harvesters are fairly common but according to Eddowes (1962) when used in a standing crop of maize, a high proportion was left on the ground. Considerable improvement in efficiency was obtained with this type of forage harvester after the crop had been "rolled" to 1·5–2 feet, by driving a tractor down the rows and harvesting in the reverse direction. One serious drawback with this method of harvesting is the high degree of soil contamination.

The chopper-type forage harvester, fitted with maize attachments under normal circumstances will result in a more efficient harvesting technique and at the same time will result in finely chopped material which is so important from many points of view. Eddowes (1962) confirmed this and pointed out that with the ordinary flail mower, attempts to improve the proportion of the crop which is picked up were successful but unfortunately resulted in much soil contamination which could lead to serious digestive upsets with stock.

In recent years precision chop harvesters have been introduced, mainly from USA and although they cost about £700, they do the best job in securing the maize crop for silage.

The lacerated cobs, leaves and stalks can either be blown into in-line towed trailers or when additional equipment, tractors and man-power are available the material can be delivered to separately towed self emptying trailers drawn alongside. Successful silage can then be made in pits, clamps or sealed concrete and stave or metal silos according to the type of feeding system required. Maize silage is high in starch and there is normally sufficient carbohydrate present for the correct fermentation process at low temperature provided there are plenty of well-formed cobs present. Only when the crop is particularly immature would consideration be given to the possible use of additives, but under normal circumstances these will not be necessary. Maize silage made from fairly mature plants will have a dry matter of about 20%. This figure is low compared with medium to good grass silages but it must be remembered that with this "arable silage", wilting is not practical and there tends to be quite a large volume of effluent. Tower silos such as "Harvestore" or "Crop Store" are ideally suited to maize silage and on average tend to produce higher quality feed than the pits or clamps on account of reduced nutrient losses—usually associated with their air-tightness. They are also specially designed to incorporate the mechanical unloading equipment which is nearly automatic and thus press button livestock feeding is a possibility. However, this is only obtained at a high capital cost and when maize silage feeding is associated with a beef rather than a dairy enterprise, it is often extremely difficult to support the case for these excellent but costly storage towers and ancillary equipment.

C. Grain Maize

Under British conditions maize grown for grain seldom dries out well before harvest, even when it is left until mid October or even later. Milbourn (1968) suggests that harvest can start during the

Fig. 10. A fully ripe maize grain crop being harvested. Photograph by courtesy of the Maize Development Association (N.F.U. Photo Service).

second half of October once the grain moisture content falls below 45%. Early autumn frosts actually help in two ways. They tend to hasten rather than delay maturity and once the maize leaves have been frosted they cease to create hold-ups by blocking the harvesting machinery. These can be of two types. Firstly, there are harvesters known as cob-pickers. They proceed through the crop, taking off the cobs, stripping them of the surrounding husk and finally deposit them in side-following trailers or in their own tanks. The naked cobs are dried in open wire cribs in many parts of the world, but in Britain one of the simpler cereal grain drying systems is used. When the cob is picked rather than combined, higher moisture content crops can be managed and hence the high figure of 45% moisture suggested earlier. It is said that cob-pickers can be used up to a fortnight before combines for gathering maize and that the grain continues to mature on the cob whilst in the drying crib.

Grain to be combine-harvested should be at least 10% lower in moisture content (i.e. 35%) and even then there is considerable chance of damaged and broken grains. This may be unimportant where the crop is to be home-fed, but when the crop is to be sold to merchants or compounders, a good, undamaged sample will be expected.

Combine harvesters suitable for the other cereals need considerable modification before they will successfully deal with maize. The first way in which this can be done is by fitting two-row or three-row maize attachments to the cutter bar (similar to the modification on forage harvesters for silage maize crops) to guide the crop in and a specially fitted drum will be needed to replace the standard one. Another method to achieve the same ultimate result is to use a cob-picking attachment. This removes the cobs from the maize plant and feeds them into the combine, the drum of which requires slight modification with additional plates. Milbourn (1968) indicates that this second method is likely to be more expensive, but when a good saleable pro-duct is required, it is preferred. One of the leading machinery distri-butors in East Anglia is offering for hire a number of Matador combine harvesters which have been adapted for use in maize crops at an inclusive rate of £7–8 per acre. With first ventures into maize growing or where small acreages are involved it would obviously be more econo-mical to hire one of these machines rather than embark on the quite costly pieces of equipment required to alter existing combines.

One of the major problems associated with maize for grain is its relatively high moisture content compared with the other cereals. It will usually leave the combine at 35% moisture or above and when it is to be sold, drying must take place quickly. With 20% or more moisture to be removed the only satisfactory driers are of the continuous-flow type and even with these the grain will often have to go through three times to bring it to a safe moisture level for storage.

When cob pickers are used to harvest the crop, drying cribs will be necessary or simple on-the-floor farm driers can be employed to remove much of the moisture. Once dried, the cob can be sold as it is for grinding for pig feed or should the grain only be required for sale then it can be separated using the modified combine harvester.

If the crop is to be home fed to livestock, then most storage of the grain or chopped cob using butyl-rubber containers within wire frameworks may be the answer. Milbourn (1968) rightly pointed out that this method of "ensiling" grain or chopped cob leaves behind the stover can easily amount to half the weight and consideration therefore should be given to whole crop ensilage at an earlier stage to exploit fully maize yields where home feeding is the outlet.

XVIII. YIELD AND CROP QUALITY

A. Grain Maize

Yield trials carried out by the National Agricultural Advisory Service in Kent in the early and mid 1960's, in conjunction with

TABLE 6

Grain Maize Variety Trial Yields

Variety	Yield of maize grain at 15% moisture content (cwt/acre)							
	1961	1962	1963	1964	1965	1966	1967	Variety Mean
Kelvedon 59	67·2	49·3	42·3	70·9	—	—	—	57·4
Inra 200	77·5	41·1	46·8	74·5	—	—	—	59·9
Annual mean	72·3	45·2	44·5	72·7	—	—	—	—
Kelvedon 59A	—	—	43·0	74·3	33·5	46·7	60·4	51·6
Inra 200	—	—	46·7	75·8	25·0	50·2	66·2	52·8
Annual mean	—	—	44·8	75·0	29·2	48·4	63·3	—

From Milbourn (1968).

N.I.A.B. and the A.R.C. Unit of Experimental Agronomy at Oxford, have been summarised by Milbourn (1966, 1968) and are reproduced above.

These results are from small scale, hand-harvested trials and although they are likely to be higher than fully mechanised commercial crops they indicate the potential yield of grain maize in Britain. Under adverse weather conditions (1965) only 1·5 tons/acre were obtained, whereas in good years (1961 and 1965) 3·5–3·75 tons of grain were harvested. It would appear that on average, without significant varietal improvement, approximately 2·5 tons of grain could be expected. This is perhaps not very high by the standards set by wheat with good husbandry under fertile conditions, but is acceptable when compared with barley and when one considers maize as a break crop when yields from these two cereals are on the decline.

In respect of quality, maize has a starch equivalent of 77·6 compared with barley at 71·4 and the corresponding protein equivalents are 7·6 and 6·5 for maize and barley respectively (Evans, 1960). It is looked upon as a high energy feed and a considerable amount is used in flaked form to improve the utilisation by young stock but it is lower in some of the essential amino acids. This point has to be considered when all or most of the food for pigs is in the form of maize.

B.   Silage and Green Fodder Maize

Estimates of dry matter yield from whole crop maize for silage or green feed have been reported earlier in Table 2 and from the data obtained during the 1960's it would appear that under favourable conditions something like 4–5 tons of dry matter/acre may be expected from green crops weighing between 20 and 30 tons. Eddowes (1962) reports silage dry matter percentages ranging from 17 to 22%, with 54–56% starch equivalent and 6–8% digestible crude protein in the dry matter from crops in the West Midlands. In South-east England Bunting (1962) showed maize capable of producing 4–7 tons of dry matter/acre and he suggested that this yield would easily exceed that obtained for kale or fodder roots in that area.

C.   Sweet Corn

Assuming that one sets out to obtain 8 plants/square yard using Table 4, approximately 6 of these will eventually produce strong growth and come to fruition. This will mean about 30,000 cob-bearing plants/acre and each will produce between 0·5 and 1·5 marketable cobs/plant depending mainly on the growing season experienced and the cultivar chosen. This results in a saleable crop of between 15,000 and 45,000 cobs/acre, a range which will be used later in assessing the gross financial output from this cash cropping enterprise.

XIX. PRODUCTION ECONOMICS, GROSS OUTPUT AND GROSS MARGINS

A. Grain Maize

With such a small acreage of grain maize (as yet) in Britain, the economic aspects either as enterprise costs or gross margins suffer through the lack of sufficient data to give a very accurate picture; however the latter has received attention from Nix (1966, 1967, 1968), the Maize Development Association (M.D.A.) (1967) and Milbourn (1968) although the data presented by these three authorities has much in common. Nix (1968), co-operating with Milbourn, has assessed the

TABLE 7

Gross Margins for Maize compared with Barley

Item	Low	Maize average	High	Barley average
Yield (cwt/acre)	25	35	45	30
Output/acre (25 shillings/cwt)	£31·2	£43·7	£56·2	35·5[a]
Variable Costs				
Seed		3·3		2·1
Fertilisers		9·0		4·5
Sprays and miscellaneous items		5·5		0·8
Total variable costs		17·8		7·4
Gross margin (crop dried on farm)	13·4	25·9	38·4	28·1
Contract drying (£3·5 per ton of grain at 35% moisture)	4·4	6·2	7·9	
Gross Margin (contract dried)	9·0	19·7	30·5	

[a] Barley valued at 21 shillings/cwt and deficiency payment assumed to be £4/acre.

(i) NOTES ON DATA FOR MAIZE
    (1) Seed: 33 lb at 2 shillings/lb.
    (2) Fertilisers: 120 units N., 80 P., 80 K.
    (3) Spray: 2·5 lb atrazine at 36 shillings/lb.

(ii) EXTRA FIXED COSTS
    Combine attachment for cob picking: £1,400. Depreciation and interest approximately £300 a year, i.e. £6/acre on 50 acres a year. Fuel costs with own drier will be high—possibly £1/ton.

gross margins for maize at low, medium and high levels of yield and these are compared with corresponding figures for barley.

From the figures given in Table 7 and looking at average yields of barley and maize it will be seen that the gross margin for barley exceeds that for maize by an amount which is approximately equal to the deficiency payment on barley. The gross output from maize is considerably higher but so too are the variable and fixed costs and therefore maize growing can only be considered as a serious alternative to barley where yields are higher than 35 cwt/acre. It is illogical to subsidise via guaranteed prices the other four cereals wheat, barley, oats and rye when 3·6 million tons of maize are being imported into Britain annually at a cost of £85 million. The areas in which grain can be produced are limited but any reduction in the amount imported would help our National Balance of Payments and in the future farmers could look for government support for maize.

B. Silage Maize

Again the number of costings available concerning silage maize are limited but the early assessments by Eddowes and Mortimer (1961) and Eddowes (1962) with additions to bring them into line with present-day prices will serve as a reasonable guide.

Assuming the green crop weight to be 25 tons/acre and a 20% loss in silage making, then 20 tons of silage would appear to cost about £45 or £2·25/ton. This is slightly higher than the figure of £2/ton quoted by Eddowes and Mortimer (1961) and Eddowes (1962) but is still below the cost of producing grass silage on account of the high yields obtained.

TABLE 8

Estimated Cost of Growing and Ensiling Maize (£/acre)

Cultivation including ploughing and drilling	5
Seed	4
Fertilisers	9
Spraying	
Herbicide	5
Pesticide	3
Bird Control	1
Harvest	8
Machinery depreciation and repair	2
Overheads	3
Rent	5
Total	£45/acre

C. Sweet Corn

The basic costs up to harvesting with sweet corn are the same as grain or silage maize but with hand harvesting, packing, transport and marketing, the overall cost/acre runs into hundreds of pounds. Figures of 2*d.* to 4*d.* as production costs/marketable cob have been tentatively suggested, but up to going to press these could not be verified by reliable economic data.

One grower in South-east England ventured into a small acreage of sweet corn in 1968. He kept accurate records and found that the total cost in growing and marketing a cob was of the order of 5*d.* and since he planned to continue with this crop one is tempted to assume a reasonable profit margin, either obtained or in view.

Costs and Returns—Grain Maize 1970–71

	£/acre	
Seed	4·2	Crop valued at £32/ton
Fertiliser	7·2	Break-even yield 25 cwt/acre
Sprays	4·2	Average yields 39 cwt/acre
Other materials	0·9	Average gross output £62·4/acre
Combining	7·5	
Drying	3·5	
Drilling and haulage	5·0	
Labour and miscellaneous costs	7·5	
Total direct costs	40·0	

Appendix
EXTRACT FROM N.I.A.B FARMER'S LEAFLET NO. 2
1971. VARIETIES OF GREEN FODDER CROPS†

MAIZE

Small acreages of maize have been grown for many years in southern and eastern counties and interest in silage maize has increased with the introduction of forage harvesters, suitable herbicides and hybrid varieties better adapted to British conditions. But there is still the problem of protection from damage by birds to which the crop is particularly liable as the seedlings emerge.

Maize is almost as susceptible to spring frost damage as potatoes and growth is very slow under cold conditions. Experiments have shown that the most satisfactory sowing period is generally from the last week in April to the first week in May. A plant population of 45,000 to 60,000 /acre is satisfactory for fodder maize and 30,000/acre for grain maize. If conditions are favourable after sowing the crop may emerge rapidly. Subsequent growth in late May and June is usually slow and the crop is very sensitive to weed competition. But growth can be exceptionally rapid during warm weather in July and August.

TYPES AND VARIETIES

The N.I.A.B.‡ has tested over 300 varieties of maize in recent years, and data for the recommended varieties are given in Table 9.

The varieties may be grouped according to their time of maturity as: early, medium and late. Choice of the appropriate group will depend on whether the crop is intended for silage, green feed or grain production. The suitability of the environment for development of the crop should also be considered, as temperature is generally the limiting factor under British conditions.

FODDER

The stover (leaves and stem) of maize, with a crude protein content of about 7% and crude fibre content of about 27%, changes relatively

† Reproduced by kind permission of the National Institution of Agricultural Botany, Cambridge.
‡ In collaboration with the Agricultural Research Council Unit of Experimental Agronomy at Oxford.

TABLE 9

Recommended Varieties of Silage Maize

Variety	Origin	Maturity class 9 = very early 0 = very late	Relative yield of dry matter	Dry matter content % at harvest	Early vigour 9 = good 0 = poor	Resistance to lodging 9 = good 0 = poor
EARLY GROUP						
Caldera 433						
(= SM92)	Netherlands	9	93	25	8	8
Dekalb 202	France	9	100	25	7	5
Inra 200	France	9	95	25	7	8
Kelvedon 59A	Britain	9	92	25	8	7
Asgrow 88	Italy	8	93	27	7	8
Anjou 196	France	7	100	23	8	6
Anjou 210	France	7	99	24	8	8
Caldera 535	Netherlands	7	107	24	8	8
Inra 258	France	7	94	24	8	8
MEDIUM GROUP						
Orlagold						
(= Orla 270)	Switzerland	5	101	23	7	5
United 352	France	5	101	23	7	8
Asgrow 77	Italy	4	107	23	6	8
Austria 290	Austria	4	105	23	7	8
Kelvedon 33	Britain	4	101	22	7	7
Inra 321	France	2	106	20	6	6
Mean		—	100	24	7	7

No variety of the Late Group is Recommended.

In the past a number of late maturing varieties have been listed and in 1970 these included Pioneer 383, Kingscroft KS2 and White Horse Tooth.

little in feeding value during the later stages of growth. On the other hand, the developing ear increases rapidly in feeding value. It consists largely of easily digestible carbohydrates together with a rather higher crude protein content and about half the crude fibre content of the stover. Thus, unlike most fodder crops, as the maize crop matures the increasing contribution made by the ears results in improved feeding value.

For good quality silage, with minimum silo losses, the crop should have the ears well formed and the grain in the soft dough stage. If the dry matter content is less than 18% it is unlikely that good quality silage will be produced and seepage may be excessive. For tower silos an even higher dry matter content is probably desirable. Delay in harvesting will increase both the yield and the quality of the silage but it is advisable to avoid the risk of frost damage.

Varieties of North European origin are often superior in vigour, during May and June, to those bred in warmer climates. They tend to do relatively better in adverse seasons and are generally more consistent in performance. In Southern England, varieties of the early and medium maturity groups are likely to give the best results and should be ready for harvesting in late September to early October. Where the site is particularly suitable for maize, varieties in the medium group may be chosen for their higher yields, although in adverse seasons the quality of the silage may not be so high. Conversely, varieties in the early group should be used where highest quality or early harvesting is required.

Silage maize has been successfully grown in the midland counties. While the same general considerations apply to choice of variety, the medium varieties may prove too late maturing. Further north it is unlikely that the crop will prove generally satisfactory. If maize is planted for use as green feed in late July and August, varieties of the later maturity groups are likely to give the greatest yields. In Table 9 yields are given in terms of dry matter, as varieties vary little in D-values about the range 67–70.

GRAIN

In southern England the earliest varieties will produce mature grain, but the yield of grain will exceed those of the more common cereals only on more suitable sites in favourable seasons. The crop ripens late and the moisture content of the grain is high, which may add considerably to costs where the grain has to be dried artificially. There are

nearly forty hybrids in current trials and the three most successful
varieties in the grain maize trials have been:

	Average grain yield[a] at 15% moisture content (tons /acre) (metric tons/hectare)	Average grain moisture content[b] % at harvest	Resistance to lodging 9 = good 0 = poor
Anjou 210	2·5 (6·3)	40	8
Inra 200	2·4 (6·0)	38	6
Kelvedon 59A	2·3 (5·8)	37	4

[a] Mostly from hand harvested trials, and commercial crops may be expected to
be rather lower yielding.

[b] Even apparently small differences in the average moisture content of varieties
are very important as these may indicate that under adverse conditions a variety
will not produce mature grain.

References

Anderson, E. (1947). "Corn before Columbus". Pioneer Hi-bred Corn Co.

Anderson, E. and Brown, W. L. (1952). "Origin of Corn Belt Maize and its
Genetic Significance. Heterosis". Iowa State College Press.

Anderson, E. and Cutler, H. C. (1942). "Races of Zea mays. 1. Their
Recognition and Classification". *Mo. Bot. Gard. Ann.* **26,** 69–88.

Beadle, G. W. (1930). *Cornell Univ. Agric. Exp. Sta.* **12,** 9.

Becker, W. R. 1952. "Hybrid Maize in Holland". *World Crops* **4,** 304.

Berger, J. (1962). "Maize Production and the Manuring of Maize". Centre
D'Etude de L'Azote 5, Geneva.

Bland, B. F. (1953). "Maize as a Silage Crop in Eastern England". *J. Min.
Agric.* **60**(7), 311–314.

Brown, W. L. and Anderson, E. (1947). "The Northern Flint Corns". *Mo.
Bot. Gard. Ann.* **34,** 1–28.

Bunting, E. S. (1962). "Maize as a Silage Crop in South-east England".
J.R.A.S.E. **123,** 46–54.

Bunting, E. S. and Blackman, G. E. (1951). "Assessment of the Factors Controlling the Productivity of Maize in England". *J. agric. Sci.* **41,** 271–281.

Burtt-Davy, J. (1914). "Maize. Its History, Cultivation, Handling and
Uses". Longmans Green & Co., London.

Castle, M. E., Foot, A. S. and Rowland, S. J. (1951). "American Hybrid
Maize for Silage in the South of England". *J. agric. Sci.* **41,** 3, 282–287.

Chesneau, J.-C. (1966). "Les semis du mais. Le mais en France". Association
generale des producteurs de mais.

Cobley, L. S. (1956). "Botany of Tropical Crops". Longmans, Green &
Co., London.

Collins, G. N. (1912). "The Origin of Maize". *J. Wash. Acad. Sci.* **2,** 520–530.

Cutler, H. C. (1946). "Races of Maize in South America". Harvard University Bot. Mus. L. **12.**

Darlington, C. D. and Upcott, M. B. (1941). *J. Genet.* **41,** 275.

Darlington, C. D. and Wylie, A. P. (1945). "Chromosome Atlas of Flowering Plants", p. 417. Allen & Unwin, Ltd, London.

De Candolle, A. (1884). "Origin of Cultivated Plants". Kegan Paul, Trench & Co., London.

Eddowes, M. (1962). "Maize as a Silage Crop in the West Midlands". *J.R.A.S.E.* **123**, 55–62.

Eddowes, M. and Mortimer, R. G. (1961). *J. Inst. Corn Agric. Merchants* **9**(2).

Emmerson, R. A., Beadle, G. W. and Fraser, A. C. (1935). "A Summary of Linkage Studies in Maize". *Cornell Agric. Exp. Sta. Mem.*, 180.

Evans, R. E. (1960). "Rations for Livestock". M.A.F.F. Bulletin No. 48.

Hall, Sir A. D. (1945). "The Soil. An Introduction to the Scientific Study of the Growth of Crops" (fifth edition). John Murray, London.

Hanna, W. F. 1924. "Growth of Corn and Sunflowers in Relation to Climatic Conditions". *Bot. Gaz.* **78**, 200–214.

Harshberger, J. W. (1893). "Maize: A Botanical and Economic Study". *Contrib. Bot. Lab. Univ. Pennsylvania.* **1**(2), 75–202.

Harshberger, J. W. (1896). "Fertile Crosses of Teosinte and Maize". *Garden and Forest.* **9**, 522–523.

Hayes, H. K., Immer, F. R. and Smith, D. C. (1955). "Methods of Plant Breeding". McGraw-Hill Book Co., New York.

Jenkins, M. T. (1937). "Corn Improvement". *U.S. Dept. Agric. Yearbook*, 455–522.

Jugenheimer, R. W. (1958). "Hybrid Maize Breeding and Seed Production". F.A.O. Agric. Development Paper No. 62. F.A.O. of the U.N., Rome.

Kappen, H. (1929). "Die Bodenazidität nach agrikultur-chemischen Gesicht-spunkten dargestellt.", p. 237. Berlin.

Kempton, J. H. (1931). "Maize, the Plant Breeding and Achievement of the American Indian". *Smithsn. Inst. Series.* **11**(7), 319–349.

Kempton, J. H. (1938). "Maize—our Heritage from the Indians". *Smithsn. Inst. Rpt. 1937*, 385–408.

Kuleshov, N. N. (1933). "World's Diversity of Phenotypes of Maize". *J. Am Soc. Agron.* **25**, 688–700.

"La Culture de Mais" (1966). Association generale des producteurs de mais. Institute technique des cereales et des fourrages.

Linnaeus (1737). Genera Plantarum.

M.A.F.F. (1968). "List of Approved Products and their uses for Farmers and Growers". Agric. Chem. Approval Organisation. Plant Pathology Lab., Harpenden, Herts.

M.A.F.F. Bulletin No 138. (1958). "Irrigation". First published June, 1947. Reprinted with amendments, May, 1958. H.M.S.O.

M.A.F.F. Short Term Leaflet No. 71. (1967). "Irrigation Guide". M.A.F.F., Whitehall Place, London.

Mangelsdorf, A. J. (1947). "The Origin and Evolution of Maize". *Adv. Genet.* **1**, 161–207.

Mangelsdorf, A. J. (1948). "The Role of Pod Corn in the Origin and Evolution of Maize". *Mo. Bot. Gard. Ann.* **35**, 377–406.

Mangelsdorf, A. J. (1950). "The Mystery of Corn". *Sci. Am.* July.

Mangelsdorf, A. J. (1952). "Hybridization in the Evolution of Maize Heterosis". Iowa State College Press.

Mangelsdorf, P. C. (1965). "Essays on Crop Plant Evolution" (Sir Joseph Hutchinson, ed.). Cambridge University Press, London.

Mangelsdorf, P. C. and Oliver, D. L. (1951). "Whence Came Maize to Asia?" Harvard University Bot. Mus. L. **14,** 263–291.

Mangelsdorf, P. C. and Reeves, R. G. (1938). "The Origin of Maize". *Natl. Acad. Sci. Prod.* **24,** 303–312.

Mangelsdorf, P. C. and Reeves, R. G. (1939). "The Origin of Indian Corn and its Relatives". *Tex. Agr. Exp. Sta. Bull.* 574.

Mangelsdorf, P. C. and Smith, C. E., Jnr. (1949a). "Harvard University Bot. Mus. L. **18**(8), 213–247.

Manglesdorf, P. C. and Smith, C. E., Jnr. (1949b). *J. Hered.* **40,** 39–43.

Martin, H. (1965). "Insecticide and Fungicide Handbook for Crop Protection". Issued by the British Insecticide and Fungicide Council. Black Scientific Publications, Oxford.

M.D.A. (1967). "Maize as a Grain Crop". Catalogue presented by the Maize Development Association. October–November, 1967

Milbourn, G. M. (1966). "Maize for Grain—a Break Crop in Cereal Growing?" *Agriculture,* 32–36.

Milbourn, G. M. (1968). "Maize for Grain". A grower's handbook. Department of Agriculture, Wye College, Ashford, Kent.

Moore, W. C. (1948). "Fungus, Bacterial and other Diseases of Crops in England and Wales—1943–1946". M.A.F.F. Bulletin No. 139.

Moore, W. C. and Moore, F. J. (1961). "Cereal Diseases". M.A.F.F. Bulletin No. 129. H.M.S.O.

National Institute of Agricultural Botany (1967, 1968). "Guide to Varieties under Trial, Observation and Multiplication".

Nix, J. (1966, 1967 and 1968). "Farm Management Pocket Book" (second edition). Department of Agriculture Economics, Wye College (University of London).

Pizer, N. H. (1963). Technical Report of the Agricultural Land Service. **8.**

Randolph, L. F. (1928). *Cornell Univ. Exp. Sta. (Ithaca) Mem.* **117,** 1.

Randolph, L. F. (1932). *P.N.A.S.* **18,** 222.

Rhoades, M. M. (1950). *J. Hered.* **41,** 58.

Rhoades, M. M. (1955). "The Cytogenetics of Maize". "Corn and Corn Improvement". Academic Press, New York.

Salter, P. J. and Williams, J. B. (1967). *J. Soil Sci.* **18,** 1.

Stoner, C. R. and Anderson, E. (1949). "Maize among the Hill Peoples of Assam". *Mo. Bot. Gard. Ann.* **36,** 355–404.

Vavilov, N. I. (1931). "Mexico and Central America as the Principal Centre of Origin of Cultivated Plants of the New World". *Appl. Bot., Genet., Plant Breed. Bull.* **26,** 179–199.

Weatherwax, P. (1935). "The Phylogeny of *Zea mays*". *Am. Midl. Nat.* **16,** 1–71.

Weatherwax, P. (1950). "The History of Corn". *Sci. Mo.* **71,** 50–60.

Weatherwax, P. (1954). "Indian Corn in Old America". Macmillan, London.

Weatherwax, P. and Randolph, L. F. (1955). "History and Origin of Corn. Corn and Corn Improvement". Academic Press, New York.

Weijer, J. (1952). "A Catalogue of Genetic Maize Types Together with a Maize Bibliography". *Bibliog. Genetica.* **14,** 189–425.

Wellhausen, E. J. 1952. "Heterosis in a New Population. Heterosis. Iowa State College Press.

Williams, J. R. Parry. (1960). "Frit Fly in Maize". *Pl. Pathology* **10,** 3.

Woodman, H. E. and Amos, A. (1924). "Maize Silage I". *J. agric. Sci.* **14,** 461–468.

Woodman, H. E. and Amos, A. (1928). "Maize Silage II". *J. agric. Sci.* **18,** 194–199.

CHAPTER 6

Beans

I. Introduction 265

PART 1. HORSE OR FIELD BEANS

II. Winter Beans 269
    A. Soils 269
    B. Climate 270
    C. Place in Crop Rotation 271
    D. Manuring 271
    E. Cultivations 272
    F. Seeding 273
    G. Depth of Sowing 274
    H. Row Width 274
    I. Method of Seeding 275
    J. Seeding Rate 276
    K. Seed Dressing 276
    L. Varieties of Winter Beans 277
    M. Breeding System with Beans 277
III. Practical Considerations in Pollination 281
    A. Recommendations 282
IV. Weed Control 283
    A. Control of Wild Oats 283
    B. Simazine and Dinoseb for Weed Control in Winter Beans . . 284
V. Pest Control 285
    A. Wood Pigeons 285
    B. Weevils 285
    C. Aphids 286
VI. Spring Beans 287
    A. Soil, Climate and Manuring 287
    B. Time of Sowing 288
    C. Depth of Sowing 288
    D. Establishment by Ploughing-in 288
    E. Seed Rate 288
    F. Seed Dressing 289
    G. Varieties of Spring Beans 289
    H. Weed Control 290
    I. Pests 291
VII. Harvest 293
    A. Binder Stage 293

    B. Combine Stage 293
    C. Time of Harvesting and Moisture Content . . . 294
    D. Storage 295
  VIII. Yield 297
   IX. Grain Quality and Use of Field Beans in Livestock Feeding . . 298
    A. Chemical Composition 298
    X. Production Costs, Output and Gross Margins with Winter and
       Spring Beans 300
   XI. Diseases of Field Beans 302
    A. Chocolate Spot 302
    B. Leaf Spot 302
    C. Stem Rot 303
    D. Virus Diseases 303

PART 2. FRENCH BEANS (KIDNEY)

  XII. Introduction 303
 XIII. Soils 306
 XIV. Place in Cropping Sequence 306
  XV. Cultivations 306
 XVI. Manuring 307
XVII. Plant Spacing, Row Width and Seed Requirements . . 309
XVIII. Sowing 310
 XIX. Varieties 311
  XX. Cultural Weed Control 312
 XXI. Chemical Weed Control 312
XXII. Pest Control 314
XXIII. Irrigation 315
XXIV. Harvest 315
 XXV. Yield and Quality 317
    A. Yield 317
    B. Quality 318
XXVI. Crop Value, Output and Gross Margin 319
    Appendix 320
    References 323

I. INTRODUCTION

Beans have been grown for a long time in British agriculture and the bulk of them under cultivation at present may be classified as "Field" or "Stockfeed" types. According to Gill and Vear (1966) beans belonging to *Vicia faba* have been cultivated in Europe since prehistoric times and they have not been found in the wild state. They suggest that present-day types may be derived from the small species *Vicia narbonensis* L., a Mediterranean annual. A number of other species have developed on a small scale in horticulture as a vegetable for human consumption and with the advent of complete mechanisation in recent years, dwarf beans have been adopted as an agricultural crop.

TABLE 1

Beans in British Agricultural and Horticulture

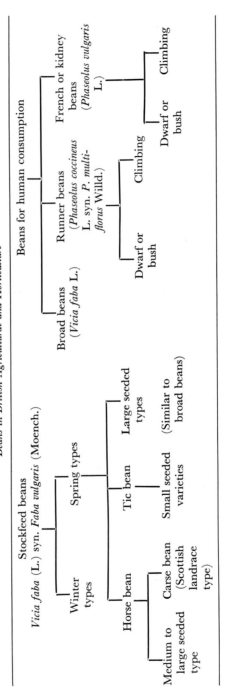

The various types of beans which are important in British agriculture and horticulture are shown in Table 1.

In their book entitled "The Chromosome Atlas of Flowering Plants", Darlington and Wylie (1955) enumerate the beans grown throughout the world and from this exhaustive list those of importance in Britain are shown below with their appropriate chromosome number.

Species	Common name	Chromosome number	Reference	Remarks
Vicia faba (X = 6 or 7)	Broad, horse or field beans	12 (24)	McLeish (1953) Rybin (1939)	Cultivated for fodder and as a vegetable in Mediterranean regions.
	Broad, horse or field beans	12	Hirayoshi and Matsumura (1952)	European strains.
	Broad, horse or field beans	14	Hirayoshi and Matsumura (1952)	Japanese strains.
Phaseolus coccineus (X = 11)	Scarlet runner bean	22	Karpechenko (1925)	Cultivated vegetable.
Phaseolus vulgaris (X = 11)	French, kidney or dwarf bean	22	Thomas (1945)	Cultivated both for grain and as a vegetable.

At the end of the nineteenth century, a quarter of a million acres of beans were grown but by the beginning of the Second World War the U.K. total amounted to only 135,000 acres. During the war most arable crops increased in acreage and beans were no exception. The U.K. acreages during the period 1939–69 are represented graphically in Fig. 1 and the major trends during these thirty years can easily be picked out. The figures for 1966, 1967 and 1968 have shown an increase over the immediately previous year of 20, 35 and 64% respectively and the big increase in 1968 was largely due to the government support (£5/acre grant) which was introduced in 1968. The recent increased acreage is also due to a number of other technological improvements in the fields of weed and pest control and also the possibilities of contracts and assured markets for some growers using new cultivars. Many arable farms are dominated by the cereal crops. These are easily managed, have a low labour requirement and less capital is invested compared with nearly all other farming systems. Frequent or continuous cereal cropping can be satisfactory for a period, and simple farming systems on these lines were developed during the 1960's. However, once the major cereal diseases such as eye-spot and take-all have appeared, reduced yields are almost certain to follow. Break crops which keep the land free from the two main cereals wheat and barley are now important features of these simple farming systems. Field beans come under this category and are considered by many to be one of the most useful break crops. They can be managed with the existing machinery and equipment on cereal farms, are

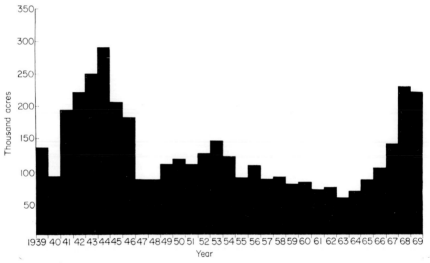

Fig. 1. Field bean acreage statistics (U.K.) 1939–69.

likely to leave the land in a better state of fertility and the produce finds a ready market for pigeon and livestock feeding. Much of the renewed interest in field beans can be attributed to their suitability as break crops.

Greenwood (1958) made a study of bean growing in North-west Europe and restricting his observations to field beans compared acres and productivity in Britain with four of our close Continental neighbours.

From Greenwood's figures, reproduced in Table 2, it can be seen that field beans occupied a bigger acreage in U.K. than any of our continental neighbours and the yield per acre was similar to that obtained in Holland, West Germany and France. In Belgium the relatively small acreage was confined to fertile soils and the field beans were almost exclusively cash crops, of the small seeded varieties, grown to satisfy the demands of pigeon keepers. This explains in part the relatively high yield obtained in Belgium compared with the other four.

Broad beans, runner beans and French beans are important vegetable crops in Britain with virtually all the acreage listed under England

TABLE 2

Field Beans in Five European Countries

Country	Field beans Acreage	Field beans Yield (cwt/acre)	Total arable land (million acres)	Acres of beans/ 1000 acres arable land
United Kingdom	123,000	15·4	18	6·8
Holland	3,500	16–17	2·6	2·4
West Germany	45,000	12–16	21·5	2·1
Belgium	7,500	19–23	2·5	3·0
France	67,000	10–14	52·6	1·3

After Greenwood (1958).

and Wales and in practice most of the crop is grown in England. The Ministry statistics showed that 30,000 acres were being grown in 1968, with approximately one-third broad beans and two-thirds runner and French beans. The dwarf types of runner and French beans have become important on quite a large scale in agriculture and this chapter will deal with these in Part 2, following the detailed account of horse or field beans grown for livestock feed.

PART 1.   HORSE OR FIELD BEANS

II.   WINTER BEANS

A.   Soils

Winter beans are usually associated with soils which are heavy by nature and provided that they are properly drained, high yields will result. Clays, silts and heavy loams come into this category of suitable soils although each of them necessitates a slightly different approach to seed-bed preparation in the autumn. Winter crops of beans will stand low temperatures provided they are not required to endure prolonged waterlogging. They will not tolerate excess moisture for long during the late autumn and winter period and a good free draining seed-bed is essential to combat heavy winter rain. Being a member of the

Leguminoseae, beans are capable of fixing atmospheric nitrogen through their association with species of bacteria (Rhizobia), provided the soil conditions are suitable. This usually infers a near neutral pH as one of the main requirements for the bacteria and thus for the bean crop an optimum pH range of between 6 and 7 has been suggested (Small, 1946). Most legumes produce better growth if they can derive their nitrogen both from the soil and through fixation and a ratio of 1:2 has been suggested for the amount of nitrogen from these two sources respectively. Soils for the bean crop should therefore contain good reserves of organic matter from which the plants may derive part of their nitrogen as a result of mineralisation. In the past the application of farmyard manure has been advocated specifically for this crop but where this is available for an arable rotation it would produce higher financial returns from crops like potatoes or vegetables.

B. Climate

The climatic conditions most suitable for winter sown crops are mild winters followed by warm dry springs and summers. Very wet winters are responsible for loss of brairds and so too are very severe frosty conditions as many of the cultivars are somewhat less winter hardy than is desired for extremes of cold which can sometimes occur. Very low temperatures following very wet periods or occurring without protective snow cover can be particularly damaging to the early brairds of beans.

The winter crop is mainly confined to the Midlands and Southeast England and although the winters there can sometimes be harder than is desired, spring and summer are usually good. Some cultivation is attempted in the western districts of England and in Wales and following the introduction of the cultivar Daffa, from the Welsh Plant Breeding Station, and with the renewed interest in this crop, farmers in other areas are considering winter bean growing. According to the N.I.A.B. Farmers' Leaflet No. 15, 1968—Recommended Varieties of Field Beans—which is included in the Appendix, this new cultivar has shown good winter survival in the West.

Excessive rainfall during the growing season usually results in crops with too much vegetative growth and unless the weather is warm and dry at seed setting time the resulting yields will be poor. Very hot dry summers on the other hand lead to restricted growth and this is usually accompanied by epidemic levels of black aphids. The multiplication rate of these insects under these conditions can be very high and before the introduction of systemic insecticides they presented a real hazard to bean growing. The spring-sown beans suffer more than the winter ones, since they are usually at a much earlier growth stage when the attack occurs.

C. Place in Crop Rotation

By far the largest amount of our arable land is in cereals and both the winter and spring sown bean crops are almost certain to follow either wheat or barley in cropping sequence. They are two of the most useful and easily managed crops to the mainly cereal grower and are considered to be satisfactory break crops in close cereal cropping. Winter beans will follow the earliest harvested cereal, in order that the land may be ploughed in the autumn, in time to produce a good seed-bed. It is possible, but unlikely, that beans will go in after grassland but good crops can be obtained following potatoes on medium loams and silts where ploughing may not be necessary in seed-bed preparation. With this cropping sequence the beans can make use of the residues particularly when F.Y.M. and liberal quantities of inorganic fertilisers have been applied. Satisfactory crops of beans can usually be obtained without further application of fertiliser, thus resulting in low production costs.

D. Manuring

When land destined for beans is tested and found to be on the acid side, the appropriate amount of limestone or chalk should be applied to the soil and worked in, well in advance, in order to raise the pH value to approximately 6·5. Farmyard manure, according to Soper (1958), will benefit the bean crop except where the soils are of an organic nature or are high in fertility. It is likely to be most useful on the lighter soils, but since its application is costly and the value of the yield increase only moderate, it can no longer be recommended without reference to these relatively unfavourable economic aspects. Strong land and soils of good fertility will grow high yielding crops of beans without farmyard manure and these are the obvious places for this crop where production costs can be kept to a minimum.

Mineralisation of the soil organic matter plus residues from previous cropping and fertilising will usually contribute sufficient nitrogen for beans until such times as they develop nodules in association with the bacteria and utilise the atmospheric nitrogen which is fixed. Inorganic fertiliser applications will therefore usually consist of a phosphate and potash compound or a mixture of the straight fertilisers superphosphate and muriate of potash. Experiments have shown moderate to poor responses in yield to these two major nutrients but when used they also provide some degree of protection against very severe attacks of the fungus which causes chocolate spot (see later for full description). Cooke (1964) demonstrated the value of fertiliser placement for legumes such as peas and beans when he showed that 3 cwt/acre of a P/K compound placed 2 inches to the side of the seed and 3 inches below the soil surface gave higher yields than double this quantity of fertiliser broadcast. However, where the phosphate and potash status of the soil is reasonably high, this degree of benefit from placement will seldom be achieved.

Fertiliser placement is suggested for bean crops which are grown on wide rows which facilitate this operation but when the crop is drilled using narrow coulter spacings as for cereals (4–7 inches) then the phosphate and potash will have to be broadcast. Should this be the case then the required quantity of fertiliser should be applied to the seed-bed and worked into the top 3–4 inches of the soil, before sowing.

Recommended Levels of Phosphate and Potash for Beans
(units/acre) Related to Soil Analysis

	Low	Medium	High
Phosphate (P_2O_5)	40–50	20–30	Nil
Potash (K_2O)	80–100	40–60	Nil

Farmers may be tempted to combine-drill their fertiliser and seed, when grown on narrow rows, but even the lowest level of potash given above is likely to damage germination and this practice *cannot* be recommended. Compound fertilisers with the nutrient ratio 0:1:2 (N:P:K) will supply the crop needs without modification but where 0:1:1 (N:P:K) are the only ones available which do not contain nitrogen extra potash in muriate form will be needed. Using a 0% N:10%P_2O_5: 20% K_2O compound fertiliser 2–3 cwt/acre would be required when soil analysis shows P and K levels to be medium and 4–5 cwt/acre when very deficient.

E. Cultivations

For winter beans to be sown in mid October, ploughing to a depth of 6 inches in September should be followed by minimal cultivations to obtain a medium seed-bed in the top 3–4 inches with a firm "bottom" below this. Some degree of clod should be left on the surface of a number of soil types, e.g. silts and loams, when there is a known danger of surface capping following winter rain. This will avoid some plant losses which occur as a result of frost when the young plants may be "heaved" out at thawing and will help to eliminate some of the physical damage which occurs at ground level following low temperatures.

F. Seeding

TIME OF SOWING

In the past, the time of sowing of winter beans was an important consideration which was linked with establishment and winter proudness of the crop. With very early sowings, or when the autumn and early winter temperatures are considerably higher than usual, excessive vegetative growth results which can be seriously damaged by severe winter frosts. The same is true today, but the growth stage in the spring is much more important now as it can often determine whether or not chemical weed control can be practised. Crops which are too far forward in the spring may have passed the critical stage and chemical weed control using some of the most useful herbicides may not be safe. On the other hand late sowings can result in very thin brairds due to the attention of birds, attacks from soil fungi or simply via poor field germination on account of lower soil temperature. Soper (1958) pointed out that early to mid October drilling could result in a yield of several more cwt of grain/acre compared with early November sowing and that winter hazards were greatly increased after mid October. The average results from eight time of sowing trials, reported by Greenwood (1959) are reproduced below and demonstrate clearly these yield reductions from late sowings. Only at abnormally high

TABLE 3

The Effect of Time of Sowing at Different Seed Rates on the Yield of Winter Beans. Difference in cwt/acre compared with the Corresponding mid October Sowing

Seed Rate (lb/acre)	Sowing date	
	Early (mid September)	Late (mid November)
50	+1·5	−2·2
100	+0·7	−4·3
200	−0·3	−5·1
300	−1·6	−3·4
400	−2·3	−2·3
500	−2·1	−3·3

seeding rates were there significant reductions in yield with early sowings compared with mid October.

If it is impossible to get winter beans in during October then it may be a wise decision to switch to the spring sown crop, since a February drilling is often possible with these and successful control of the black aphid can now be obtained.

G.   Depth of Sowing

The ideal depth of sowing for beans is approximately 3 inches and most authorities are in agreement on this subject. Planting below this level gives added bird protection, but brairding is later and can be quite irregular. It is important to have 3 inches of soil above winter beans so that they obtain a firm establishment and are not "heaved" out by frost. Heavy winter rains can often flatten the ground on some soil types, in spite of leaving a reasonable amount of surface clod and some of the seeds may just lie below the surface if shallower planting is practised.

H.   Row Width

According to Soper (1958) the essential factor for obtaining a high yield with beans is to see that a large number of plants are established per square yard. The actual arrangement of these plants is relatively unimportant. Row width would therefore be determined according to convenience of growing the crop and would fall naturally into three groups, narrow, medium and wide rows.

Narrow	*Medium*	*Wide*
6–7 inches	12–14 inches	18–21 inches

In respect of the narrow rows, equivalent to normal cereal coulter row spacing, weed control would be obtained by either chemical means or should the land be fairly clean and initial crop growth fast, then weeds could well be suppressed by the crop itself. Where farmers may be growing another legume (e.g. dried peas), they may consider the medium row width of 12–14 inches as suitable, with the idea of inter-row cultivation for eliminating weeds during the initial stages of growth. It is unlikely that the majority of growers would choose this row width due to the difficulties experienced in cultivations at 12–14 inches and to the fact that only one or two hoeings may be possible as damage at later stages would result.

Wide rows 18–21 inches apart afford the best opportunity of mechanical weed control and at the same time provide access for tractors when systemic insecticides are used for the control of blackfly, but this of course is more important with spring beans. Where full advantage from both wild and honey bee populations is considered desirable, the wide rows afford these insects easier access to the crop and pods are set to within 9 inches of the ground and are better distributed along the stalks compared with plants from narrow rows (see Fig. 3).

I. Method of Seeding

Some time ago it was traditional to "plough beans in". This old method of establishment eliminated a separate seeding operation since the seed was distributed from a bean box which was fitted to the plough. A good covering of soil resulted and the seed-bed was firm below but the main disadvantage with this method was in the fact that emergence was irregular and so too were the rows, making mechanical

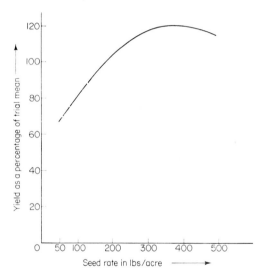

FIG. 2. The influence of seeding rate on the yield of field beans. After Greenwood (1959).

cultivation extremely difficult, if not impossible. Exceptional circumstances may still lead some farmers to plough beans in, but under most circumstances the crop is better drilled. This ensures that the beans go in 3–4 inches deep to avoid damage from a residual herbicide such as Simazine. The older cup-feed drills, usually associated with shoe coulters, are probably the most suitable for the bean crop since they pick up and deposit the seed with the least abrasion. Force feed drills whilst being more accurate in relation to the cereals sometimes crack and break the larger seeded bean varieties and this should be checked before drilling commences. Disc coulter drills can be satisfactory, provided due attention is paid to evenness in seed-bed preparation. Single disc drills can be useful when the conditions of drilling are far from ideal but the double disc drills tend to ride out of the soil when the tilth is inadequate and coverage may well be insufficient.

J. Seeding Rate

Although there is evidence to show tillering with bean plants at low seeding rates this compensation is not sufficient to equal the increase in yield obtained at higher plant densities. Greenwood (1959) has summarised the effect of seeding rate on yield from thirteen experiments carried out at Cambridge with both winter and spring cultivars. The mean results (ignoring time of sowing) are shown in Fig. 2 and demonstrate that 400 lb/acre will produce the maximum yield although 200–300 lb/acre appears to be the economic optimum range. The rates recommended by Soper (1958) of 3·5 bushels for good conditions and 4 bushels/acre for less satisfactory seed-beds (equal to 220–250 lb/acre) and the recent Ministry of Agriculture suggestion of at least 220 lbs/acre fall within the economic optimum range suggested by Greenwood. With lower production costs in mind some farmers are sowing 168 lb (12 stone) thus saving 0·5 cwt of seed/acre. A saving of slightly less than £1/acre results but should there be any plant losses due to winter kill, bird or insect activity or through disease then yield reductions will follow. It is important to obtain a good plant stand and low seed rates of the order of 12 stones/acre can only be recommended for small seeded cultivars and in general 2 cwt of seed will be needed for the medium–large seeded varieties.

K. Seed Dressing

Fungus diseases causing damping-off and foot rot can be responsible for some poor thin braids particularly under cold wet conditions when germination and initial growth is very slow. Seed dressings based on *Captan* (Orthocide) or *Thiram* (Fernasan A) will aid germination particularly under adverse conditions and should be used as a routine measure for winter beans for all sowings except possibly the very early ones. These dry fungicidal dressings can reduce the rate of flow of bean seed in some drills and where prior calibration is envisaged a sample which truly represents the bulk must be used. For details of seed dressings see M.A.F.F. list of Approved Chemicals.

L.   Varieties of Winter Beans

Winter bean cultivation is restricted to the Southern part of Britain and it is not a widely grown crop on the Continent. As a result there has been little variation in the varieties which were available and they have been few in number. Greenwood (1959) summarised the variety trials carried out by N.I.A.B. during the period 1951–56 and Smith and Aldrich (1967) reported on the work carried out between 1952 and 1964. The National Institute of Agricultural Botany has issued a descriptive list each year since 1960—"Varieties of Field Beans", Farmers' Leaflet No. 15. The most important agronomic characters were summarised for many cultivars, seed of which was available through the trade, and it was up to individual farmers to make their own choice. In 1968 the Council of this National Institute approved for the first time a recommended list of cultivars, having obtained much information during the 1960's in view of the increased interest in this crop. This has eliminated many of the old and poorer varieties and the choice is now much simplified for growers. Full details of the 1970 "Recommended List of Varieties of Field Beans" can be obtained from the Appendix.

In previous years the variety Gartons S.Q. has been included in the descriptive lists and has been one of the most consistent yielders over many years. It is not on the present recommended list being some 6–10% lower yielding than Daffa, Maris Beaver or Throws M.S.

M.   Breeding System with Beans

The economic yield from crop plants is usually derived from vegetative growth or through sexual reproduction and in the case of field beans it results from sexual reproduction within the flowers. Fertilisation usually occurs either as selfing (e.g. wheat, barley or oats) or by cross-pollination (e.g. sugar beet or brassicas) but in the case of beans both self-pollination and cross-pollination occurs. The results from cytological examination of individual plants from field bean populations strongly suggest that the bean crop is inherently a cross-pollinated species which is forced to inbreed under British conditions through the lack of suitable insect vectors (Rowlands, 1955). It has been established that under normal conditions winter beans produce 30% fertile seed by cross-pollination and 70% from self-pollination. Where conditions are severe, with low survival rates, almost half of the plants are cross-pollinated (Fyfe and Bailey, 1951). Sirks (1923) had previously indicated that some 20–40% of seed from field bean populations was cross-fertilised due to insect activity (mainly bumble and honey bees) and this was confirmed later by Picard (1953).

K

Cross-pollinated seed gives rise to more vigorous and higher yielding plants (Drayner, 1959), and the proportion of seed obtained by out-pollination should therefore have a marked effect upon the yield. It is almost certain that the stocks of beans maintained by growers in Britain in the past have not shown any trend towards increasing yield because they have suffered from an "in-breeding depression", being mostly self-fertilised. On the other hand if the crop relied entirely upon pollinating insects for its existence and these were not forth-coming, or were in short supply, the yields would be very poor. Thus with field beans, which are partly self- and partly cross-fertilising, complete failure to set seed is not likely.

The value of introducing honey bees to beans at flowering has been questioned in recent years although many field observations have demonstrated significant improvement by their presence. As a result there are two schools of thought on this subject holding views which are diametrically opposite.

(i) OTHER EVIDENCE IN FAVOUR OF MAINTAINING OR INCREASING POLLINATING INSECTS

Scriven *et al.* (1961), in a survey of bean growing in Lincolnshire, found that honey and bumble bees were numerous in districts where the crop was good, but few were found in those areas where the crop was regarded as poor or unprofitable. In following up this observation, these authors set out to ascertain the effect on yield of self-pollination and cross-pollination and also to see whether introducing honey bees would be beneficial. In areas with few pollinating bees, self-fertilised crops yielded below 15 cwt/acre whereas cross-pollinated crops in the open yielded over 35 cwt/acre. Controlled experiments carried out in later years using caged plots revealed a similar pattern and the results obtained over a 3-year period are reproduced in Table 4.

These results clearly indicate the value of bees and their effect on yield via cross-pollination and confirm the view expressed by Riedol and Wort (1960) that insect pollination is required for maximum yield.

Drayner (1959) showed that when hybrid bean plants (produced by cross-pollination) were kept in isolation, they produced about 70% of the seed that came from mechanically self- or cross-pollinated ones. Also that little or no seed was produced by autofertilising on inbred plants (i.e. those produced by self-pollination) and these needed mechani-cal disturbance (sometimes termed tripping in this context) for seed to be set. Thus if seed from both cross- and self-fertilisation needs external influences for maximum seed production, pollinating insects such as bees will be required by virtually all stocks of beans.

Free (1966) demonstrated that caged field bean plants had more seeds per pod and more seeds per plant when bees were introduced but the caging itself also had a very marked effect on yield.

TABLE 4

The Effect of Method of Pollination on Yields of Winter Beans

Treatment	1957	1958	1960	Mean
1. Self-pollination (bees excluded from cages)	22·8	12·3	23·0	19·4
2. Bee pollination (bees included in the cages)	43·6	27·4	37·5	36·1
3. Reduced light (free-flying bees)	40·1	27·4	40·4	36·0
4. Control (free-flying bees)	38·5	32·4	39·7	36·8
5. Open fields (free-flying bees)	38·0[a]	27·5[a]	—	—
Mean	36·2	24·9	35·1	32·1

[a] Excluded from means and statistical analysis. S.E. of general mean $\pm 1\cdot 4$. After Scriven *et al.* (1961).

Riedol and Wort (1960) indicated that when insect pollination of beans is prevented, pod production on the lowest inflorescences, which normally produce the most pods, is reduced. A compensatory increase on the upper inflorescences results giving even pod formation up the stem and moderate yields [Fig. 3(b)]. With adequate insect pollination, more pods are produced and their arrangement on the bean plant is illustrated in Fig. 3(a).

From a practical point of view the harvesting of plants of type (a) presents less of a problem with lower percentage seed losses due to shattering than those of (b) and are much to be preferred. They are also higher yielding. Bond (1969) maintained that insect pollination was beneficial even without yield increases since the insects brought about a quicker pod setting which resulted in an improved uniformity in ripening.

In 1962, Bond and Fyfe, in reviewing bean breeding work, emphasised the reality and extent of heterosis by comparing hybrids and inbred bean stocks against their respective controls and this can only be brought about by cross-fertilisation (Table 5).

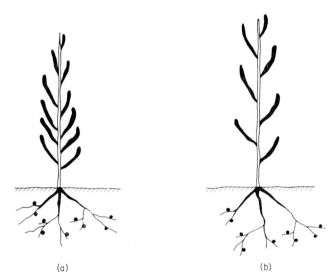

(a)   (b)

FIG. 3(a). Field bean plant with sufficient pollination. (b) Field bean with insufficient pollination.

Evidence not supporting the view that an increase in pollinating insects (i.e. via the introduction of honey-bees) will inhance bean yields.

The controlled experiments carried out by Scriven *et al.*, figures from which are reproduced in Table 4, were not able to demonstrate an increase in bean yields on the introduction of honey bees to caged plots compared with the open field and if these critical observations were done in areas where the total bee population was poor, it seems that cross-pollination by whatever insects are available, plus self-pollination, appears to give reasonably good yields.

Bond and Hawkins (1967), studying bee behaviour in male-sterile and male-fertile field beans in 1965 and 1966, report little bean pollen collection in 1965 by a colony of honey bees placed beside the crop and none of these insects were seen pollinating either of these two bean

TABLE 5

Yields of Hybrid and Inbred Stock of Beans Expressed as a Percentage of their Respective Controls

	1956–57	1957–58	1958–59
Hybrids	152	180	113
Inbreds	83	43	97

After Bond and Fyfe (1962).

stocks but in the following year large amounts of pollen were gathered. However, in a private communication Bond (1968) again records little pollen gathering in the years 1967 and 1968.

On their experimental bean plants, Riedel and Wort (1960) could not produce mature pods on more than one-third of the flowers although more than one-third were pollinated. They concluded that such a low level of pod production does not necessarily indicate insufficient pollination and Hodgson and Blackman (1956) and (1957) recorded even lower proportions of pods to flowers. A few years previously Soper (1952) had reported that a large proportion of flowers and young pods were shed from the plants during the growing season and he suggested that inadequate nutrition, crop light intensity and cross-fertilisation may limit pod setting.

Pollination of field beans can be effected by a number of insects but the most important of these are bees. Colonies of honey bees may be introduced round the periphery of the crop or they may occur close by, quite fortuitously, and will be attracted to the beans. Wild bees, known collectively as bumble bees, are very important pollinators of beans. They are much bigger and can trip the flowers much more easily and with a longer proboscis (tongue) they can easily reach to the sexual parts of bean flowers. *Bombus agrorum*, *B. hortorum*, *B. lapidarius*, *B. ruderatus* and *B. terrestris* are the important species of bumble bees which can pollinate legumes but quite frequently *B. terrestris*, which has a shorter tongue than the others, makes holes at the base of the corolla of the bean flower from the outside. Quite a large proportion of honey bees visiting beans are nectar-gatherers and they can frequently be found robbing the flowers from the punctures left by *B. terrestris*. The honey bees which are pollen gatherers are attracted to hedgerow flowers from which pollen is more easily obtained and many will continue to visit the wild hedgerow species even though they may be introduced to the bean crop.

Finally, and by no means least, honey bees which are introduced to caged plots of beans are bound to visit the front of the flowers and any benefit which may result from this may not be repeatable in open fields since they have the puncture marks left by *B. terrestris* as an alternative entry.

III. PRACTICAL CONSIDERATIONS IN POLLINATION

In order to obtain maximum yield with field beans, cross-fertilisation should be exploited and this has seldom been questioned. It is also better to end up with a bean plant of type (a), Fig. 3, where a high proportion of the pods are at the lower end of the stem. Thus in order to make use of whatever cross-pollinators are available (pollen beetles, honey bees or bumble bees), the crop is best grown on wide rows so that these insects have easy access to the bean flowers right down the stem on both sides of the plants. It is well known that honey bees dislike foraging in dense stands of closely spaced beans particularly when the crop is being blown about and the wider row widths would therefore be more suitable.

Pollen-gathering bees are more valuable pollinators than those in search of nectar (Free, 1958, 1960, 1962) and Free (1965) also demonstrated that feeding sugar syrup to the colonies makes them more efficient pollinators and many more actually pollinate the bean flowers. He further suggested that feeding honey bee colonies with sugar syrup containing the scent of the flowers would direct their activities to the crop and thus increase efficiency. Thus the honey bees need to be properly managed so that they are collecting plenty of pollen in order to be effective in successful pollination. It is important that they are introduced to beans at flowering otherwise they may only collect nectar from the crop.

One of the major problems in the past concerning honey bees in bean crops has been the control of blackfly. When spraying is carried out, this should be done before the bees are introduced or whilst they are shut in but in order to protect both the bumble and honey bees, the more expensive granular systemic insecticides will be necessary should blackfly colonies develop after the onset of flowering.

A. Recommendations

Working backwards from the number of flowers/acre of beans, the number which need to be visited to obtain a honey bee load, and the pollen requirements of a good stock of bees, Scriven et al. (1961) arrive at the figure of 1 hive/4 acres of beans. In view of the fact that not all the bees are pollen gatherers, that many other plants have pollen which is easier to collect and that in bad weather the bees only forage for short periods, these authors recommend one hive of honey bees for each acre of beans.

It is unlikely that fields will be without some wild bees and in some areas the bumble bee and other insect pollinators may be sufficient and therefore one cannot be dogmatic and recommend 1 hive of honey bees/acre of beans for all circumstances. Even with a surfeit of bees only 30–40% of beans are cross-pollinated and if a reasonable percentage of out-pollination is obtained through the wild bees, then the introduction of honey bees will have only a marginal effect, plus the fact that the bean grower would usually incur an extra cost of about £3/acre. When the population of wild bees is known to be very low and the pollination of field beans recognised as poor in a district, then this additional outlay in bean growing can be justified and can also lead to higher profit margins through increased yield. The maximum benefit which the introduction of honey bees can bring has been estimated by Cooper (1967) at 11 cwt/acre, which valued at £14 can leave an additional profit of about £11 per acre.

When the acreage of field beans is small, the wild insect vectors will probably be sufficient in most districts and the introduction of honey bees will not be necessary. When large acreages are involved, however, it will certainly be worthwhile under most circumstances to employ colonies for pollination.

IV. WEED CONTROL

The difficult perennial weeds such as couch grass or thistles should be tackled by cultivations in the autumn before the crop is sown and by inter-row cultivation in the spring where the rows are placed at wide intervals. Many broad-leaved weeds can also be dealt with at the same time, however for crops drilled with narrow coulter spacings or when cultivation is either impractical or undesirable, chemical control of these annuals can be obtained using the appropriate herbicides. These are listed in the "Weed Control Handbook", Volume II—Recommendations", and are reviewed frequently, so it is important that the most up-to-date edition is consulted. At the time of writing, this was the fifth edition, by Fryer and Evans (1968) from which the following (Table 6) recommendations have been extracted.

A. Control of Wild Oats

The recommended levels of Simazine for the control of weeds in beans are not sufficiently high for effective control of wild oats (*Avena ludoviciana* and *Avena fatua*). *Barban* at 8–10 oz active ingredient/acre applied when the wild oats have 1–2 leaves is effective and whilst this is a fairly expensive treatment (£4–£4·50 when sprayed overall) the cost can be reduced by band spraying. If the area designated for beans has a previous history of wild oat infestations, and chemical control is certainly required, then *Triallate* at 1·5 lb active ingredient/acre applied to the soil in 20–30 gallons of water can be recommended. It can be applied pre- or post-drilling, needs incorporating into the soil by harrowing and is a cheaper method of wild oat control than

TABLE 6

Herbicides for Field Beans

Type of treatment	Herbicides	Weeds controlled
Pre-emergence contact	Diquat plus paraquat Paraquat Cresylic acids Pentachlorophenol Sulphuric acid	All seedling weeds
Pre-emergence contact with some residual	Dinoseb in oil Dinoseb-amine	Germinating and seedling broad-leaved weeds
Pre-emergence residual	Simazine	Germinating annual grasses and broad-leaved weeds
	Tri-allate	Germinating wild oats plus annual grasses
	Chloropropham plus diuron Chloropropham plus fenuron	Germinating weeds
Post-emergence	Simazine	Spring germinating weeds in winter beans
	Dinoseb-amine	Seedling broad-leaved weeds
	Barban	Wild oat species at a young stage

After Fryer and Evans (1968).

Barban. Triallate will also control blackgrass (*Alopecurus myosuroides*) and its application does not interfere with other soil-applied herbicides such as Simazine.

B. Simazine and Dinoseb for Weed Control in Winter Beans

The spraying of winter beans for the control of broad-leaved weeds using *Simazine* can be done soon after drilling or it may be left until January or February provided that spring growth has not really started and the plants are still in the "hard stage", having just come through the winter. Rate of application will range from 8 to 16 oz active ingredient/acre, with the higher level associated with heavy soils and to avoid damage to the crop, the seed should be at an even depth, 3–4 inches below the surface.

For farmers who wish to leave the decision of chemical weed control to the spring, and therefore assuming that no soil acting herbicide has been applied, then the eradication of broad-leaved annual weeds can be obtained using Dinoseb compounds. *Dinoseb-amine* is recommended at 1·5 lb active incredient/acre for winter beans provided there is not active spring growth and this will mean that the plants are about 3–6 inches in height. When the crop is much higher than this and the beans have passed the "hard stage", damage will follow spraying and thus late applications cannot be recommended. Optimum air temperature range for best results with dinoseb spraying is 44–64°F (6·7–17·8°C) (M.A.F.F. Short Term Leaflet No. 60); below this temperature the herbicide is less effective and above, the resistance of the crop to damage is much reduced.

Dinoseb acetate at 2·5 lb/acre can also be used on winter beans in the spring, applied in medium volume before the crop reaches 8 inches in height.

Dinoseb acetate plus monolinuron is being marketed as either a pre- or post-emergence spray for beans. The monolinuron has been added because dinoseb acetate does not control *Polygonum* spp. well, being less effective than dinoseb-amine, and spraying should take place before there is significant growth past the two leaf stage.

V. PEST CONTROL

A. Wood Pigeons

These can occasionally be troublesome following the drilling of bean crops into poor seed-beds, where uneven coverage and shallow planting has resulted in the attention of this pest. Once the crop has brairded, pigeon activity on bean crops diminishes and thus protection during the early germinating stages is vital. Bird-scaring devices of all descriptions plus frequent visits to the field with shotguns can usually be employed to protect the crop during this vulnerable period. If possible slightly higher seeding rates than those generally recommended can be usefully employed when bird damage is certain, i.e. fields in the close proximity to woodland occupied by pigeons.

B. Weevils

Pea and bean weevils (*Sitona* spp.), in the adult form, feed on the foliage of a wide range of leguminous crops, including field beans, and the larvae of these species feed on the root nodules. The adult weevil is one of the earliest insects to become active in the spring and the damage they cause is easily identified, being U-shaped notches eaten out of the leaf edges. Damage is usually severe round the headlands of crops, particularly when surrounded by thick hedges, which make ideal overwintering sites for this insect. It is often severe if peas, clover or

other legumes have been grown on the land recently or may be found
in close proximity to the beans. Nearly all crops of beans suffer some
damage from weevil activity but it is seldom of economic significance.
Bean plants can usually grow away satisfactorily even though they have
been attacked, however should the infestation occur at a very young
stage in the development of the crop or be unusually severe in extent
then insecticidal control measures can be employed.

Good control of weevils can be obtained using DDT sprays or DDT
dust and the following are recommendations which appear in "The
Insecticide and Fungicide Handbook" (Martin, 1969):

DDT sprays at 10 oz active ingredient/acre, either as 20 oz 50%
wettable powder or 2 pints emulsifiable concentrate; or DDT dust at
2·5 lb active ingredient/acre (50 lb 5% dust); or gamma-BHC dust
at 6 oz active ingredient (56 lb 0·65% dust)/acre.

The DDT emulsion will persist longer on the bean foliage than the
DDT wettable powder and both these will last longer than the DDT
dust.

Photographs of typical weevil damage and further details of this pest
may be obtained from M.A.F.F. Advisory Leaflet No. 61.

C. Aphids

The black bean aphid, often simply termed "blackfly" (*Aphis
fabae*), overwinters on the Spindle-Tree (*Eunonymus europaeus*) or the
Sterile Guelder rose (*Viburnum opulus* var. roseum) and during the
summer can be found in large numbers (colonies) on many crop
plants including beans, sugar beet and mangels and also on weeds
such as docks and fat-hen. It is a serious pest of spring sown crops
but damage is less severe and less frequent on winter beans. Full details
of the life-cycle of this pest have been given by Jones and Jones (1964)
showing the various stages in development throughout the year and the
form of reproduction. In the spring and early summer the winged
female aphids (alatae) migrate to the beans and their colonies are
formed at the top of the plant amongst the newest growth. The females
are able to reproduce without the intervention of the male (asexually)
and when the weather is hot and dry this is extremely rapid and the
whole plant can soon be covered and affected by large numbers of these
insects. Significant reductions in yield will follow if preventive *or*
control measures are not carried out and recommendations for each
of these appear below.

Prevention: in areas where black bean aphids are serious pests each
year, prevention of aphid build-up can be effected by the use of

systemic insecticides, provided they are put on well before the first flowers appear.

(a) *Menazon*, available as a liquid, is recommended at 4 oz active ingredient/acre (10 fluid oz of 40% conc), applied at low–medium volume (20–40 gal/water) just prior to flowering.

(b) *Phorate*† and *Disulfoton*† are granular systemic insecticides which are applied in small quantities (10–20 lb commercial product) either over-all or band placed over the rows.

Control measures may be required in some years, where aphids are not normally a problem and since Menazon, Phorate and Disulfoton give unsatisfactory control if applied after the aphids are present, other insecticides based on one of the following are recommended for a speedy kill: dimethoate, formothion, malathion, mevinphos†, nicotine†, oxydemeton-methyl†, phosphamidon†, or schradan.

Recommended eradication treatments where preventive measures have not been taken:

Disulfoton† at 12 oz active ingredient (10 lb 7·5% granules)/acre or phorate† at 1 lb active ingredient (10 lb 10% granules)/acre or demeton-S-methyl† at 3·4 fl oz active ingredient (6 fl oz 58% e.c.) in 40 gal/acre; or oxydemeton-methyl† at 3·4 fl oz or active ingredient (6 fl oz 57% e.c.) in 40 gal/acre.

After Martin (1965, 1969).

VI. SPRING BEANS

A. Soil, Climate and Manuring

Soil, climate and manurial requirements of spring beans are the same as those described for winter crops. In general, the spring crop is lower yielding but there are several advantages which in some districts and seasons make them preferable to winter beans. Firstly, the amount of frost damage is less and there is a much lower "complete" failure rate with spring sowings, plus the fact that chocolate spot infections, caused by species of the *Botrytis* fungus, are less severe. From a cropping sequence point of view spring beans are more easily

---

† Granular insecticidal applications are not normally harmful to bees, neither are insecticide sprays applied before the beans are in flower but where control measure spraying is carried out whilst the beans are in flower, then local bee-keepers should be warned and where bees are employed for pollination, these should be shut in whilst the spraying is in operation.

Demeton-S-methyl, Disulfoton, Mevinphos, Nicotine, Oxydemeton-methyl, Phorate, Phosphamidon, Schradan are chemicals included in the Agriculture (Poisonous Substances) Regulations and the necessary precautions whilst handling and using them must be taken. Users are strongly advised to read the M.A.F.F. Leaflet A.P.S. 1 "The Safe Use of Poisonous Chemicals on the Farm".

accommodated and more time can be spent in ploughing and cultivating in seed-bed preparation in spite of early spring sowing being
required. On the other hand, blackfly infestations are more severe
and the growing season is very long, particularly in Northern England
and Scotland and as a result harvesting often takes place under
relatively poor conditions.

B. Time of Sowing

In view of what has just been stated above, to obtain a mature crop,
which is ready to harvest under reasonable weather conditions, sowing
should take place in February or March as soon as ground conditions
permit. In fact spring beans should be the first crop to be drilled in the
spring.

C. Depth of Sowing

The optimum depth of sowing is similar to that for the winter crop,
i.e. 3–4 inches below the surface and this is very important when
residual soil applied herbicides are to be used.

D. Establishment by Ploughing-in

In the past, ploughing-in of the seed from a bean box fitted to the
plough was a common method of establishing bean crops. It meant
that ploughing and sowing were completed in one operation, the seeds
were well protected from birds and they went into a firm seed-bed.
One serious disadvantage with this method is the fact that inter-row
cultivation is either very difficult or impossible to achieve due to
irregularity of row-width.

A number of farmers still prefer this method of sowing, but many
of them have modified their operations slightly. Seed is broadcast onto
the surface with a fertiliser spinner and this is followed by shallow
ploughing. Reasonable plant stands can be obtained in this way, but
emergence may be irregular and higher than average seed-rates may
be necessary under some circumstances. Weed control is obtained either
by suppression from the crop itself (not often satisfactory) or by spraying and where pre-emergence herbicides are selected, the wheel-mark
damage to the crop can be minimised compared with post-emergence
applications.

E. Seed Rate

The tillering capacity of spring varieties is much less than the corresponding winter ones, but since the field germination is usually higher,
the amount of seed required per acre is about the same.

F. Seed Dressing

Soil temperatures are normally low when spring beans are sown in early (February and early March) and to avoid losses due to disease, the seed should be dressed with anti damping-off preparations based on thiram or captan. Seed dressings are less important when drilling takes place towards the end of March or into April, since the soil temperatures are normally rising fairly quickly and with speedy germination and initial growth, the young braird is less affected by soil inhabiting pathogens.

G. Varieties of Spring Beans

Recommended varieties of spring beans for sowing in England and Wales are published annually by The National Institute of Agricultural Botany (N.I.A.B.) in their Farmers' Leaflet No. 15 and the 1970 one is reproduced in full in the Appendix. Previously the N.I.A.B. had listed and described the most important cultivars in this leaflet and according to the size of the seed, they were classified as tic or horse beans.

(i) TIC BEANS

These are round and small, with a 1000-seed weight between 10 and 20 oz. They are normally grown as a cash crop since quite a large proportion is sold for pigeon feed at home or on the Continent. They command a slightly higher price than horse beans and together with slightly higher yields, a significantly greater financial return can be expected per acre.

Tarvin (formerly Garton's Pedigree), Blue Rock, Maris Bead and Minor are presently the highest yielding tic beans and one of these cultivars should be chosen unless there is strong local evidence to support the adoption of other varieties.

In this connection Herz Freya may be one such variety. It is very early maturing, has a short straw and could therefore better suit the requirements in Northern England, and possibly Scotland, although in the South it has given some poor yields, mainly due to its susceptibility to bean leaf roll virus.

(ii) HORSE BEANS

These usually have flat oval seeds, are much bigger than tic beans, with a 1000-seed weight between 20 and 30 oz and apart from those which go for further multiplication, they are all used for livestock feed.

Strubes is the only variety presently recommended for general use in England and Wales. It is very early maturing, has short straw and with resistance to bean leaf roll virus can be grown in the South as well as in the Northern parts.

Suffolk Red has been grown for many years in the South-east. Long straw and relatively late maturity make it unsuitable for the North, but it has given some good yields in East Anglia where it originated as a farmer's stock. Most of the seeds are red, but a significant proportion may turn out to be white or buff coloured.

Granton, a Scottish landrace variety was grown for many years as the control in variety trials and was cultivated quite widely in Scotland. This short strawed early maturing cultivar was useful in the North, but with susceptibility to bean leaf roll virus, it suffered reductions in yield when grown in the South.

(iii) CARSE BEANS

These are specific to Scotland and are stocks of beans which have developed and been maintained in the Carse land areas (heavy clay) notably Stirlingshire. They are small round to oval beans, similar to the Tics in size and yield well in their "own" environment. Quite often, they outyield the recommended varieties, when grown alongside in trials, but they cannot be grown with success in the South where the bean leaf roll virus can be responsible for 30–40% yield losses with these susceptible stocks.

(iv) LARGE-SEEDED TYPES

Interest in these is now largely academic since their cultivation on a field scale ceased several years ago. Mazagan is a representative of this group and being similar to the broad bean, was quite often found growing in gardens. The plants of this variety have poor field characters (i.e. slender stems which are prone to lodging) but the grain is large and of good quality. The seed is approaching one inch in length (1000-seed weight over 30 oz) and presented some difficulty in drilling, being so large. It is unlikely that these large-seeded cultivars will become popular in the future, even if suitable ones were available, partly on account of drilling difficulties and also due to the fact that higher than normal seeding rates are required.

H. Weed Control

Inter-row cultivation for spring bean crops grown on wide rows can be successful in controlling the annual weeds on some soil types, but where the crop is grown on heavy land, and in the very dry areas, post-emergence cultivation becomes difficult and on many occasions undesirable. Herbicides for weed control in bean crops have been listed in Table 6, and of these the following are more widely used on the spring sown crop in practice.

Simazine at 0·5–1 lb active ingredient in 20–30 gallons of water/acre applied post-drilling well before the crop emerges, will usually give satisfactory weed control, provided that the crop has been evenly drilled at 3–4 inches below the surface and that some rain follows the application of the herbicide. The lower dosage rate can be used except where difficult weeds like Knotgrass and Black Bindweed (*Polygonum* spp.) predominate and then 0·75–1 lb active ingredient/acre will be required.

In the absence of soil-acting pre-emergence herbicides, Dinoseb Acetate at rates up to 2·5 lb/acre can be used post-emergence on spring beans, in a manner similar to that described for the winter sown crop. Dinoseb-amine at 1·5 lb/acre in high volume also applied as a post-emergence herbicide may be used on spring beans but the amount of scorch suffered by the crop is likely to be greater than with winter beans, since growth is almost certain to be fairly active at the time of application. This crop damage can be minimised by lower application rates but not without a reduction in the degree of weed control obtained.

Tri-allate applied pre-sowing or Barban post-emergence can be used for wild oat control as described under winter beans.

I. Pests

The control of wood pigeons and weevils has already been outlined under winter beans and measures suggested apply equally well to the spring crop. Aphids on the other hand are more important related to spring beans and will require further attention.

(i) APHID CONTROL

The Black Bean aphid (*Aphis fabae*) is a much more serious pest of spring sown crops than with the winter ones and control measures already discussed under winter beans are particularly important. Aphid control is therefore much more widely practised and is more often needed with spring beans, a factor which should prompt most farmers to consider the use of wide rows, if insecticides are used as remedial rather than routine control measures. Green aphids are also to be found on beans and they arrive slightly earlier than the black aphids and may introduce virus diseases to the crop. Black aphids usually appear in early June and they may be responsible for spreading the viruses already present and they cause considerable physical damage to the plant especially when multiplication rates are high (East Midlands N.A.A.S Circ., 1968). Control of both types of aphid can be obtained by a protective treatment to the field margins only and this less costly method can be useful in the Midlands and North of England where the problem is less severe. In the South and East of England where aphid numbers are high and build-up is rapid, treatment of field margins is inadequate and spraying, dusting or application of granular insecticide to the whole field will be required. Where fields are treated early, they can be subject to aphid attack later on since most of the insecticides have a relatively short persistency and a second treatment may be required in exceptionally bad years.

(ii) MINIMISING INSECTICIDAL SPRAY DAMAGE TO HONEY BEES

Having seen that honey bees can be of use in the pollination of field beans and in the general interest of insect protection, due consideration should be given to bees and beekeepers when using insecticides for

TABLE 7

Main Chemical Dangers to Bees Classified According to Avoidance Measures (after Cooper, 1964)

Problem		Safe		Unsafe
Crop in flower	Pest	At all times	For evening spraying or 24 hour hive closure	Fairly safe with hive closure on day of spraying and next day
Beans: broad, field and runners	Aphids (Blackfly and green-fly)	Menazon G Nicotine DS Schradan Disulfoton, G, ABB Phorate G, ABB	Malathion Demeton-methyl Oxydemeton-methyl Menazon S Mevinphos Derris Pyrethrins Disulfoton G, ADB Phorate G, ADB TEPP	Dimethoate Formothion Phosphamidon

Code: in the above table, application is by spray except where otherwise specified.

ABB = applied before bloom;   D = dust;   S = spray.
ADB = applied during bloom;   G = granules;

aphid control. The following points suggested by Cooper (1964) should be borne in mind:

(a) Where possible the highly toxic materials should be applied pre-flowering;

(b) If at all possible the less toxic insecticides should be chosen;

(c) Insecticidal application confined to early morning or late evening will do less damage since fewer bees are flying at these times;

(d) Warn local beekeepers to close their hives on the day of spraying and this will apply particularly to those who have placed hives around or within the bean crop itself.

This author lists the various available insecticides for black- and green-fly control into two categories, safe and unsafe for bees and the pertinent part of this classification is thought of sufficient importance to be reproduced for the guide of both farmers and beekeepers (Table 7).

VII. HARVEST

A. Binder Stage

In the past a large proportion of the crop was cut by binder and although most of the acreage today is combine-harvested, it was felt that binder stage should be mentioned and briefly described. Plants at this point of maturity will still be green at the top although senescence will have occurred to a large extent at the bottom. There the leaves will have turned yellow-brown, a few will have dropped and the pods at the bottom and in the middle of the plant will have changed from green to brown. At this stage, binding can go ahead without any fear of serious seed losses due to shattering and the stooks will have to remain in the field for two or three weeks before stacking takes place. It was often said that beans should only be bindered in the morning when there was still some dew about in order to minimise the losses due to shattering. This infers that the crop is almost dead ripe and that once the moisture has evaporated the pods will split and seeds will fall to the ground. This statement therefore really refers to crops for binding which are past the optimum stage. It has been shown that maturation is more uneven without pollinating bees compared with crops where sufficient insects visit the plants, thus giving additional support for the introduction of honey bees to beans.

B. Combine Stage

When beans are to be combine-harvested, the crop should be fully mature with the seeds ripe and as dry as possible. The whole crop looks black. Much of the leaf will have fallen and those remaining are black, as are the pods. The point of attachment of the bean seeds to the pod, known as the hilum, is black on most varieties, when the crop is ripe. However, since some cultivars have a white hilum this is not a foolproof test. When bean crops are tall, combine-harvesting them is a much more efficient method of securing the grain and often the reel is removed so that the cut plants fall directly onto the combine platform. This will eliminate the seed loss which can occur when the pods are brittle because the beans tend to drop immediately the bottom ripest pods are touched. Combine-harvesting a dead-ripe crop of beans is probably best tackled on damp or dull days rather than bright sunny ones to minimise grain losses. If, however, the crop is late in being harvested, then on no account should farmers delay combining, since the weather conditions can quickly deteriorate and the grain moisture content is often excessive.

Several modifications to combine harvesters will be required when moving into beans from cereals and most if not all of these listed below need to be considered.

1. Reduced drum speed.
2. Wider concave clearance.
3. Bottom riddle will need to be changed.
4. Top sieve requires to be wide open.
5. Straw elevator chain will probably need slackening (bean straw wrapped round one which is tight can be responsible for it snapping).
6. An increase in the fan speed will almost certainly be necessary as the amount of trash with beans is usually greater than with cereals.

C. Time of Harvesting and Moisture Content

Moisture contents of combine-harvested beans are usually higher than those of cereal crops particularly in Northern England and

	Winter bean harvest	*Spring bean harvest*
South and Eastern England	August	Early September
Midlands	Early September	Late September–early October
North of England	—	October
Scotland	—	October–November (or even later)

Scotland as harvest often takes place in October or November and it is not uncommon to see combines in the field later than this following a difficult, late season. An approximate "harvest calendar" has been attempted for the various regions, but it must be stressed that the harvest date for beans, particularly spring sown, is less predictable than with cereal crops. In spite of having relatively high moisture contents, the grain can be secured without damage, providing the

Low	*Medium*	*High*	*Very high (Bean moisture %)*
15–20%	21–25%	26–30%	31–40% and even higher on occasions

modifications, previously listed, are carried out. Drying of the wet grain should follow immediately and virtually all types of farm drying systems will cope with beans satisfactorily. Where the produce is destined to be sold for seed, the drying temperatures should be kept as low as possible and with high moisture contents the process should not be hurried, otherwise the outer layers become very hard and although the grain appears to be properly dried the inside will still be soft and wet and the passage of moisture from the centre to the periphery will take some time. Excessive heat applied to wet beans can cause the seeds to split. With large seeded crops like peas and beans, two stage drying is often recommended to ensure that satisfactory moisture reduction is achieved and to keep the seed whole. Beans should be dried to 15% moisture and stocks which start with high or very high moisture contents (see above), are better moved or agitated during drying othewise they tend to cake together and set quite hard. Continuous or batch driers are useful in this respect also the in-bin drying systems where the automatic transfer from drying to holding bin and back can be achieved.

Because the beans are larger than cereals, have a thicker skin and are usually at higher moisture levels, recommended temperatures for drying are somewhat lower. According to the M.A.F.F. Short Term Leaflet 60, to obtain relatively uniform stress-free drying, unheated air or air with a 7–10°F (3–4·5°C) temperature lift (cf. ambient) should pass through the beans stored at uniform depth. On-the-floor drying systems or radially ventilated bins are well suited to this "slow but sure" type of moisture removal whereas tray or sack driers, with shallower layers may be operated at higher temperatures, i.e. 20–25°F (9–11°C) above ambient temperature.

D. Storage

Once the moisture content of field beans has been reduced to 15%, they can be stored satisfactorily where they have been dried or transferred to holding bins awaiting disposal either to be sold for grinding or for livestock feed on the farm. In the past, and to a limited extent today, old beans are preferred to newly harvested stocks for livestock feed during the period September to December. It is thought that the feeding of newly harvested beans to stock can result in digestive upsets and it is possible that this may be linked with the alkaloid content. Beans offered for sale will usually be classified as ordinary or old stocks and the latter often command a small premium.

Pea and bean beetles, known as Bruchids, appear to attack the grain whilst in store. Although most species of this pest are tropical or sub-tropical in origin and therefore only of limited importance when they appear on imported legumes, some are capable of completing their full life cycle under our cool temperate conditions and thus can assume economic importance. These small beetles (3–4 mm long) *Bruchus rufinamus* Bohem. (bean beetle) and *Bruchus pisorum* L. (pea beetle) lay eggs on the growing crop, and the resulting larvae bore into the seed. They develop rapidly eating the endosperm and on occasions the embryo and once formed the adult beetle bites its way out. These adults spend the winter in the stored beans or somewhere sheltered indoors, to emerge in the late spring to lay their eggs on developing bean pods and thus complete their life cycle. Full details of this can be found in M.A.F.F. Advisory Leaflet No. 126 (1968) together with some excellent photographs of the damage which these insects can inflict. Bean stocks to be used for seed should be examined to ascertain the extent of the damage once these beetles are noticed and to prevent attacks to the next crop, heavily infested seed should be fumigated. For large quantities of seed, methyl bromide, aluminium phosphide or a mixture of ethylene chloride and carbon tetrachloride can be used but for small quantities to be fumigated on farms a mixture of 3 parts of ethylene dichloride to 1 part of carbon tetrachloride applied evenly at 6 pints per ton is the Ministry's recommendation. This mixture is poisonous and inhalation of the fumes must be avoided. Respirators should be worn when fumigation is done under confined conditions and the safety precautions as laid down by the Ministry must be observed.

ACID TREATMENT FOR THE STORAGE OF MOIST BEANS

At the time of writing a few farmers and several experimental workers are undertaking the acid treatment of stockfeed beans, with the idea of storage without drying. Propionic acid is being used at about 1% (higher rates are being tried at high–very high moisture levels) and reports so far indicate that this level of application appears to reduce the pH sufficiently to inhibit unwanted fungal and bacterial growth.

At an approximate cost of £1·25/ton this is not high by the standards set by contract drying, particularly considering the amount of moisture to be removed from many loads of beans. Problems associated with this new technique appear to be in an even application of these small quantities of acid and the applicators seem to be in short supply. The latter position however is not permanent and could quickly be reversed should it be confirmed from the 1968 crop that this new method of storage was completely successful.

VIII.   YIELD

Yield of both winter and spring sown beans has been very poor in the past. The national averages show that only 16–17 cwt were produced/acre between 1939 and 1959, except in Scotland towards the end of that period, where over 1 ton/acre was being obtained from a relatively small acreage. No distinction is made between autumn and spring sown crops in these National Statistics, but trials carried out by the N.I.A.B. in the Midlands and Eastern Counties in recent years showed that winter beans gave about 5 cwt more grain/acre than spring cultivars under directly comparable conditions (N.I.A.B. Farmers' Leaflet No. 15, 1968). Smith and Aldrich (1967) indicated that, from National Agricultural Advisory Service Surveys, about one-quarter of the national bean acreage was sown with winter varieties, almost all of which could be located in the eastern counties. Soper (1958) suggested that 25–30 cwt/acre was a good yield and yields exceeding 30 cwt/acre are obtainable with new cultivars and practices according to M.A.F.F. Short Term Leaflet No. 60 (1967). In calculating gross margins for spring and winter sowings Nix (1968) used the following yields (cwt/acre) and these would appear suitable estimates for the late 1960's. Having accepted these as average standards in

	Low	Average	High
Spring beans	20	25	30
Winter beans	25	30	35

present-day bean growing, the extremes will complete the picture in terms of productivity. Under poor conditions of land and season spring bean yields may be as low as 10–15 cwt/acre and the reverse conditions with winter beans could easily result in 40 cwt/acre and 50 cwt would appear to be the ceiling yield with autumn sowing.

IX. GRAIN QUALITY AND USE OF FIELD BEANS IN LIVESTOCK FEEDING

A. Chemical Composition

With an increasing acreage and interest in the bean crop during the late 1960's, the aspects of quality were scrutinised and Eden (1968) surveyed the available data related to analytical composition. Mean values are given in Table 8, for both winter and spring sown crops and for those who are especially interested in bean quality, the ranges given by this author in the original text will prove useful.

TABLE 8

Dry Matter Composition of Field Beans (%)

	Spring sown (104 samples)	Winter sown (28 samples)
Crude protein	31·4	26·5
True protein	28·2	24·0
Ether extract	1·5	1·5
Crude fibre	8·0	9·0
Nitrogen-free extractives	55·2	59·0
Total ash	4·0	4·0
Silicious matter	0·1	0·1
Calcium	0·16	0·19
Phosphorus	0·66	0·68
Magnesium	0·13	0·13
Potassium	1·17	1·22
Sodium	0·01	0·02
Chlorine	Trace	Trace
Manganese (ppm)	14	14

After Eden (1968).

From the mean values, it will be noted that spring beans have nearly 5% more crude protein and contain 1% less fibre than winter beans and are thus of significantly higher quality. Previously a crude protein of 25·5% of the total weight or 29·7% of the dry matter was assumed for all beans (Evans, 1960).

This difference in protein between winter and spring beans was also reported by Smith and Aldrich (1967) whose figures are reproduced below.

Spring beans (34 samples)   31·6% protein (N × 6·25)
Winter beans (10 samples)   27·7% protein (N × 6·25)

Bond and Toynbee-Clark (1968) examined the protein content of winter and spring beans, and by sowing the winter cultivars in the spring, they were able to demonstrate that the spring varieties still showed a high albuminoid level. From this they concluded that the differences were inherent within the varieties and that date of sowing or growing season had no effect. The average results obtained by these workers in respect of beans grown in South-west and South-east England appear beneath and suggest a difference of approximately 3% in crude protein levels.

TABLE 9

*Mean Percentage Crude Protein (N × 6·25) of Bean Dry-matter
All Spring Sown*

South-west England (1967)			South-east England	
Spring cultivars (3) 30·58 ± 0·71	Winter cultivars (7) 27·46 ± 0·46	1966	Spring cultivars (1) 28·9	Winter cultivars (3) 26·1 ± 0·4
		1967	Spring cultivars (3) 29·69 ± 0·20	Winter cultivars (2) 26·81 ± 0·24

After Bond and Toynbee-Clark (1968).

The spring sowing of winter cultivars, often practised when the autumn and winter weather is exceptionally wet and drilling has to be delayed until the turn of the year, is therefore unlikely to improve the protein content and spring varieties should be selected for spring sowings. Carpenter and Johnson (1968) also report on the chemical analyses and metabolisable energy (M.E.) of field bean types, and they confirm the protein difference between winter and spring stocks, but not the crude fibre variation which emerged from Eden's survey. From the livestock feeding viewpoint, beans can be considered as having a starch equivalent of 66 and protein equivalent of 20, thus making bean meal a mixed starchy-protein concentrate. By comparison with other feeds, beans have 2·5–3 times the protein of cereals but only 0·5–0·67 of that of imported cakes such as linseed, ground-nut or soya. According to the M.A.F.F. Short Term Leaflet No. 55, entitled "The Use of Beans in Livestock Feeding", 1 cwt of beans should be looked upon as equal to 0·5 cwt barley plus 0·5 cwt soya bean cake. This assessment based on the nutritional value of beans enables farmers and livestock feeders to make economic comparisons between beans and other protein sources without forgetting the amount of carbohydrate which the beans contain. Beans can be fed to all classes of stock with safety and practical suggestions and recommendations can be found in Short Term Leaflet No. 55 or textbooks relating to animal feeding.

One final point in respect of the nutritive value of beans and that

TABLE 10

Chemical Analyses and Energy Values of Winter and Spring Field Bean Dry-matter

| | Field bean types | | |
| | Throws M.S. | Minor | Strubes |
Analyses on 90% dry-matter basis	(Winter)	(Spring tic)	(Spring horse)
Crude protein (%)	23·4	27·1	26·7
Ether extract (%)	1·4	1·5	1·3
Crude fibre (%)	5·9	6·1	6·1
Ash (%)	3·5	3·1	3·3
Nitrogen-free extractives (%) (by diff.)	55·8	52·2	52·6
M.E. (Metabolisable energy— kcals/g at 90% D.M.)			
(a) Determined by chicks	2·52	2·68	2·80
(b) As predicted from analyses	2·52	2·78	2·59

After Carpenter and Johnson (1968).

is the amino acid lysine, which is important in pig and poultry feeds. With both these classes of stock, lysine is an essential amino acid which must be present in the diet, because unlike the ruminants, they are unable to synthesize it. Beans have a lysine content of about 1·8%, compared with 1·5% in groundnuts and 2·5–3·0% in soya beans (M.A.F.F., 1966a) and thus at the same level of protein intake, the amount of lysine consumed when beans are fed is significantly greater than with these two other protein sources.

X. PRODUCTION COSTS, OUTPUT AND GROSS MARGINS WITH WINTER AND SPRING BEANS

The following details of production costs, associated with average yields, are given by Nix (1968) to arrive at gross margins for field beans.

Average yielding crops of spring beans have a lower gross margin than barley whereas similar crops winter sown are greater in this respect, but it must be remembered that in some years bean yields can

TABLE 11

Gross Margins for Spring and Winter Field Beans

	Spring beans £/acre	Winter beans £/acre
Variable costs		
Seed	4·3	5·0
Fertiliser	2·2	2·2
Sprays	5·5	3·3
Total variable cost	12·0	10·5
Output 25 cwt at 26 shillings/cwt plus £5/acre grant	37·5	44·0 (30 cwt/acre at 26 shillings/cwt plus £5/acre grant)
Gross margin	25·5	33·5

be very low due to poor season, insufficient insect activity, pests and diseases. For those who consider gross margins inadequate in assessing the contribution which beans can make towards the general farm economy or in assessing profits, full enterprise costs have been attempted and appear below.

TABLE 12

Specimen Production Costs for Field Beans

	Spring beans £	Winter beans £	
Labour (13 man hours @ 6s. 6d.)	4·225	3·90	(12 hours)
Tractor (8 hours @ 5s.)	2·00	1·75	(7 hours)
Machinery depreciation and repairs	4·00	4·00	
Fuel (other than for tractor running)	0·90	0·90	
Materials			
Seed	4·00	5·00	
Fertiliser	2·25	2·25	
Sprays	5·50	3·30	
Rent	4·50	4·50	
Total Direct Costs	27·375	25·65	
Adjustment for manurial residues (−)	1·00	1·00	
Removal of straw (+)	1·00	1·00	
Share of general Farm expenses (Overheads)	2·50	2·50	
Total Cost/Acre	29·875	28·15	

With the £5/acre grant (1968, 1969 and 1970) this means a net cost of production of £25 and £23/acre for spring and winter sowings respectively and thus with field beans for stockfeed (£27/ton) the break-even yield is in the region of 17–18 cwt/acre. Should production costs be higher than indicated or the price lower, then 1 ton/acre would be more like the break-even point. On the other hand small spring tic beans for the pigeon trade, although of limited sale, usually command a higher price than the ordinary beans and at £32/ton the break-even yield is of the order 15–16 cwt/acre. These figures are approximations only, but will serve as a guide to indicate possible profit margins when stockfeed or tic beans are grown as a cash crop.

XI.   DISEASES OF FIELD BEANS

Chocolate spot, leaf spot, stem rot, mosaic and leaf roll viruses are the diseases associated with field beans.

A.   Chocolate Spot

This disease is caused by the fungi *Botrytis fabae* and *Botrytis cinerea* and is the most important fungal disease of the bean crop. Severe infections causing complete crop failure may occur with winter sown crops but chocolate spot is seldom of economic importance on spring beans. The spores of botrytis species are carried over from one season to the next on bean straw (haulm), on self-sown volunteer plants and occasionally on the seed. Brown spots appear on the leaves of infected plants in the winter and warm wet springs will induce these brown lesions to increase markedly in size and they will eventually appear on the stem. The flowers and young developing pods drop off severely infected plants thus producing drastic yield reductions. There are no fungicides available for the control of chocolate spot but there are two ways of reducing the incidence of this disease in winter beans. Firstly, it is important that haulm from previous bean crops is properly disposed of by burning or removal followed by good ploughing, particularly where the next crop is grown in close proximity. Secondly, it has been shown that lack of potash can aggravate this disease and so for winter beans it is advisable to see that plenty of this nutrient is available.

B.   Leaf Spot

Leaf spot, due to the fungus *Ascochyta fabae*, occurs on both winter and spring beans. Greyish brown lesions occur on the leaves, sometimes circular in form, and these symptoms can readily be confused with those caused by *Botrytis* species. Later developing lesions of leaf spot on the pods reach the seed, causing discoloration and death. The disease is mainly seed-borne and it is important not to use heavily infected seed. Farmers wishing to use their own stocks for seed are advised to have them tested for the presence of *Ascochyta* along with the germination test and this service is offered by the Official Seed Testing Stations. It is recommended that stocks with more than 5% leaf spot infection should not be used (M.A.F.F. Short Term Leaflet No. 60, 1967a).

C.   Stem Rot

Mainly caused by the fungus *Sclerotinia trifolium* var. fabae although clover rot (*S. trifolium*) may also be present on bean crops. This disease is associated with frequent bean cropping or beans following red clover, but since this herbage legume is less popular now than in previous years, stem rot is less common amongst beans. Affected plants begin to rot at ground level and eventually they collapse and patches appear in the field where the crop has died off. When significant quantities of stem rot appear in beans, freedom from this crop and red clover for four to five years will usually eliminate the fungus resting bodies (sclerotia) from the soil and it is from these that the crop contracts the disease.

D.   Virus Diseases

Mosaic and leaf roll viruses are known to attack field beans. They are probably introduced by aphids from other legumes and thus the early use of aphicides on the beans will help to control the spread even though the viruses may be present. Bean leaf roll virus, as the name suggests, causes an inward rolling of the leaf margins and they become yellow and brittle. Photosynthetic activity is restricted and severe yield reductions can follow. Differences in varietal susceptibility have been recorded and up-to-date information concerning this can be obtained from the most recent recommended lists (N.I.A.B. Farmers' Leaflet No. 15). Amongst the spring beans Strubes is resistant, whilst Herz Freya and Scottish Carse are susceptible and these last two varieties should not be chosen for cultivation in the South and East on account of the high aphid populations there.

PART 2. FRENCH BEANS (KIDNEY)

XII.   INTRODUCTION

The French or Kidney bean, *Phaseolus vulgaris* L., is a self-pollinated annual, probably of sub-tropical origin, which is cultivated today in many parts of the world under a wide variety of climatic conditions.

According to M.A.F.F. Bulletin No. 87 (1962) the cultivars are most conveniently divided into eight groups although difficulty is found in giving examples of two of these (Table 13).

The most important group for marketing fresh was the dwarf form of green string bean. These have been superseded by runner beans as they rapidly deteriorated into something very coarse and stringy and had a very limited pulling period. For canning, quick-freezing and more recently for dehydration, the green stringless varieties of the dwarf type are being used. The acreage statistics relating to French

TABLE 13

Types of French Beans

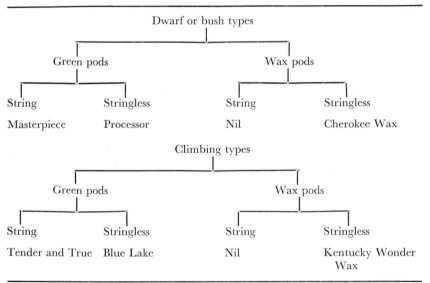

beans in England and Wales over the past fifteen years, shown in Table 14, demonstrate the increasing importance of this crop.

During the 10-year period 1953–62 there was a steady increase in French bean growing from 1620 to 4820 acres. As the standard of living rose, more people were interested in higher class vegetables such as dwarf beans and peas and this is reflected by the acreage increase in the case of beans. A much greater rise took place after 1962 with almost double the quantity being grown the next year. This is accounted for mainly by the expansion of beans for processing, i.e. canning and quick-freezing. Since 1963 new methods of dehydration have proved successful on a commercial scale and this has also helped to maintain the large acreage reached in 1964 (nearly 10,000 acres). It was estimated that at least 90% of the French bean production was used for processing (M.A.F.F., 1967a), and a detailed study of the acreages in 1963 and 1964 revealed that 93% and 95% respectively were dwarf

TABLE 14

French Bean Acreage (June returns) England and Wales

Year	1953	1954	1955	1956	1957	1958	1959	1960
Acreage	1620	1978	1800	2477	2686	3076	3948	4180
Year	1961	1962	1963	1964	1965	1966	1967	
Acreage	4044	4820	8136	9906	7706	9047	9800	

beans. Developments in the mechanisation of this crop, particularly in respect of complete harvesters has also helped to stimulate widespread interest and nowadays dwarf beans are grown on a large scale in agriculture (see Fig. 4). Processing companies are also keen to see

FIG. 4. Typical of the extent and uniformity of dwarf bean crops today. Chemical weed control employed between the rows using a mixture of diquat and paraquat (Pregtone Extra). Photograph by courtesy of Plant Protection Ltd.

dwarf bean growing on a large scale, where the full benefits of mechanisation are felt, where economies of large scale production can be obtained and small numbers of growers or farmers are involved.

In 1962, over three-quarters of the dwarf beans were grown in Eastern England and the main counties involved were Norfolk and Suffolk. From the same survey, nearly 60% of the holdings where the crop was grown had over 10 acres each and 1·7% of the total number of holdings had 30 acres or more and were cultivating 31·8% of the area devoted to this crop.

XIII. SOILS

Soils for the cultivation of dwarf beans should be as near to neutral as possible (pH 6·5) and where lime is required to raise the pH, this should be applied well in advance of the crop and worked in. They will grow satisfactorily on a wide range of soils, provided that the drainage is good. Clay soils and those which remain cold and wet, however, are really unsuited on account of poor slow growth and many seeds will rot in them, failing to braird. Soils with high organic status like the peats will only produce moderate yielding crops on account of an excess of vegetative growth and the light, sharp sands, which tend to be deficient in many of the plant nutrients, can seldom be brought to a high enough fertility level to support beans. The ideal soil texture for this crop is a deep loam or light silt, which has a good organic matter status from the manuring and residues of previous crops.

XIV. PLACE IN CROPPING SEQUENCE

It is unlikely that dwarf beans will present a problem from a rotational point of view, especially in agriculture. They can follow roots, potatoes or sugar beet and will do well if these previous crops have received applications of farmyard manure. It is possible to grow them after cereals, but the state of fertility is usually fairly low and as a result crops could well be on the light side. In horticulture they are readily incorporated into the cropping sequence following such crops as roots, brassicas, lettuce or potatoes.

XV. CULTIVATIONS

Where the land is to receive farmyard manure or compost, up to 30 tons/acre can be applied in the autumn or early winter and the land should be ploughed to about 6–8 inches as soon as possible. Where the organic status of the soil is good and only inorganic fertilisers are to be supplied for the crop, then provided there is no plough pan or large quantities of surface trash to be buried, then the depth of ploughing can be reduced to 4–6 inches. Dwarf beans are normally grown on wide rows and to facilitate precision work at drilling and with inter-row cultivations, one-way ploughing is recommended to leave the land uniform and flat. As far as possible, frost action should be employed to facilitate seed-bed preparation and this means early ploughing particularly on the stronger type of land. Dwarf beans are sown fairly late, on account of their susceptibility to frost and a large measure of cultural weed control can thus be obtained by exploiting the stale seed-bed technique. This simply means the preparation of a rough seed-bed well before the crop is due to be sown and once the annual weeds have germinated they can be quickly killed off by cultivation, before the proper crop seed-bed is considered. Seed-beds for dwarf beans are similar to those already described under field beans. A good deep tilth should be obtained and this need not necessarily be fine because the seeds are quite large. Since the drilling will take place in the late spring, the desired seed-bed should be arrived at via the least possible number of cultural operations in order to conserve soil moisture. Where contact herbicides are to be used on the crop, they are much more effective where a good fine tilth has been obtained. Early germination will be encouraged and these weed-killers can prove very useful provided the land is free from large clods.

XVI.   MANURING

Until recently there have been few precise experiments to ascertain the optimum level of fertiliser application for dwarf beans. Previous recommendations through Ministry publications in 1961 and 1962 suggested the following rates in the presence and absence of farmyard manure.

	N	P_2O_5	K_2O
		Units/acre	
F.Y.M.	—	60	60–120
No F.Y.M.	20	60	60–120

Where farmyard manure was available this rate of fertiliser application was achieved by using 6 cwt/acre of the compound 0% N: 10% P_2O_5; 10 or 20% K_2O, or 3 cwt/acre of 0% N: 20% P_2O_5:20% K_2O

alone or with 1 cwt of muriate of potash. It was further suggested that the crop may respond to nitrogen when grown on poor soils or without farmyard manure, through the addition of 20 units of nitrogen. Presently farmers and growers are using significantly higher quantities of fertiliser nitrogen in their manuring of this crop mainly because it does not receive farmyard manure, and from the limited evidence of critical experiments carried out by the Pea Growing Research Organisation this practice can be justified. Significant responses to applications of nitrogen were recorded, whilst small and very small ones resulted from phosphate and potash applications respectively. It was suggested by Gent (1966) that dwarf beans responded to nitrogen because of their relatively short growing season or through lack of suitable nodule forming bacteria in the soil. Soil analysis should be helpful in assessing the correct amount to use and in the light of results from present practice and judging by the up-to-date fertiliser trials with dwarf beans, the following levels would seem reasonable where farmyard manure is not available.

Fertilisers for Dwarf Beans, Grown without Farmyard Manure
(units/acre)

	Low	Medium	High
Nitrogen (N)	120 (150)	60 (90)	—
Phosphorus (P_2O_5)	60	30	—
Potash (K_2O)	60	30	—

Level of major nutrient as assessed or measured by soil analyses; i.e. where soil analysis is low for all major nutrients, 6 cwt/acre of a 20% N:10% P_2O_5 and 10% K_2O compound fertiliser would be required.
Nitrogen levels in brackets indicate upper limits where soil nitrogen reserves are low or medium.

Experiments using ordinary and inoculated dwarf bean seed have shown how beneficial this practice can be in respect of nodulation. The root nodule bacteria, in symbiosis with the bean plant, can fix large quantities of atmospheric nitrogen. Inoculation may thus become standard practice since it can provide a much cheaper source of nitrogen for the crop. Where an inoculation service is available, farmers would be advised to have their seed treated, especially where the crop is being grown on land which has never carried beans before.

Trials where the nitrogen fertiliser has been split, applying half as a basal dressing and half as a top dressing, have not shown any clear-cut advantage over ones where the same total amount was applied to the seed-bed and since an additional operation is required, this practice cannot receive general support. However, if heavy rain follows seeding, or for a variety of reasons the crop is yellow and slow in growth, then a top dressing of 20–30 units of nitrogen, as nitro-chalk or nitra-shell will normally prove advantageous.

XVII. PLANT SPACING, ROW WIDTH AND SEED REQUIREMENTS

Previously, dwarf beans on a small scale were sown by hand in rows 1·5–2·5 feet wide. A continuous line of seeds was put in, with the idea of thinning after brairding to about 6–9 inches. This method was not widely adopted, being wasteful in seed and it was more usual to put in 2 seeds at 6 inch intervals. Field germination with dwarf beans has always been low compared with laboratory germination and thus singling where 2 seeds were placed together would not be a time-consuming operation. In order to arrive at reasonable estimates of seed rate/acre some knowledge of seed size, plant population requirements and field germination must first be ascertained. In the following table it

TABLE 15

Calculated Seed Rates for Dwarf Beans for Two Stated Final Plant Populations
(lb/acre)

Plant population	Row width	Plant spacing	Seeds/lb					
			1000	1200	1400	1600	1800	2000
43,560	18 inches	8 inches	131	109	93	82	73	65
	24 inches	6 inches						
52,272	30 inches	4 inches	157	131	112	98	87	78

Laboratory germination assumed to be 82%, percentage germination of viable seeds 40% and therefore true field germination only 33%.

is assumed that a plant population of 43,000 or 52,000 is required and that field germination is 33%.

With the exception of the very small-seeded varieties, these assumed plant populations will require something like 0·75–1·5 cwt seed/acre and this is in line with earlier recommendations made in Ministry of Agriculture Leaflets and Bulletins. Work carried out in the mid 1960's at the Pea Growing Research Organisation (P.G.R.O.) (then at Yaxley, now at Thornhaugh-Peterborough) and referred to by Gent

L

(1966) suggested that higher plant populations than those assumed in Table 15, are required to maximise yield. Greater profitably through higher overall yields was obtained using 1·875 cwt/acre (15 stones) linked with narrow row widths of 8–12 inches. In 1966 it was considered impossible to harvest dwarf beans mechanically on rows less than 15 inches apart and these high yields obtained from higher than average seed-rate using closer planting led to some new thinking amongst machinery designers and manufacturers. It also stimulated further plant population studies at P.G.R.O. to confirm or reject these new ideas and the latest information on this and other topics related to peas or beans can be obtained from them. In 1966 Gane suggested that row widths of not over 16 inches should be used with seed rates which result in not more than 4 inches between plants. Using varieties with 1400 seeds/pound and assuming a true field germination of 33%, then 210 lb of seed is required/acre.

XVIII. SOWING

Dwarf beans are very susceptible to frost damage and early spring sowing has to be discouraged. When the soil temperatures are low a large proportion of the seed will rot and even without actual frost damage following emergence, the brairds would be very thin and re-sowing would have to be considered. Germination only begins around 49–50°F (10°C) and it is thus necessary for something approaching this temperature to be reached before the seed goes in. Actual

Growth Temperatures for Kidney Beans

Minimum	Optimum	Maximum
49°F (9–10°C)	92·6°F (34°C)	115°F (46°C)

(After Hall, 1945.)

temperature requirements for growth are shown above. Care should be taken when handling dwarf bean seed as it is very fragile and if bags are dropped onto concrete surfaces, cracks occur in the seed coat and germination is adversely affected. Fungicidal seed dressings should be used as a routine measure to protect the young seedlings from soil-borne fungi and as with field beans the flow rates of drills should be checked with dressed rather than ordinary seed to ensure correct seed rates. Where trouble from pests such as the bean seed-fly is anticipated, a combined dressing incorporating both fungicide and insecticide should be used.

The earliest sowing, in frost free areas, will take place from the middle to the end of April, but the bulk of the crop should go in during May or when late planted in early June.

Field germination, predators and chemical weed control are the three main factors which determine depth of sowing and the compromise usually reached is for drilling to take place at about 2–2·5 inches.

XIX. VARIETIES

Where dwarf beans are grown for the green market, varietal choice is made by the grower or farmer but when grown for processing the company concerned usually stipulates which cultivars are to be used. As indicated earlier in the classification of French beans, the varieties presently cultivated are bush types belonging to the stringless group having green pods. A list of these appears in the Ministry's Bulletin

TABLE 16

Dwarf Bean Varieties Under Consideration in 1967 and 1968

N.I.A.B. 1968 Stringless dwarf french bean trials	P.G.R.O. Ltd. 1967 Dwarf bean variety studies
(a) *Main trial* at Headquarters Trial Ground —Cambridge, at Sprowston, Norfolk and the Scottish Horticultural Research Institute, Mylnefield, Dundee.	(a) *Main trial* at Thornhaugh.
1. Glamis 9. Early Gallatin 2. Bonvert 10. Executive 3. Idelight 11. Tenderette 4. Slenderwhite 12. Tendercrop 5. *Harvester* 13. Encore 6. *Processor* 14. Gallatin 50 7. Tenderwhite 15. Cascade 8. Slim Green 16. Bush Blue Lake 274	1. D.6 VIII 2. Earligreen 3. Glamis 4. Green Pod 60209 5. *Harvester* 6. *Processor* 7. Sprite 8. Tender Crop
(b) *Preliminary trial* at Headquarters Trial Ground, Cambridge A total of 54 varieties.	(b) *Preliminary trial* at Thornhaugh. A total of 31 varieties.

No. 87 (1962) but as with varieties of most other crops they are being continually reviewed and replaced when something better comes along. In 1966 Gent listed the most popular varieties in current use. These included Tendercrop, Harvester, Processor, Gallatin 50 and Prelude and it will be noted that these are being used as standards in the present day variety trials (Table 16).

With a much greater interest in dwarf beans, particularly for processing, variety testing of established and newly introduced cultivars receives considerable attention and the bulk of it is carried out by the National Institute of Agricultural Botany and the Pea Growing Research Organisation. Up-to-date information on the best varieties to grow can be obtained from these two bodies or from the processing companies. Some indication of the interest in this crop can be gleaned by looking at the present list of varieties under consideration in main and preliminary trials in Table 16.

Once a variety has shown promise in the preliminary trials, it automatically goes for further testing in the main trials and is compared with established ones (i.e. Harvester and Processor) which are used as controls. In the 1967 N.I.A.B. main trial all varieties produced high yields of acceptable quality, with Glamis again showing its rapid maturity compared with the others (N.I.A.B. 48th Report, 1967).

Having demonstrated the value of high plant populations in obtaining good yields, the trials at P.G.R.O. are conducted on a 14-inch row width and the seed is sown with an intra-row spacing of approximately 2 inches.

XX. CULTURAL WEED CONTROL

When dwarf beans were grown on wide rows (18–30 inches) inter-row cultivation for weed control was widely practised. This could be done as soon as the crop brairded and again before the plants began to meet in the row either once or twice. Two or three cultivations were sufficient to maintain weed-free land between the rows, but weeds developing between plants within the row became a nuisance and hand work was necessary. It was often the practice in the past to push up soil on either side of the plants to see that they obtained a firm ground hold but this could not be recommended for today's crop which is mostly mechanically harvested. Any form of cultivation after sowing is usually frowned upon firstly because of the loss of soil moisture and secondly because the dwarf bean harvester must be raised to avoid picking up soil and stones during operation. Chemical weed control has been widely adopted following the introduction of a number of satisfactory herbicides and dwarf bean growing without post emergence cultivations is much preferred especially by the processing companies.

XXI. CHEMICAL WEED CONTROL

The contact pre-emergence sprays as listed in Table 6, can be used to kill seedling weeds which emerge before the crop and of these *Paraquat* or *Diquat plus Paraquat* are the most useful.

Contact pre-emergence herbicides such as *Dinoseb-amine* (3–4 lb/acre) or *Dinoseb in oil* (2·25 lb/acre) applied in medium to high volume will also give a good control of weeds which are present and they also exhibit some residual value in the control of later germinating annuals (Fryer and Evans, 1968).

Of all the herbicides "screened" by P.G.R.O. in recent years mono-linuron stood out as the most selective pre-emergence chemical for dwarf beans. Control of a wide range of weeds can be obtained and this herbicide is well tolerated by the crop (P.G.R.O., 1968). Weed control with monolinuron is mainly through root uptake and although there is some contact action, this is not sufficient for some of the more resistant weeds. As a result a mixture of monolinuron and the contact herbicide dinoseb acetate has been tried.

Monolinuron plus dinoseb acetate, marketed as Ivorin by Farbwerke Hoechst A.G., is approved under the Ministry's Agricultural Chemicals Approval Scheme and Fryer and Evans (1968) indicate, for informa-tion only, the rates of application.

Soil type	Dinoseb acetate		Monolinuron
Light	24 oz	plus	8 oz per acre
Medium	30 oz	plus	10 oz per acre
Heavy	36 oz	plus	12 oz per acre

Ivorin is recommended by the P.G.R.O. (1968) for use on dwarf beans and the rates of application related to soil type are shown below. With early sown crops, best results are obtained by delaying spraying until 3–4 days before crop emergence but with later sown crops which germinate more quickly, spraying should follow soon after drilling. Provided a fine tilth is secured without large clods and seed-beds are moist at the time of application or rain soon follows, weed control will usually be good enough to dispense with inter-row cultivations.

Soil Type	Rate of Ivorin (lb product/acre)
Heavy: clays, heavy loams, silts, Fen and Peat	6
Medium: medium silts, loams, clay loams, sandy soils with high organic matter	5
Light[a]: Sands, light silts	4

[a] Ivorin is not recommended for very light sands containing a high proportion of coarse sand and with a low humus content, except at grower's own risk.

Propachlor applied at 4 lb/acre in medium to high volume, within 2 days of drilling, has given good control of weeds by residual action but later applications lead to crop damage. As with Ivorin, best results are achieved from fine, clod-free seed-beds and although this herbicide looks promising, further tests are deemed necessary before recommendations are made.

XXII. PEST CONTROL

The two major pests, excluding birds, are the bean seed fly (*Delia cilicrura*) and black aphids (*Aphis fabae*).

Bean seed fly maggots tunnel into the germinating seeds, destroying them or injuring them so that the plants produced are poor and stunted. Damage can largely be prevented by using a combined insecticide/fungicide seed dressing. Dieldrin with Thiram is approved for this purpose in spite of the former being one of the highly persistent chlorinated hydrocarbons. Better germinations are obtained using this combined seed dressing. It is thought that there exists a positive interaction between insecticide and fungicide components and even when bean seed fly alone appears to be the trouble, the combined dressing is recommended.

Blackfly (black bean aphis) can be very troublesome on spring sown field and broad beans, but attacks on winter sown field beans and dwarf beans are usually less severe. In the case of dwarf beans, the crop is normally at a very young stage when the first generation of this pest appears in late May–early June and it usually escapes infestation. Attacks will normally come towards the end of June, in July and August and insecticides for the control of this pest should be chosen with the persistency of the chemical in mind.

The 1968 Agricultural Chemicals Approval Scheme list the following insecticides as being suitable for aphid control in French and runner beans:

Demeton-S-methyl	(21)[a]	Malathion	(1)
Dichlorvos	(1)	Mevinphos	(3)
Dimethoate	(7)	Nicotine	(2)
Disulfoton	(42)	Oxydemeton-methyl	(21)
Formothion	(7)		

[a] Twenty-one days between final spraying and harvest—see text.

Martin (1965) indicated the minimum periods in days which are required between spraying and harvest when these various aphicides are used on beans for human consumption. These figures are given opposite the appropriate chemical and must be strictly complied with and to avoid tainting even longer periods may be required. When blackfly attack on dwarf beans comes early, Demeton-S-methyl or Oxydemeton-methyl can be used but if these pests appear just before harvest then Malathion, Mevinphos or Nicotine should be chosen. Processing companies are likely to reject crops which contain this pest and even when the infestation is late in appearing control measures will often be necessary.

XXIII. IRRIGATION

Dwarf French beans are widely grown in the drier, arable areas of Britain and when soil moisture deficits arise, and irrigation water and equipment are available, the application of 1–2 inches of water will give substantial yield increase. It is suggested in the Ministry's Irrigation Guide, Short Term Leaflet No. 71 (1967b) that this crop would do well when drilled into a soil which is near field capacity and water could well be needed in April or May. However, later applications during the crop response period (after green bud stage has been reached) will normally be of greater benefit and at that time it pays to keep the soil near to field capacity.

Recommended Levels of Irrigation for French Dwarf or Stringless Beans

Time of application		Soil available water capacity (AWC)		
Crop stage	Period	Low	Medium	High
After green bud stage	June and July	1 inch at S.M.D. 1 inch	2 inches at S.M.D. 2 inches	2 inches at S.M.D. 3 inches

M.A.F.F. Short Term Leaflet No. 71.

XXIV. HARVEST

With over 90% of the dwarf bean acreage for processing and large acreages involved per grower, harvest is almost invariably a completely mechanised operation. Single-row harvesting machines move down the row removing leaves, stalk and pods. These are elevated to a position where the leaves and stalks can be removed by blowing out of the rear, leaving the pods to be collected in sacks on a platform, in a manner similar to cereal grain collection on a bagger combine harvester. For illustration of these harvesters and the crop during this period see Figs 5, 6 and 7. The harvested beans are quickly removed to the factory where they receive immediate attention and are processed for canning, quick-freezing or freeze-drying.

Fig. 5. Dwarf bean harvester (Model ST5 made by Mather and Platt Ltd.). Photograph by courtesy of Mather and Platt Ltd.

Fig. 6. Dwarf bean harvester with covering plates removed to show harvesting mechanism. Photograph by courtesy of Mather and Platt Ltd.

Fig. 7. Dwarf bean crop before (*right*) and after (*left*) mechanical harvest. Photograph by courtesy of Mather and Platt Ltd.

The relatively small acreage for the fresh green trade is hand picked into shallow baskets and should be removed to a cool packing shed to retain the quality. Dwarf beans plants are small with many pods close to or touching the ground and hence soil contamination by rain splashing can often be fairly extensive. Produce like this may need washing before presentation at market and a thorough drying of these is necessary before packing to enhance their appearance and to ensure reasonable keeping quality.

XXV. YIELD AND QUALITY

A. Yield

In 1949, when row widths of 28 inches were being used and seeds placed at 4–6 inch intervals, a yield of 2 tons/acre was regarded as a good crop and would be obtained by several picks by hand during the Season (Hoare, 1949). Information collected by the Ministry of Agriculture in England and Wales during the period 1954–64 showed that 2 tons/acre was about the average yield at the beginning and except in unfavourable seasons, it would be reasonable to look for 3 tons/acre.

Progress in cultivations, chemical weed control and higher yielding varieties have undoubtedly helped to improve on the last figure quoted and yields between 3 and 4 tons can be obtained with ease in the 1970's.

Variety trials carried out by the P.G.R.O. in 1967 showed present cultivars to be capable of producing 4–5 tons/acre and 6 tons/acre is not considered unobtainable with high seeding rates and a 14 inch row width (Gent, 1966).

B. Quality

A large number of individual characteristics are taken into account when assessing quality in dwarf beans and with the crop produce going for canning, quick-freezing or dehydration to a number of companies, quality definitions in respect of one method of processing may not fit either of the other two. The following factors, however, are taken into account when judging new varieties.

(a) Yield compared with standard varieties.
(b) Plant growth habit and vigour.
(c) Maturity rating compared with standard varieties used as controls.
(d) Number or proportion of pods which touch the ground.
(e) Resistance to pest and disease attack.
(f) Pod characteristics:
    (1) Colour (indicated by a number which relates to R.H.S. Colour Chart).
    (2) Overall assessment of shape, extremes are very curved to straight.
    (3) Average or maximum length.
    (4) Cross-section.
    (5) Factors contributing to loss of quality, i.e. seed development, stringiness and pod wall lignin formation.

The agronomic characteristics (a) to (e) are mainly the concern of the farmer or grower but manufacturing companies are also involved to see that varieties are suitable for growing in order that their contracts are readily acceptable to producers. They are particularly interested in the number of pods touching the soil as this will usually determine the degree of soil-contamination and these characteristics are of prime importance. Having looked at these factors, which, when integrated constitute crop quality, the final assessment is made once a reasonable quantity has passed the processing test, i.e. canning, quick-freezing or dehydration.

XXVI. CROP VALUE, OUTPUT AND GROSS MARGIN

During the period 1954–64, growers' net prices varied between £40 and £80/ton and gross output/acre ranged from £120 to £180. Average yields of 2–3 tons with a price of £60/ton would give this range and it can be seen how important the yield/acre is in obtaining high gross outputs. With the vast majority of dwarf beans grown on contract for processing, the price/ton will be fixed and to maximise the profit obtained from this crop, high yields must be produced at reasonable cost.

In recent years dwarf bean crops have become completely mechanised. Acreages/grower have risen sharply and prices have tended to remain low by previous standards. Output and gross margin are assessed on low, average and high yields and these are shown below.

Output and Gross Margin from Dwarf Beans (per acre)

Output	2 tons	3 tons	4 tons
Value of output (at £40/ton)	£80	£120	£160
Variable costs	£30	£40	£50
Gross margin	£50	£80	£110

Appendix

EXTRACT FROM N.I.A.B. FARMERS' LEAFLET NO. 15
"RECOMMENDED VARIETIES OF FIELD BEANS"†

Field beans provide a useful break crop for farmers specialising in cereal production, as they require no additional equipment and have a low labour requirement. Beans are a valuable source of protein, and can be used with cereals to make balanced rations for most classes of livestock.

This leaflet lists the main characteristics of the recommended varieties of winter beans and spring beans. The National Institute of Agricultural Botany tests all promising new varieties of field beans at several centres over a number of years. No variety is recommended on less than three years' trial. The varieties are classified as follows:

G   Recommended for General use

S   Recommended for Special use

Figures in the tables are based on accumulated results in the last five years of N.I.A.B. trials, and varieties are listed in order of expected yields of beans. The percentage yield is shown for each variety in relation to the control variety which is shown as 100. Numerical scales indicate the relative value of other characters such as straw length, earliness of ripening and seed size; these are based on a 0–9 scale with a high figure indicating that the variety shows the character to a high degree. The scales for winter beans, however, do not necessarily correspond with those for spring beans.

WINTER AND SPRING BEANS

Direct comparisons between winter and spring beans, made mostly in the Midlands and Eastern counties, have shown an average yield difference of about 5 cwt/acre in favour of winter beans. Winter beans ripen two to four weeks earlier than spring beans and usually have longer straw. They are, however, more liable to chocolate spot disease than spring beans, and in areas where this disease is common spring beans are more reliable.

The average crude protein content (N × 6·25) for spring beans is normally higher than for winter beans.

For information on the growing and harvesting of field beans see "Field Beans", M.A.F.F. Short Term Leaflet No. 60 (1968).

† Reproduced by kind permission of the National Institute of Agricultural Botany, Cambridge.

RECOMMENDED LIST OF WINTER BEAN VARIETIES 1970

Yield comparisons other than with the control are not strictly valid, and differences of 4% or less should be treated with reserve.

Varieties classified for General Use G, Special Use S	Throws M.S. G	Daffa S	Maris Beaver G
Agricultural Characters:			
Yield as % of Throws M.S.[a]	100	100	94
Shortness of straw	6	6	6
Earliness of ripening	6	6	6
Seed Characters:			
(greatly influenced by conditions of growth)			
1000 seed weight (ounces)	25	24	25
Crude protein % at 15% moisture	24	24	24

A high figure indicates that the variety shows the character to a high degree.

[a] Changes in the yields for 1970 reflect not only the addition of 1969 trial data, but also the decision to quote each year only mean trial yields for the previous five years.

VARIETIES FOR GENERAL USE

THROWS M.S.

A synthetic variety produced from four lines which are multiplied separately and combined before marketing. This gives the variety a certain amount of hybrid vigour, particularly in the first one to two years after the lines are brought together.

Hasler & Co. Ltd, Dunmow, Essex.

MARIS BEAVER

Selected from a Cambridgeshire local stock of winter beans. Provides a reliable source of seed for the type of winter bean commonly grown in the Eastern counties for many years.

Plant Breeding Institute, Cambridge.

VARIETY FOR SPECIAL USE

DAFFA

A variety bred in Wales for western conditions. Although results have been variable, it has tended to show good winter survival in the West, where it has outyielded Throws M.S. It has usually been inferior to Throws M.S. in the Eastern counties.

Welsh Plant Breeding Station, Aberystwyth.

SPRING BEANS

The two main types of spring beans, tic beans and horse beans, are characterised by their seed size and shape. All varieties currently recommended are of the tic type.

Tic beans are small (between 10 and 20 ounces/1000 seeds) and rounded. Some are sold for the pigeon trade at home and on the Continent and the remainder used for livestock feed.

Horse beans are similar in size to winter beans (20 to 30 ounces/1000 seeds) and are flatter and larger than tic beans. They are grown entirely for livestock feed.

RECOMMENDED LIST OF SPRING BEAN VARIETIES 1970

Yield comparisons other than with the control are not strictly valid, and differences of 4% or less should be treated with reserve.

VARIETIES FOR GENERAL USE

Varieties classified for General Use G, Special Use S	Maris Bead G	Tarvin G	Herz Freya S	Minor G	Blue Rock G	Francks Ackerperle G
Agricultural Characters:						
Yield as % of Minor[a]	104	103	101	100	97	98
Shortness of straw	7	5	8	6	5	7
Earliness of ripening	7	4	9	6	5	8
Seed Characters: (greatly influenced by conditions of growth) 1000 seed weight						
(ounces)	15	15	15	13	15	13
Crude protein % at 15% moisture	27	28	25	29	27	26

A high figure indicates that the variety shows the character to a high degree.

[a] Changes in the yields for 1970 reflect not only the addition of 1969 trial data, but also the decision to quote each year only mean trial yields for the previous five years.

MARIS BEAD

High yielding. Moderately early and short   Plant Breeding Institute, strawed. Seed size medium small.   Cambridge.

TARVIN

High yielding. Late maturing with long straw. Seed size medium small.

Gartons Ltd, Warrington, Lancs.

MINOR

Medium late maturing with medium straw length. Seeds are small and round. Basic seed from the breeder is not available in 1970

Gembloux, Belgium.

FRANCKS ACKERPERLE

An early maturing tic bean with short straw and small round seed.

Franck, Germany.

BLUE ROCK

A late maturing, long strawed variety with small round seed.

David Miln & Co., Ltd, Chester, Cheshire.

VARIETY FOR SPECIAL USE

HERZ FREYA

A very early variety with short straw. Susceptible to bean leaf roll virus and gives poor yields in Southern England in some seasons. Recommended for use in the North where early maturity is particularly important.

Herz, Germany.

CHOCOLATE SPOT

Although spring beans are usually less severely attacked by chocolate spot than are winter beans, infections can occasionally be severe. In 1968 Francks Ackerperle and Herz Freya were more severely infected than other spring varieties.

References

Bond, D. A. (1968). Private Communication.

Bond, D. A. (1969). "Field Beans—A Manual for Farmers and Advisors". Fison's Cambridge Division, Harston, Cambridge.

Bond, D. A. and Fyfe, J. L. (1962). Plant Breeding Institute, Cambridge. Annual Report, pp. 4–26.

Bond, D. A. and Hawkins, R. P. (1967). *J. agric. Sci. Camb.* **68,** 243–247.

Bond, D. A. and Toynbee-Clark, Gillian (1968). *J. agric. Sci. Camb.* **70,** 403–404.

Carpenter, K. J. and Johnson, C. L. (1968). *J. agric. Sci. Camb.* **70,** 391–392.

Cooke, G. W. (1964). "Fertilizers and Profitable Farming" (second edition). Crosby Lockwood & Son, Ltd., London.

Cooper, B. A. (1964). East Midlands N.A.A.S. Circular. ENT/5/1/5. M.A.F.F.

Cooper, B. A. (1967). East Midlands N.A.A.S. Circular. ENT/5/1/21 D. M.A.F.F.

Darlington, C. D. and Wylie, A. P. (1955). "Chromosome Atlas of Flowering Plants" (second edition). Allen and Unwin Ltd., London.

Drayner, Jean M. (1959). *J. agric. Sci. Camb.* **53,** 387–403.

East Midlands N.A.A.S. Circular (1968). "Protecting Beans from Blackfly". ENT/5/1/22 D. M.A.F.F.

Eden, A. (1968). *J. agric. Sci. Camb.* **70,** 299–301.

Evans, R. E. (1960). "Rations for Livestock". M.A.F.F. Bulletin No. 48. H.M.S.O.

Free, J. B. (1958). *Bee World* **39,** 221.

Free, J. B. (1960). *J. Anim. Ecol.* **29,** 385.

Free, J. B. (1962). *J. Anim. Ecol.* **31,** 497.

Free, J. B. (1965). *J. agric. Sci. Camb.* **64,** 167–168.

Free, J. B. (1966). *J. agric. Sci. Camb.* **66,** 395–398.

Fryer, J. D. and Evans, S. A. (1968). "Weed Control Handbook. Vol. 11: Recommendations", p. 37. Blackwell Scientific Publications, Oxford and Edinburgh.

Fyfe, J. L. and Bailey, N. T. J. (1951). *J. agric. Sci. Camb.* **41,** 371–382.

Gane, A. J. (1966). *Agriculture.* **73,** 250–254.

Gent, G. P. (1966). "Developments in Dwarf Beans". *The Grower* **66,** 5, 143–144.

Gill, N. T. and Vear, V. C. (1966). "Agricultural Botany". Duckworth and Co., London.

Greenwood, H. N. (1958). *J. Roy. agric. Soc. Lond.* **119,** 52–56.

Greenwood, H. N. (1959). *J. Roy. agric. Soc. Lond.* **120,** 70–77.

Hall, A. D. Sir (1945). "The Soil. An Introduction to the Scientific Study of the Growth of Crops", p. 119. John Murray, London.

Hirayoshi, I. and Matsumura, M. (1952). *Jap. J. Breed.* **1,** 219.

Hoare, A. H. (1949). "Vegetable Crops for Market and Processing". Crosby Lockwood and Son, Ltd., London.

Hodgson, G. L. and Blackman, G. E. (1956). *J. exp. Bot.* **7,** 147–165.

Hodgson, G. L. and Blackman, G. E. (1957). *J. exp. Bot.* **8,** 195–219.

Jones, F. G. W. and Jones, Margaret G. (1964). "Pests of Field Crops", p. 52. Edward Arnold Ltd., London.

Karpechenko, G. D. (1925). *Bull. appl. Bot. Pl. Breed.* **14,** 143.

McLeish, J. (1953). *Heredity.* **6** (Suppl.), 125.

M.A.F.F. (1962). Bulletin No. 87, "Beans". H.M.S.O.

M.A.F.F. (1966a). Short Term Leaflet No. 55, "The Use of Beans in Livestock Feeding". M.A.F.F., London.

M.A.F.F. (1966b). "Fumigation with the Liquid Fumigants Carbon Tetrachloride, Ethylene Dichloride and Ethylene Dibromide. Precautionary Measures 1966". H.M.S.O.

M.A.F.F. (1967a). Short Term Leaflet No. 60, "Field Beans." M.A.F.F., Pinner, Middlesex.

M.A.F.F. (1967b). "Horticulture in Britain. Part 1: Vegetables." H.M.S.O.

M.A.F.F. (1967c). Short Term Leaflet No. 71, "Irrigation Guide." M.A.F.F., London.

M.A.F.F. (1968). Advisory Leaflet No. 126, "Pea and Bean Beetles". M.A.F.F. Pinner, Middlesex.

M.A.F.F. (1968). Advisory Leaflet No. 61, "Pea and Bean Weevils".

Martin, H. (1965), (1969). "Insecticide and Fungicide Handbook for Crop Protection", second and third editions. Issued by the Brit. Insect and Fung. Council. Blackwell Scientific Publications, Oxford.

National Institute of Agricultural Botany. (1967). Forty-eighth Report and Accounts, 28.

National Institute of Agricultural Botany. (1968, 1970). Farmers' Leaflet No. 15, "Recommended Varieties of Field Beans".

Nix, J. (1968). "Farm Management Pocketbook". Department of Agricultural Economics, Wye College.

P.G.R.O. (1968). Pea Growing Research Organisation. Information Sheet No. 1, March 1968. Thornhaugh, Peterborough.

Picard, J. (1953). *Ann. Amel. Plantes.* **3,** 57.

Riedol, I. B. M. and Wort, D. A. (1960). *Ann. Applied Biol.* 48, 121–124.

Rowlands, D. G. (1955). *Agric. Progress* **30,** 137–47.

Rybin, V. A. (1939). *C.R. Acad. Sci., U.R.S.S.* **24,** 368, 483.

Scriven, W. A., Cooper, B. A. and Allen, H. (1961). *Outlook on Agriculture* **3**(2), 69–75.

Sirks, M. J. (1923). *Meded. Landhoogesch., Wageningen.* **26,** 40.

Small, J. (1946). "pH and Plants. An Introduction for Beginners", p. 146. Baillière, Tindall & Cox, London.

Smith, B. F. and Aldrich, D. T. A. (1967). *J. nat. Inst. agric. Bot.* **11,** 133–146.

Soper, M. H. R. (1952). *J. agric. Sci. Camb.* **42,** 335–346.

Soper, M. H. R. (1958). "Field Beans". M.A.F.F. Publication. H.M.S.O.

Thomas, P. T. (1945). Unpublished data quoted by Darlington, C. D. and Wylie, A. P. in "Chromosome Atlas of Flowering Plants" (2nd edition, 1955). Allen and Unwin, London.

Wallace, T. (1951). "The Diagnosis of Mineral Deficiencies in Plants by Visual Symptoms". Plates 180–187 inc. H.M.S.O.

CHAPTER 7

Peas

    I. General Introduction 327
   II. Taxonomy of Peas 330
  III. Soils and Climate for Peas 332

PART 1. PEAS FOR HARVESTING DRY

   IV. Introduction 333
       A. Major Production Areas 334
    V. Varieties 335
       A. Marrowfat Peas 335
       B. Blue Peas 336
   VI. Seed-bed Preparation 337
  VII. Manuring 337
       A. Seed Rates 339
 VIII. Seed Dressings 340
   IX. Drilling and Spatial Arrangements 341
    X. Time of Sowing 342
   XI. Weed Control 343
       A. Cultural 343
       B. Chemical 343
  XII. Control of Pests 347
       A. Wood Pigeon 348
       B. Pea Weevil 348
       C. Aphids 348
       D. Pea Moth 348
 XIII. Harvest 349
       A. Method 1 349
       B. Method 2 351
       C. Method 3 352
  XIV. Warning and Information on Desiccants 353
   XV. Quality of Samples 353
  XVI. Yield 356
 XVII. Production Economics 356

PART 2. CONTRACT VINING PEAS FOR CANNING, QUICK-FREEZING AND DEHYDRATION

XVIII. Introduction 357
  XIX. Major Production Areas. 358

XX. Varieties 358
XXI. Seed-bed Preparation 361
XXII. Manuring 362
XXIII. Seed Rate 362
XXIV. Seed Dressing 364
    A. Drilling, Row Widths and Time of Sowing 364
XXV. Time of Sowing 364
    A. Accumulated Heat Units (AHU) 364
XXVI. Chemical Weed Control 365
    A. Post-emergence Herbicides Relating to Peas to be Harvested .
    Green 365
XXVII. Control of Pests 366
XXVIII. Harvest 366
    A. Method 1 366
    B. Method 2 369
    C. Method 3 369
XXIX. Yields with Vining Peas for Freezing and Canning . . . 373
XXX. Economic Aspects of Vining Peas 373

PART 3. PEAS FOR MARKETING GREEN—"PULLING PEAS"

XXXI. Introduction 375
XXXII. Production Areas 376
    A. Size of Green Pea Crops 377
    B. Production of Green Peas 377
XXXIII. Sowing Dates 377
XXXIV. Varieties 378
XXXV. Harvesting and Marketing 378
XXXVI. Yield 378
XXXVII. Crop Protection 379
    A. Narcotic Bait 379
    References 380
    Addendum 382

I. GENERAL INTRODUCTION

Peas are leguminous plants of the sub-family Papilionoidaea and belong to the general class of Dicotyledons. Cultivated and wild forms are assumed to have come from countries of Southern Europe, probably bordering the Mediterranean, and Italy or the nearby islands have been suggested as possible points of origin (Bell, 1948; Percival, 1947). European cultivation goes back at least to the Bronze Age and peas became important on account of their "soil ameliorating properties". Their ability to fix the atmospheric nitrogen was only discovered towards the end of the nineteenth century, but this crop had already established itself in Britain as a soil improver. With the production and widespread use of cheap forms of inorganic nitrogen through straight or compound fertilisers, the nitrogen-fixing ability of peas, although of value, is not regarded as highly today as it was in the past. Nevertheless peas make a significant contribution to British agriculture. They are grown as a vegetable crop for the green market, for canning, quick-freezing and dehydration, as a pulse crop for human or stockfeed or as a component of arable silage mixtures.

Statistics collected by the Board of Agriculture (1905), indicated that quarter of a million acres of peas were being grown in the United Kingdom at the end of the nineteenth century and with average yields stipulated at 25 bushels/acre (14 cwt) one can assume that peas were grown mainly as a pulse crop, to be harvested dry for human and animal feed.

In the early 1950's, 30–40 thousand acres of stockfeed peas were being grown in the United Kingdom, although in practice the bulk of these would be found in England. Yields of only 0·5–0·75 ton/acre were being obtained and this fact alone was probably responsible for the rapid decline in popularity which was to follow. By the end of the 1950's only one-third of the early 1950 acreage was left and by 1963 so few were being cultivated that stockfeed peas ceased to merit individual attention in the June acreage statistics.

Similar trends could be observed in the dried pea crop grown for human consumption. In the early 1950's, 120–130 thousand acres were grown, producing up to 0·75 ton/acre, but in spite of a significant rise in the output, only 25% of the former acreage was left by the early 1960's. This was also against a background of increasing imports of dried peas which during that period carried an import tariff of 10% or 7/6d. per cwt, whichever was the higher.

During the 1950's the National Farmers' Union was campaigning for higher tariffs on imported peas in an attempt to raise the price and status of the home-grown crop, but unfortunately they were unsuccessful. By 1962, only 22,600 acres were being grown but since then there has been a steady increase in the amount under cultivation. Several factors have been responsible for this revival. Processing companies are now offering contracts to farmers for the supply of both marrowfat and blue peas and higher yields are being obtained through better varieties and improved growing techniques.

Peas grown for the green market have declined in acreage since 1950, but the drop has not been of the same magnitude as that in the groups previously described. The reduction in demand can be attributed mainly to higher living standards which in turn have meant that more people can afford the better quality frozen peas which do not need time in preparation. The shelling of fresh peas can be very time-consuming and where crops have been left too long in the field or samples have been in the shops for some time before purchase there is a big drop in quality. Pulling peas on a farm scale requires a large labour force (casual workers) at harvest and with the exception of farming areas on the fringe of large urban populations, like the West Riding of Yorkshire, these large labour squads are becoming increasingly difficult to muster. For this and other reasons the pulling pea acreage is unlikely to rise in the future.

Over the last one and a half decades, peas for canning and freezing have increased in importance. In spite of the acreage having trebled during this period, large quantities were being imported, notably from the Scandinavian countries and America.

Year	Total acreage (England, Wales and Scotland)	Estimated imports of frozen peas (cwt)
1956	51,500	4510
1957	55,900	97,663
1958	59,200	123,188
1959	66,600	202,659
1960	73,500	286,856

Farming Express (1961).

In England and Wales the increase in acreage has been steady but in Scotland a peak was reached in 1960 and since then the allocation to most farms holding viners has dropped slightly. In 1961 separate acreages were obtained for the canning and quick-freezing peas and in 1965 the acreage associated with dehydration was first published.

	1961	1962	1963	1964	1965
Peas, green for canning	33,707	29,900	35,294	36,010	31,402
Peas, green for quick-freezing	34,244	36,090	43,498	45,388	40,162
Peas, green for dehydration	N.R.	N.R.	N.R.	N.R.	7,431
Total	—	—	—	—	78,995
	1966	1967	1968	1969	1970
Total acreage for canning quick-freezing and dehydration	87,100	97,200	104,500	112,974	127,500

N.R. denotes no return in the statistics.

A new, improved dehydration process introduced in 1960 has in recent years led to a substantial acreage of peas being grown on contract for this purpose and explains the rise in the above figures, particularly 1963 compared with the previous year. Prior to 1965 peas for dehydration were included in the acreage designated "quick-freezing".

II. TAXONOMY OF PEAS

Class: Dicotyledon
    Family: Leguminosae
        Sub-family: Papilionoidaea
            Tribe: *Vicieae*
                Genus: *Pisum*   (Willis, 1966)
                Species: *abyssinicum*
                                      arvense
                                      elatius
                                      fulvum   (Darlington and
                                                      Wylie, 1945)
                                      sativum   (Bell, 1948)

Species of the genus *Pisum* have a basic chromosome number $X = 7$, and Darlington and Wylie (1945) list them all as diploids.

Species	Chromosome number	Reference	Comments
abyssinicum	14	Fedotov (1935)	Grown as a vegetable and pulse crop–Abyssinia
elatius	14	Fedotov (1935)	Grown as a vegetable and pulse crop–Mediterranean and South-west Asia
fulvum	14	Fedotov (1935)	Found in Asia Minor and Syria
sativum (Garden pea)	14	Sansome (1933)	Grown for fodder, as a vegetable and pulse crop

In Britain the cultivated forms of peas belong to either *arvense* or *sativum*. The former species includes the red and purple flowered forms which are grown principally as a pulse crop for stock feeding, whilst the latter species embraces all the white flowered and white or green seeded forms which are primarily used for culinary purposes (Bell, 1948). In the past "field peas" referred only to the arvense forms, since these alone were grown on a field scale whilst the sativa forms were known simply as "garden peas" being cultivated in market gardens,

TABLE 1

Types of Stockfeed and Culinary Peas

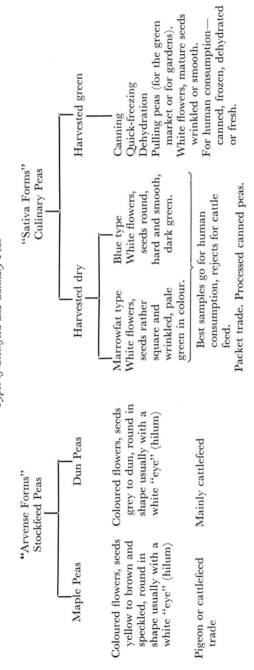

"Arvense Forms" Stockfeed Peas

Maple Peas
Coloured flowers, seeds yellow to brown and speckled, round in shape usually with a white "eye" (hilum)

Pigeon or cattlefeed trade

Dun Peas
Coloured flowers, seeds grey to dun, round in shape usually with a white "eye" (hilum)

Mainly cattlefeed

"Sativa Forms" Culinary Peas

Harvested dry

Marrowfat type
White flowers, seeds rather square and wrinkled, pale green in colour.

Blue type
White flowers, seeds round, hard and smooth, dark green.

Best samples go for human consumption, rejects for cattle feed.
Packet trade. Processed canned peas.

Harvested green

Canning
Quick-freezing
Dehydration
Pulling peas (for the green market or for gardens).
White flowers, mature seeds wrinkled or smooth.
For human consumption—canned, frozen, dehydrated or fresh.

allotments and private gardens but not on a farm scale. Today there are very few "field peas" under cultivation whilst the "garden pea" is probably the most important vegetable crop associated with Agriculture.

It is now customary to use *Pisum sativum* for all edible forms of peas and to differentiate between "garden" and "field types", a sub-species classification has been adopted. Garden peas are recognised as cultivars of *P. sativum hortense* and varieties of field peas as *P. sativum arvense.*

Although there have been references from time to time about winter peas, the majority if not all of the present cultivars are not winter hardy and the crop is normally spring sown. All forms can be described as weak-strawed annuals which climb by means of their leaf tendrils in a manner similar to the vetch plant.

The principal types of peas are listed and briefly described in Table 1 and where significant acreages are involved in present-day farming, production details will be considered at length later in the chapter.

Stockfeed peas are now of historical importance only and it is therefore proposed to concentrate on the various forms grown for human consumption in three parts, namely Part 1—Peas for harvesting dry; Part 2—Contract peas (Canning, quick-freezing and dehydration) and Part 3—Pulling peas for the green market.

The requirements of soil and climate are almost the same for each of these groups and thus consideration will be given to these aspects once at the beginning of this chapter and crop protection will be dealt with briefly at the end.

III. SOILS AND CLIMATE FOR PEAS

The most suitable soil type for peas is a medium calcareous loam. Very light soils, provided they are not acid, will grow early crops of peas, but except in years of higher than normal rainfall, the yields will tend to be below average. Heavy soils in general are not suited to pea growing on two accounts. Firstly, they generally hold more water and thus tend to be late and secondly the deep tilth required by this legume is often impossible to obtain. If these heavier soils are really well drained, with a high lime content and a good soil structure then peas can be grown successfully. On the light to medium soils which facilitate the growing of this legume, the fertility or organic status should be high. Although this crop is a nitrogen fixer, through its association with rhizobial bacteria, the initial establishment and early growth is better if the peas can obtain some of their nitrogen nutrition from the soil.

The average optimum pH value for peas has been assessed at 6 by

Small (1946) having consulted many of the early authorities on this subject. The range 5·5–6·5 was suggested by the same author although in light of farm practices and since Trénel (1927) reported injury at or below 6, it would be wise to consider the range 6–7 as optimal, since considerable variation can occur within one field. Olsen (1925) reported some alka-tolerant varieties of peas, pH 6·9–7·5, and there is no doubt that they can be grown on chalky or limestone soils in Britain although "lime-induced chlorosis" may occur through trace element deficiencies.

In general the drier, sunnier regions of Britain are preferred for pea growing. Dull wet conditions promote excessive haulm growth resulting in low yields. Quality in respect of peas harvested dry or green is poor and harvest of both groups can be difficult and costly. Peas to be harvested green can be grown successfully in the medium rainfall districts, but when the crop is for harvesting dry, it must be confined to the lower rainfall areas, to ensure the best possible chance of good harvesting conditions. As a result the yields obtained are variable and quite often limited because of lack of moisture rather than any other single factor.

PART 1. PEAS FOR HARVESTING DRY

IV. INTRODUCTION

Dried peas can be obtained loose or in packet form by housewives to be reconstituted as a vegetable or for use in soups. A large proportion of the production is sold to processing companies who market processed canned peas and significant quantities are used in catering establishments and fish and chip shops. According to the Ministry of Agriculture, Fisheries and Food (1967), the pre-1939 quantities used and the amount home-produced were as follows:

Sold loose and used in catering establishments 50,000 tons	} 70,000 tons	15,000 tons home produced	
Used for canning 20,000 tons		55,000 tons imported	

Before the war the greater part used consisted of blues in spite of three-quarters of the home production being marrowfats. The greater popularity of the marrowfats has been ascribed to the fact that the main variety of this type of pea (Harrison's Glory) could be picked green and sold for the greengrocery trade when the demand existed.

During the Second World War, importation of dried peas ceased and to stimulate home production the government provided farmers with a guaranteed price and market for their crops. This resulted in a rapid acreage increase from 19,000 acres pre-1939 to a record figure of 180,000 acres in 1948 and at that time over 90% of the acreage was on contract to industrial processors. 1949 was a very good year with record yields and production outstripped demand and the government had to pay for the surplus. In June, 1950, the guaranteed prices for peas was removed and during the next twelve years the acreage dropped steadily until in 1962 it had fallen to about 22,500 acres which was similar to the pre-war figure. Between 1950 and 1962 contracts for dried peas were drastically reduced but since the mid 1960's they have been reintroduced by the larger processing firms. During this period of decline in the 1950's and early 1960's imports fluctuated considerably but the general trend was for them to increase slightly. Since the mid 1960's, advances in husbandry techniques, better varieties and increased confidence through definite contracts have all led to increased interest in dried peas.

Dried Pea Acreage In England and Wales—Late 1960's

(Total of marrowfats and blues)

Year	1966	1967	1968	1969	1970
Acreage	35,200	39,900	49,500	55,800	73,300

International Commitments and Agreements in Respect of Dried Peas

1. *G.A.T.T.* There are no bindings on dried peas under the General Agreement on Tariffs and Trade.

2. *Commonwealth agreements.* There is a margin of preference of 10% guaranteed to Australia and New Zealand.

3. *E.F.T.A.* (European Free Trade Association). Since dried peas are included in Annex D of the Stockholm Convention, they are not subject to the progressive removal within the Association of tariffs and quantitative restrictions which applies to commodities in the industrial field.

A. Major Production Areas

The dried pea acreage is now associated with the intensive arable counties of Eastern England. The three major areas of production in order of importance are: 1, East Anglia; 2, Lincolnshire; 3, East Riding of Yorkshire. With much of the present acreage concentrated in East Anglia, average yields have shown significant improvement as can be seen by the following figures which have been taken from the Ministry's statistics.

	Marrowfats	Blues
Average yield 1955–59 (incl.) cwt/acre	15·8	16·8
Average yield 1960–64 (incl.) cwt/acre	22·5	25·4

V. VARIETIES

Peas for human consumption which are to be harvested when the seed is ripe are described in many ways. The terms "drying", "dried", "harvesting" or "threshed" are commonly used to signify this crop.

In the past three types were recognised, namely Marrowfats, Large or Dutch Blues and the Lincolnshire Small Blues but the present acreage consists only of Marrowfats and Large Blues.

A. Marrowfat Peas

The flowers and seeds of each of these two groups have been described briefly in Table 1, and the attractive pale green, rather square-shaped marrowfats find a ready market for the packet trade once the split and damaged ones have been removed. The most important variety was Harrison's Glory and present-day cultivars are either direct descendants of this one or are said to be "Harrison's Glory type". The following bifocated list of marrowfat varieties groups them roughly into past and present varieties.

Varieties of Marrowfat Peas

Older varieties, some of which are outclassed	Present cultivars and recent introductions
Emigrant	Big Ben
Harrison's Glory	Maro
Zelka	Green Golt (previously known as No. 56/773)
	No. 814

In the past marrowfats have produced larger grains of higher quality than the large blues, but they have not possessed such good field characters. Straw (haulm) lengths of 2·5 feet on average have been some 3–6 inches longer than the blues and thus there was a much greater tendency to lodge particularly where the soils contained significant nitrogen reserves. The marrowfats have usually been outyielded by the blues to the extent of 1–3 cwt/acre. In spite of this yield disadvantage, marrowfats are more popular because they command a higher price on the open market or through contracts with commercial companies and they still occupy about 80% of the dried pea acreages as the following figures show.

	1963	1964	1965
Acres of marrowfat peas	25,330	26,434	27,543
Acres of blue peas	8504	5569	6446
Total	33,834	32,003	33,989
Marrowfats as % total	75%	83%	81%

B. Blue Peas

Processing companies are the main outlet for this type of dried pea and the highest quality samples are chosen for reconstituting into processed canned peas. Higher yields and better field characters make these more acceptable to the grower but prices are always lower and thus the acreage is only small compared with the marrowfats.

Varieties of the Large Blue Type of Dried Peas

Older varieties, some of which are becoming outclassed	Present cultivars and recent introductions	
Mansholt's Gek	Alaska	Cobri
Servo	Dik Trom	Faceta
Stijfstro	Pauli	Hylgro
Unica	Rondo	Jo 06638
	Vedette	Octavus
		Riito
		No. 815

Variety trials with both types of dried peas were originally conducted and organised by the Home-grown Threshed Peas Joint Committee (set up in 1944) and later by the Pea Growing Research Organisation (1956). Results from 39 trials over a 4-year period have been summarised by Reynolds (1958) and these are reproduced in Table 2 below and additional information can be obtained from an earlier paper on this subject (Reynolds, 1957).

The best variety of blue pea outyields the highest marrowfat by over 2 cwt/acre, confirming the earlier yield estimates. The present variety testing of dried pea varieties is mainly conducted by the P.G.R.O. whose permanent headquarters and trial grounds are at Thornhaugh, nr. Peterborough.

TABLE 2

Variety Trial: Yields of Dried Peas (cwt/acre at 16% moisture)

Variety and type	Yield (cwt/acre)	Yield expressed as percentage Servo
Servo (B)	25·3	100
Rondo (B)	24·1	95
Unica (B)	23·4	92
Emigrant (MF)	23·1	91
Stijfstro (B)	23·0	91
Mansholt's Gek (B)	22·5	89
Zelka (MF)	21·5	85

VI. SEED-BED PREPARATION

Early spring drilling of the dried pea crop is essential and thus ploughing in the autumn and winter in the previous year will be necessary for all soils except the very light sands which can virtually be turned over at any time. Advantage should be taken of the frost mould which is usually left and cultivations in late February and early March should be designed to produce a deep even tilth. Where pH values are well below the optimum early application of lime is necessary. If this is applied in the late autumn or winter, the rain will probably have incorporated it into the soil, but should it be put on in early spring, just before drilling, then it must be properly worked into the top 4–6 inches by cultivations.

VII. MANURING

From field experiments it has been shown that dried peas do not normally benefit from the applications of fertiliser nitrogen. When applied it is usually responsible for greater straw length which is simply an embarrassment, causing the crop to lodge badly. The only exceptions to this general rule are when the winters have been exceptionally wet, thus leaching most of the soil nitrogen, or when the soil nitrogen status (via organic matter) is extremely low, or for exceptionally early drillings when a small quantity of nitrate nitrogen could be beneficial. Manuring will normally involve only phosphate and potash of the major nutrients and compound fertilisers containing these two should be chosen. Peas are said to respond to applications of potash more frequently than to phosphate and compound fertilisers with a phosphate : potash balance of 1:1·5 or 1:2 are usually suitable. Cooke's recommendations from "Fertilisers and Profitable Farming" (1964) are as follows:

TABLE 3

Recommended Levels of Phosphate and Potash for Dried Peas, Related to Soil Analysis (units/acre)

| | P and K status as shown by soil analysis | | |
	Low	Medium	High
Phosphate	45	20	0
Potash	55	35	0

These rates are suggested for side-band placement, pre-drilling or combine drilling although about 2 cwt/acre of a P/K compound fertiliser is about the maximum which can be combine-drilled both from practical and safety points of view. The highest levels of potash suggested above may cause a reduction in the field germination if combine drilled. If the fertiliser is to be broadcast, then double the above recommendations would have to be applied to obtain the same increase in yield and since the response is very small it is doubtful whether this practice would show a profit. Broadcast applications are however worthwhile when the soil phosphate and potash status is low on soil analysis and when the crop is to be drilled on narrow rows.

When more than 2 cwt/acre of a phosphate–potash compound is recommended it is best applied by means of a placement drill which deposits the fertiliser about 2 inches to the side and 1 inch below the seed through separate coulters. This is seldom satisfactory at coulter spacings of less than 10–12 inches and under these circumstances the fertiliser should be combine-drilled, if needed in small quantities or broadcast and worked into the seed-bed before drilling when the rates of potash may seriously injure germination.

The potash status of the soil has been suggested as being a useful guide to the amount of fertiliser which can be recommended for peas (M.A.F.F., 1960).

With such a poor yield response to fertilisers, they should be omitted altogether when the soil levels of phosphate and potash are good and also when dried peas immediately follow crops which have been heavily manured. Yield responses to potash at varying "fertility" levels were reported by the P.G.R.O. (1963) and the results from 25 experiments are summarised in Table 4. In spite of the poor responses

Soil potash status	Fertiliser recommendation (cwt/acre 0% N: 10% P_2O_5: 20% K_2O compound or equivalent placed to the side and below the seed)
Medium	2 cwt
Low	4 cwt
Very low	6 cwt
	(or 1 cwt muriate broadcast + 3 cwt fertiliser placed)

indicated below, potash remains the most important major nutrient and as far as phosphate is concerned, the experiments have shown that

TABLE 4

Yield Response to Potash at Varying Soil Fertility Levels

Readily available potash as shown by soil analysis	Number of experiments	Average increase or decrease in yield from broadcasting 2 cwt of muriate of potash/ acre (120 units K_2O)
Low	7	+2·1 cwt/acre
Medium	5	+0·3 cwt/acre
High	13	−0·1 cwt/acre

the soil must be acutely deficient before any measurable response in yield to broadcast superphosphate can be demonstrated.

A. Seed Rates

Seed rates pre-1939 were high, in the region of 2·25 cwt/acre (18 stones) but in those days field losses were higher than they are today and thus a larger quantity was needed to produce the same braird. Nowadays 1·75–2 cwt/acre (14–16 stone) is considered sufficient for the large-seeded blues and marrowfats and 1·5 cwt for the small blues. Owers and Yule (1956) reported early trials concerning many aspects of dried pea growing and amongst this report were brief details of seed-rate experiments (Table 5).

Seed costs are high with all types of pea growing, but even if the value /cwt is 0·5 or 0·3 of the cost of seed/cwt, these figures show an economic return at the highest rate—2 cwt/acre. For a number of practical reasons associated with seed losses in the field it is not worthwhile reducing seed rates and farmers who have done so in the past are convinced that it is a "false economy". This has been confirmed from field experiments and even when all reasonable precautions have been taken to minimise losses prior to brairding, 20–25% of the seed sown fails to produce plants. Recent experiments with vining peas also confirm the necessity for high seed rates and even at about £100/ton,

TABLE 5

The Effect of Seed Rate on Yield with Unica and Zelka Dried Peas (Mean of three years, 1950–52 incl.)

Seed rate	Yield as percentage 1 cwt seed rate
1 cwt/acre	100 (28·7 cwt/acre)
1·5 cwt/acre	112
2 cwt/acre	120

seed rates of 16–18 stones can be justified over rates between 8 and 16 stones/acre.

In recent years the P.G.R.O. has been concerned indirectly with seed rate experiments with peas through their research into optimum spatial arrangement. When testing various row widths and intra-row spacing this is bound to involve seed rate variations and the most recent information on this subject can always be obtained from this organisation directly or through their publications.

VIII. SEED DRESSINGS

Investigations in the 1950's into the use of seed dressings for peas showed that by using the more effective chemicals, a better and stronger braird could be obtained which in turn would produce higher yields (Home-Grown Threshed Peas Joint Committee Report, 1950–51; Blair and Copp, 1953 and Blumer and Harder, 1953). Owers and Yule (1956), however, were unable to obtain consistently better results with thiram, thiram plus gamma-BHC, 1% organo-mercury or TCBQ seed dressings compared with undressed controls over a period of 4 years. It was pointed out that these were tested on peas sown at the optimum date in good seed-bed conditions and that the seed dressing should be considered as added insurance against failure. Had they been tested with early sown peas, in far from ideal circumstances, then the results could well have shown significant advantages in field establishment and yield. It is under these conditions that seed dressings for peas are universally recommended. When soil temperatures are low in the early spring, soil-borne pathogens are likely to be much more damaging to the initial growth thus lowering considerably the percentage field germination. The present recommended dressings are based on the fungicide Captan and one of the commercial preparations, known as "Orthocide Pea Seed Dressing" is marketed by Murphy and appears in the 1969 M.A.F.F. Agricultural Chemicals Approval Scheme.

IX. DRILLING AND SPATIAL ARRANGEMENTS

In order to escape serious damage from pests, peas should be drilled at 1·5–2 inches below the surface and protection will often be required after drilling until they have fully brairded. Shallower planting than this will usually result in significantly higher incidences of bird damage and when they are planted much below 2 inches, the germination and field establishment will fall. Where the levels of phosphate and potash fertilisers to be used are low, they may be combine-drilled but it is safer to drill the seed through one coulter and place the fertiliser to the side and below, down another at medium and high application rates.

In the past most of the peas were grown on row widths about double that of the normal cereal coulter spacing and many were put in with ordinary drills where two coulter tubes were channelled into one shoe. At this distance (12–14 inches), fertiliser placement is possible and cultural weed-control through one or two inter-row cultivations can also be employed. Chemical weed control can now be employed with a high degree of success and since the dried pea crop is grown in the driest areas of the country, farmers are often pleased to be able to dispense with cultivations which tend to reduce the available soil moisture.

Narrow row widths are being employed and the P.G.R.O. is engaged in trials to ascertain the optimum spatial arrangement within these closer spaced rows, which tend to produce higher yields. Eight-inch row widths with intra-row spaces of 1·5, 2, 2·5 and 3- and 4-inch row widths with intra-row spaces of 3, 4, 5 and 6 inches are being investigated, and although mostly associated vining peas, the results may apply equally well to the dried crop in principle. From the limited evidence of these experiments where yield has been assessed as the weight of dried rather than green peas, the widest spacings 8 × 3 inches and 4 × 6 inches have resulted in the poorest yields and again the concept of high seeding rates for maximum yield is confirmed.

Where peas were grown in wide rows, comparisons involving tractor hoeing or post-emergence Dinoseb spraying for weed control showed

M

that similar yield increases could be obtained which implied similar effectiveness in the elimination of weed competition (Proctor *et al.*, 1955). The pre-emergence herbicide Prometryne gives excellent weed control and its adoption into current practice has been fairly rapid. Thus a considerable amount of the tractor wheel damage can be eliminated if pre-emergence herbicides are used and advantage can and should be taken of the yield increase using narrow rows of 4 or 8 inches. The difference in yield between 4- and 8-inch rows is not significant provided that the intra-row spacing on the 8- inch rows is halved to maintain high seeding rates. Thus within the limits set by high seeding rate and relatively narrow row widths the spatial arrangement does not appear to be important.

Other advantages with dried crops on narrow rather than wide rows, suggested by Gane (1962) are as follows:

1. More even maturation.
2. More even size of pea.
3. Better degree of cutting efficiency at harvest.
4. Greater competition against weeds.

X. TIME OF SOWING

Extensive field trials carried out over a period of 8 years, involving 60 experiments, were reported and discussed by Proctor (1957) and they emphasize the value of early sowing to obtain maximum yield with dried peas. General recommendations to emerge from these investigations are for sowing to take place as early in March as possible under normal circumstances, but when soil conditions will allow a mid to late February start, this opportunity should not be missed. Damage by frost to the common cultivars is negligible (Reynolds, 1957) and when the seed is protected against soil borne diseases these early sowings can maximise the profit from this crop.

M.A.F.F. (1960) recommendations in respect of sowing time are similar to those suggested by Proctor, namely a mid February start where conditions permit, but in North Lincolnshire and Yorkshire it would be better to wait until early March. Each week's delay in sowing after the first week of March will normally lower the final yield by 1 cwt/acre and early sown peas suffer less from pests and will normally mature under better weather conditions. They do better on land which has been ploughed early and in this connection it is said that they prefer a stale furrow. Land which has had plenty of time to weather over the winter will only need the minimal cultivation and where the surface has been left flat with a good frost mould early sowings can occasionally take place directly onto the furrow. The data below in Table 6 emphasise the value of this early drilling and confirms the P.G.R.O. results reported earlier.

TABLE 6

Early, Mid and Late Sowings with Dried Peas—Comparative Yields

Late February/Early March Sowing	Late March Sowing	Mid April Sowing
115%	100%	75%

XI.   WEED CONTROL

A.   Cultural

Peas do not exhibit vigorous initial development and thus they can easily be dominated by many fast growing weed species and hence in the past they have often been termed a "dirty" crop. Row widths of 12–14 inches have made it possible to inter-row cultivate immediately after brairding and once again before they have reached a stage when mechanical damage may result. These cultivations can be useful for the control of germinating annual dicotyledonous weeds between the rows, but within the row itself weeds can develop and seriously compete with the crop and some of the perennials and difficult grass weeds are not completely brought under control. It is likely that available soil moisture is reduced by inter-row cultivation, and for higher yields the evidence is in favour of narrow rows and therefore chemical weed control, although often more costly, is generally preferred.

B.   Chemical

With such a large volume of research currently in progress concerned with herbicides for peas, present recommendations, although worthy of note, can quickly be superceded by more effective ones and for the most up-to-date information on this rapidly changing subject readers particularly interested in peas should consult the following literature or members of their local Agricultural Advisory Service.

"Weed Control Handbook". Issued by the British Crop Protection Council (J. D. Fryer and S. A. Evans, eds); Volume I Principles; Volume II Recommendations.

Ministry of Agriculture, Fisheries and Food Advisory Leaflet No. 376. "Weed Control in Peas".

Pea Growing Research Organisation Ltd. Reports: Annual Report, Progress Reports, Technical Memoranda and Miscellaneous Publications.

(i) PRE-SOWING TREATMENTS FOR THE CONTROL OF WILD OATS (*Avena fatua*)

Wild oats have become one of the most important weeds of arable land in the low rainfall areas and they can seriously reduce the yield of peas. The chemicals TCA, Propham or Tri-allate can be used to control them. These three herbicides are worked into the seed-bed, pre-drilling, in order to kill the wild oats as they germinate, and the degree of control is usually about 80%, but on occasions with Tri-allate it may rise to as high as 95% or more (M.A.F.F., 1968).

TCA (Sodium or ammonium trichloracetate) is applied at 7 lb/acre, in medium to high volume onto a level seed-bed about a fortnight before drilling. An alternative method to spraying is to apply this herbicide mixed with the fertiliser which the peas would normally receive, but this mixing requires a special operation and should not be done on the farm (Fryer and Evans, 1968). The application of TCA will usually take place towards the end of February and to obtain the best kill of wild oats, it should be well incorporated into the top 4 inches of soil. The control of wild oats is good but since it is more damaging to the crop than the other two herbicides, TCA is now dropping out of favour. There is also evidence to show that on the highly organic soils this treatment is less effective and it reduces the waxy bloom associated with leaf surface of the peas. This makes them more susceptible to damage from applications of Dinoseb compounds which are used to control the broad-leaved weeds post-emergence. Only half the normal dose of Dinoseb-ammonium or Dinoseb-amine can safely be used (see later on in the text), following TCA applications.

Propham, although not quite as reliable in controlling wild oats, is safer to use than TCA, particularly for early drilled peas. In the past it has been recommended at 2–4·5 lb/acre, the smallest dose for light soils and the highest level for the heavy soils (M.A.F.F., 1968). Current advice in the "Weed Control Handbook" (1968) is for 3 lb/acre for medium to heavy soils, to be applied in medium to high volume, to a level seed-bed, at least 5 days before drilling. Propham is a volatile chemical and therefore must be worked into the soil immediately after spraying. It has the added advantage of killing a number of broad-leaved weeds as well as wild oats, i.e. chickweed, knotgrass and redshank but is not suitable for soils with a high organic status. The time of application of Propham is critical. If applied too early, much will be lost before the wild oats germinate and once these weeds have emerged, then they cannot be controlled with this herbicide. In many districts spraying during the first week in March will enable Propham to deal satisfactorily with the main flush of germinating wild oats.

Tri-allate is considered to be the most effective herbicide in controlling germinating wild oats and has the least effect on the pea crop. Like Propham, it is volatile and should be incorporated into the top few inches of the soil within an hour of application. Special instructions are issued by the manufacturers relating to the actual depth of mixing and depth of drilling of the peas and both of these must be noted. Spring germinating wild oats and blackgrass (*Alopecurus myosuroides*) are both effectively dealt with by Tri-allate provided it is very efficiently incorporated into the seed-bed. It is recommended at 1·25–1·5 lb/acre, in medium to high volume, for application to fine seed-beds 2–21 days before the crop is drilled. Sooner than 2 days before and more than 21 days before drilling are equally unsatisfactory and Tri-allate has given poor results on the highly organic soils (Fryer and Evans, 1968).

(ii) PRE-EMERGENCE WEED-KILLERS

Contact. There are several contact pre-emergence herbicides which can be used for weed control in peas where the weeds have emerged before the crop. It is unlikely, but not impossible, for this position to be encountered in dried pea growing, but when it occurs one of the following contact pre-emergence herbicides can be selected for weed-control.

    Diquat
    Paraquat
    Dimexan
    Cresylic acid formulations
    Pentachlorphenol formulations.

Contact plus some residual effect. Dinoseb-amine can be used as a contact pre-emergence spray up to a few days before crop emergence and when applied at 3–4 lb/acre there will be considerable residual effect. This type of spraying is not popular with farmers on two accounts. Firstly, it is more expensive than post-emergence dinoseb spraying and secondly the residual effect is not sufficiently lasting to give protection throughout the season. Recently, Dinoseb in oil has become available and at 2·25 lb/acre it is less expensive and should be used where contact pre-emergence spraying can be justified.

(iii) RESIDUAL HERBICIDES FOR PEAS

The following residual pre-emergence herbicides are recommended for the control of annual weeds:

Prometryne
Chlopropham plus Fenuron
Chlorpropham plus Diuron

Prometryne is one of the most useful herbicides for peas and it has been adopted quickly and widely in farming. Unfortunately it cannot be recommended for use on the open-textured sandy soils because of crop damage, but on medium and heavy soils 1·25–1·5 lb/acre should be applied, in medium to high volume, at any time after sowing up to 3 days before the crop is expected to emerge (Fryer and Evans, 1968). An added advantage with Prometryne lies in the fact that some contact kill is also obtained and weed seedlings appearing at the time of spraying are controlled as well as those which germinate later. Like all residual herbicides, its efficiency increases as the seed-bed gets finer and under lumpy, cloddy conditions much of its action is lost and little residual effect remains on highly organic soils.

Chlorpropham 1 lb plus Fenuron 0·25 lb/acre or *Chlorpropham 1 lb plus Diuron 0·2–0·4 lb/acre* applied in medium to high volume are recommended mixtures for the control of annual weeds. Spraying should take place after drilling and at least a week before the crop emerges. For good results the weed seedlings should not have appeared at the time of spraying. Again fine seed-beds are essential and these herbicides cannot be employed on the very light sandy soils or those without a reasonable clay or organic content.

(iv) POST-EMERGENCE CONTROL OF WILD OATS

The herbicide Barban can be used as a post-emergence spray for the control of heavy infestations of wild oats which occur unexpectedly. For best results this difficult grass weed should be at the 1–2·5 leaf stage at the time of spraying and the relatively high dosage rate of 7·5–10 oz/acre should be applied in low volume. This treatment is usually less effective than a pre-sowing treatment of Tri-allate which should be employed when it is known that land destined for peas has a previous history of wild oats. Peas drilled on narrow rows exert significant competition after spraying and it is generally agreed that Barban produces better results under these circumstances compared with crops sown on wide rows. Some scorch to the peas will be noted following spraying.

This will soon disappear but it would be wise to delay further spraying with dinoseb compounds until the crop has shown a full recovery.

(v) POST-EMERGENCE CONTROL OF ANNUAL WEEDS

TABLE 7

Chemical	Rate of application (active ingredient /acre)	Volume	Comments
MCPB salt	2 lb	Low-medium	More effective than Dinoseb compounds with some diffi-cult weed-like thistles and fat hen. Crop should have 3–6 expanded leaves. Cheap but not as good overall perform-ance as Dinoseb compounds. Some varieties are susceptible to damage—consult most up-to-date "Weed Control Hand-book".
Dinoseb-amine	1·5–2·5 lb	At least 40 gal/acre	For use when air temperature is not below 7°C (44·6°F) and when maximum for the day will not exceed 30°C (86°F). Safe on dried peas and *some* varieties for vining and pull-ing. Crop stage—1st expanded leaf until 10 inches height reached.
Dinoseb-ammonium	2 lb	High	For use when air temperature is not below 13°C (55·4°F) and the maximum for the day will not exceed 27°C (80·6°F). Dried pea crops only. Crop growth stage—2 and 6 inches.
Dinoseb-acetate	2·5 lb	40 gal/acre	More selective than Dinoseb-amine and should be chosen for those varieties of vining and pulling peas which are susceptible to Dinoseb-amine.

After Fryer and Evans (1968).

XII. CONTROL OF PESTS

A full list of pea pests can be obtained from Jones and Jones (1964) but the most serious ones are included at this stage to maintain as far as possible the continuity of operation sequence in pea growing.

A. Wood Pigeon

One of the most serious pests of nearly all pea crops is the wood pigeon. Attacks can begin at one end of a field whilst drilling is still in progress at the other and with large pigeon populations in many districts, reductions to brairds can be both rapid and severe. Flocks of homing pigeons and semi-wild groups can often be responsible for the initial attacks on pea crops and the wood pigeons' attention is focused on this acceptable source of food. It is important to begin crop protection early where pigeons are concerned and the use of carbide-operated bird scarers, scarecrows, mechanical scaring devices and periodic shooting expeditions can be employed. The examination of crops of shot birds have revealed up to 200 peas and thus when large numbers descend upon fields of newly sown peas, the reductions in the braird can be very severe.

B. Pea Weevil

The pea and bean weevil is the first insect pest to trouble pea crops. The adults feed on the leaves of newly emerged plants carving out the characteristic U-shaped notches and it is now known that the larvae feed on the roots and any early formed nodules. Fast growing crops do not usually succumb to this pest, but should the adult weevil numbers be large, resulting in nearly all the plants being affected, and growth is slow, then control measures should be used. Dusting or spraying with DDT or dusting with gamma-BHC all give excellent control of the pea and bean weevil. Derris is another approved chemical for the control of this pest.

C. Aphids

Peas may often be attacked by aphids, commonly called greenfly, and when the weather is hot and dry the reproduction rate is extremely high and resulting damage to the crop can be severe. In respect of the dried pea crop, should there be a colony of aphids on more than one plant in five before the pods are fully formed, insecticidal sprays should be employed. DDT or one of the many systemic insecticides can be used and control is usually obtained speedily and efficiently.

D. Pea Moth

Damage caused by the pea moth larvae can be troublesome in many pea growing districts and in these areas it is usual to consider DDT spraying as a routine measure. This treatment should be put into effect when the most forward headland pods are three-quarters grown in the early sown crops or 7–10 days after the beginning of flowering with late drillings (P.G.R.O., 1962). The local advisory services in pea growing areas often operate a warning system in respect of this pest of pea crops and when this is available due regard must be taken and recommendations adhered to where at all possible (M.A.F.F., 1968).

XIII. HARVEST

Most of the marrowfat varieties and some of the blues tend to have a very sprawling habit of growth and thus the dried pea crop is not the easiest to be harvested. In the past virtually the whole of the dried pea acreage was cut, tripoded and threshed out of the tripods when ready, but nowadays with the help of crop desiccants, direct combining is a practical and attractive method for harvest and is gaining in popularity with farmers and growers.

A. Method 1

Cut with pea-cutter, crop elevated to tripods, four-pole or hut-racks (Fig. 1) and threshed by combine

Dried peas are ready for cutting when the seed is easily removed from the pod and the seed stalk is left behind. The crop at this stage will have

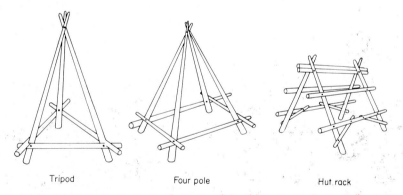

Tripod   Four pole   Hut rack

Fig. 1. Tripod, four-pole and hut rack used for pea drying prior to combining.

lost much of its green look and the pods will appear buff or parchment coloured on the outside. Specially designed pea-cutters are used which lift the haulm off the ground and over the cutter-bar and the severed plants fall onto a moving canvas (similar to that on a binder) to be deposited at the side in a narrow windrow. These (p.t.o. driven)

Fig. 2. The McBain Pea Cutter-Rower, illustrating the tractor coupling. Photograph by courtesy of P.G.R.O. Ltd.

Fig. 3. Leverton Pea Cutter. Photograph by courtesy of P.G.R.O. Ltd.

machines are carried on the three point hydraulic linkage of tractors and operate when the tractor is driven backwards, which means that the crop is cut and windrowed without wheel damage. Several pea-cutters are available and Figs 2 and 3 illustrate two of these rear-mounted backwards driven machines. The windrows are left on the ground for a short period to facilitate drying but it cannot be over-emphasised that whilst on the ground the pea crop is most susceptible to damage and loss through bad weather. The peas are loaded onto tripods, four-poles and hut-racks, in order that they will dry out and be relatively safe from bad weather damage. These are illustrated in Fig. 1 and it must be stressed that tripods are the least satisfactory and hut-racks the most efficient in drying-out and protecting the crop. Tripods without cross members towards the ground leave a large amount of the crop on the ground and with hut-racks, the smallest amount is touching the soil and the wind is able to blow through and thus drying will be most rapid. The number of "loading" or "cocking" units will depend on the size of the crop, but as a guide, using 6 ft 8 inch four-poles, approximately 16–20 sets are recommended per acre and 25 sets per acre will be required when hut-racks are employed, since they are smaller and will not carry the same weight of crop (M.A.F.F., 1960). Besides facilitating the quickest drying, hut-racks can be loaded more quickly from the windrows and they can be used when the crop is much damper, thus the danger period is usually shorter. When the crop has dried out sufficiently for threshing to proceed, combine harvesters are brought to the field and moved from set to set and the crop is hand forked into them. It is best to use a combine harvester in this manner to minimise seed losses through shattering; moving the peas to a stationary threshing mill can result in substantial losses unless extreme care is taken and providing that only a small proportion of the pods have split open.

B.   Method 2

Cutting with pea-cutter, and combine-harvesting from the swath

During exceptionally good harvesting weather, when it is hot and sunny, it is possible to use combine harvesters on the windrows. This method would be more widely used if the weather in Britain could be relied upon, as it is much quicker and less costly. One of the recognised pea-cutting machines is used to cut and windrow the crop and it is usually turned gently, often by hand, to get the produce dry with the minimum loss due to shattering. Combine-harvesting the windrows can go ahead once the haulm is dry, using pick-up reel attachments or the ordinary metal tined reel.

C. Method 3

Crop desiccated and direct combined

In recent years, the introduction of chemical desiccants to aid harvesting with a number of crops has led to quicker and easier handling methods. By using them on dried peas the crop moisture content is rapidly reduced and leaves and stems in much smaller quantities are now no longer a problem in the threshing process. Desiccants are also useful in drastically reducing the weeds which are present in the crop at harvest and which often cause harvesting difficulties as they are much greener and wetter than the crop.

Sulphuric acid, Diquat and more recently *Dimexan* have and are being assessed in P.G.R.O. trials and the results to date have been reported in the following publications by this organisation—Experiments Summary: Peas, 1965, 1966 and the 1967 Annual Report. Whilst it is possible to reduce the crop moisture content with these chemicals, there are adverse side effects and limitations upon their use.

Sulphuric acid at 15 gallons BOV/acre rapidly reduces crop moistures, gives excellent weed control to facilitate easy harvesting but has nearly always resulted in pod-splitting. In practice it would appear that about 5% pod-splitting could be anticipated and most of this would be associated with those pods which were green and immature at the time of spraying. If the crop is sprayed too early then a higher incidence of pod-splitting would almost certainly result. Acid treatment has also been associated with some loss of colour and a less even colour in the harvested peas. Fungus diseases present on the pods appear to be killed by this treatment. Diquat, tested at 3 and 4 pints/acre in medium volume, proved to be slower acting than the sulphuric acid, but eventually crop moistures were reduced and weed-control obtained without pod-splitting.

(i) RECOMMENDATION FOR PRE-HARVEST DESICCATION†

Diquat at 6–8 oz active ingredient/acre in medium volume with an added wetting agent as prescribed by the manufacturers can be used to desiccate peas for harvesting dry. Spraying should take place when the crop is fully ripe as Diquat does not hasten ripening, it simply reduces moisture levels. Harvesting should take place immediately desiccation is complete as the plants tend to be very weak and more straggly once the moisture has been taken out. Although this treatment is likely to lead to slight losses of colour, germination is not reduced.

† Fryer and Evans (1968), "Weed-control Handbook" (fifth edition),

XIV. WARNING AND INFORMATION ON DESICCANTS

Diquat-treated haulm must not be fed to any class of livestock. *Sulphuric acid as a desiccant—for information only* (Fryer and Evans, 1968).

This acid may be used as a pre-harvest desiccant on dried peas but it is of doubtful value if rain immediately follows the application. Crops to be treated with 15–20 gallons BOV (British Oil of Vitriol)/acre, which is a 77% v/v solution of sulphuric acid, should be mature and when weather conditions following spraying are good direct combine-harvesting will be possible 7–10 days later.

XV. QUALITY OF SAMPLES

When the weather at harvest is dry and sunny quality samples can easily be obtained. These are characterised by good uniformity, freedom from admixture, free from split, discoloured or faded peas and it is assumed that the moisture content when sold will be uniform and approximately 15%.

Under many circumstances it is impossible to produce direct from the combine or threshing mill a clean sample which is suitable for processing. However it is possible to dress the dried peas and take out unwanted trash with farm seed cleaners and in recent years industrial sorting machines have been made available to agriculture. One of these, namely Sortex, is capable of sorting by colour or shade differences (Figs 4 and 5). Discoloured peas and unwanted material of the same size are automatically rejected by using photo-electric cells and these are removed from the main stream by air jet (large seeded crops, e.g. beans) or by an electrostatic field (small seeded crops, e.g. rice). It is unlikely that these will appear on farms, but samples which are otherwise acceptable can be cleaned by the processing companies and under most circumstances the price paid to growers will be associated with the proportion of unwanted material which has to be removed in this way.

Processing companies vary slightly in respect of the accepted quality limits, but all concerned in this trade expect to dress ex-farm to remove waste and stained peas. Their price per ton to farmers and growers naturally depends on how much has to be removed and it is also linked with the moisture content. Prices in operation during 1968 from one of the leading manufacturers are shown in Table 7a on p. 355.

In addition the price would be reduced by 1 shilling/cwt for each 1% moisture in excess of 18% as determined on arrival at the manufacturing company's premises.

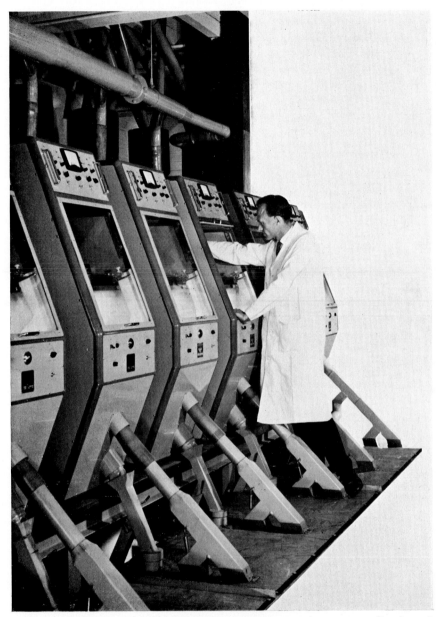

FIG. 4. A bank of electronic colour sorting machines used to separate discoloured seeds from samples of dried peas. Photograph by courtesy of Gunson's Sortex Ltd.

Acceptable sample

Rejected peas

FIG. 5. Samples of peas having passed through the electronic sorter. Photograph by courtesy of Gunson's Sortex Ltd.

TABLE 7a

Examples of 1968 Prices for Marrowfat and Blue Peas According to the Total Waste and Stain Content (£/ton)

Total waste and stain (%):	Marrowfat	Blues
Under 3	56·0	52·0
3 and under 4	55·0	51·0
7 and under 8	51·0	47·0
10 and under 11	49·0	45·0
13 and under 14	47·0	43·0
16 and under 17	42·5	38·5
19 and under 20	39·0	35·0
21 and under 22	37·0	33·0
24 and under 25	34·0	30·0

Soil contents above 0·5% will normally be deducted by weight from the deliveries and processors usually reserve the right to reject heavily soil contaminated lots (i.e. above 5%). They also have the option to accept or reject samples which contain over 25% total waste and stain.

XVI. YIELD

Average Yield of Dried Peas (cwt/acre)

5-year mean	Marrowfats	Blues
1955–59 inclusive	15·76	16·80
1960–64 inclusive	22·48	25·44

Since 1964, with increasing fertiliser placement, better varieties, good chemical weed control and reasonable possibilities of direct combining, yields have gone up further and the following estimates (cwt/acre) represent the present output on farms.

	Low	Average	High
Marrowfats	15–18	22–26	30 and over
Blues	18–20	24–28	35 and over

In respect of the marrowfats Nix (1968) uses 18, 24 and 30 cwt/acre as low, average and high yields in calculating the output and gross margin from this crop and his 1969 data is quoted again later in connection with the economics aspects of dried peas. From the price data of 1968, given earlier it will be noted that marrowfat peas were worth approximately 4s./cwt more than the blues but the latter can be expected to produce slightly higher yields. A yield differential of 2·5–3 cwt/acre with the blues is sufficient to offset the higher price obtained for the marrowfats, but the demand for blues is seldom as keen.

XVII. PRODUCTION ECONOMICS

A clear indication of the output and variable costs has been given by Nix (1969) and his data relating to Marrowfat peas are reproduced on p. 357.

These figures indicate gross margins which are higher by 3, 10 and £20/acre when compared with barley at the corresponding low, average and high production levels, but it must be remembered that harvest with dried peas is a more difficult and more costly operation. Special pea-cutters are needed, pick-up reels are often required by combines and some hard work is inevitable either in the turning of swaths prior to combining or in the erection of tripods, four-pole or hut-racks.

Pre-harvesting desiccation can eliminate this hand labour but will cost £3–£4/acre and there is a risk of some crop losses through pod-splitting. Thus, although higher gross outputs and higher gross margins can be obtained with dried peas compared with barley, it is unlikely that the

Marrowfat Peas, for Processing

Production level	Low	Average	High
Yield (cwt/acre)	20	25	30
Output (47s. per cwt)	£47	£59	£70
Variable costs:			
Seed		£10	
Fertilisers		£3	
Sprays		£4	
Miscellaneous		£1	
Total variable costs		£18	
Gross Margin	£29	£41	£52

Note: 47s. per cwt related to contract price for peas with a 12–13% wastage.

final profits would be greater where average or below average yields of peas are obtained. Where production is above average, profit margins with dried peas should be higher than those from barley and where desiccation has not been practised, the gross output can often be raised by including pea straw on the credit side.

PART 2. CONTRACT VINING PEAS FOR CANNING, QUICK-FREEZING AND DEHYDRATION

XVIII. INTRODUCTION

Processing companies who are concerned with canning, quick-freezing or dehydrating green peas offer definite contracts to farmers and growers. Although there are many regulations which have to be complied with, farmers are guaranteed a fair price according to the quality of their crop and good financial rewards can be obtained providing the yield and quality are good. During the 10-year period 1951–1960, the average yearly acreage for England, Wales and Scotland amounted to approximately 48,000 but since that period there has been a gradual rise and by 1966 about 87,000 acres were being grown in these three countries. At the beginning of the 1960's green peas harvested for canning and freezing were of equal importance. Since then the acreage for canning has remained fairly steady, but the acreage for quick-freezing has risen significantly and by 1965 approximately 7500 acres were being grown to satisfy the demands for dehydration.

XIX. MAJOR PRODUCTION AREAS

Contract vining peas tend to be localised in those areas where factories have developed and they are certain to be associated with the arable areas of England and Scotland.

In 1964, 94% of all the contract vining peas were grown in England and the Eastern, East Midland and Yorkshire–Lancashire regions were the three major production areas, as can be seen from the data on p. 359 (M.A.F.F., 1967). Norfolk was and still is the major producer in the Eastern Region and Lincolnshire the most important in the East Midlands.

In Scotland, green peas for vining are almost entirely confined to the two counties of Perthshire and Angus where they are associated with factories at Blairgowrie and Dundee respectively. The processing companies have installed small stationary viners on many of the farms where the peas are grown and contracts of around 60 acres were usually obtained. In recent years these have tended to be smaller and now stand at around 40–50 acres.

XX. VARIETIES

Peas suitable for canning, quick-freezing or for dehydration can all be obtained from the same variety. There is virtually no difference in basic requirements in respect of the crop for quick-freezing or dehydration and peas for canning are generally more mature and tend to be larger as a result. Companies with extensive interests in the crop for quick-freezing, however, have sponsored breeding work in recent years to produce new varieties especially suited for their purposes and many of these come to Britain from America and nearby European countries.

Quality assessments of vined peas include measurements of maturity which are obtained by using tenderometers. These are small machines which register the sheering force necessary to slice or puncture samples of the harvested crop, a low reading indicating young soft material and a high one is indicative that the peas are much more mature and past their best. Varieties are tested as "vining peas", firstly at a young tender stage for quick-freezing and dehydration when the tenderometer

1964 Contract Vining Pea Acreage for Great Britain
86,523

England and Wales 81,398   Scotland 5125

Eastern	South Eastern	East Midland	West Midland	South-west	Northern	Yorkshire and Lancashire	Wales	Scotland
47·8%	4·3%	30·8%	1·6%	Negligible	Nil	9·6%	Nil	5·9%

From M.A F.F. (1967).

reading usually lies between 95 and 105 and secondly they can be assessed on their suitability for canning when the tenderometer reading is above 110 but below 125/130 (usually between 115 and 125) (Martin, 1937).

Kelvedon Wonder, Gregory's Surprise and Dark Skinned Perfection are widely grown cultivars and they are also used as controls in variety trials when assessing new material. This work is carried out jointly by the following three organisations, through co-ordinated preliminary and main field trials and factory tests:

Pea Growing Research Organisation Ltd (P.G.R.O.), Thornhaugh, Peterborough.
National Institute of Agricultural Botany (N.I.A.B.), Cambridge and out-stations.
The Fruit and Vegetable Preservation Research Association, Chipping Campden.

The main trials are composed of varieties which have already gone through the preliminary trials and thus they have shown merit in these initial tests. Although this variety testing is done by three government-sponsored organisations, they have not, as yet, produced a recommended list, as with many of the other crops such as cereals, potatoes or sugar beet. The reason for this lies partly in the fact that varietal choice is not available to farmers and growers, who are in the hands of the processing companies. Thus it would appear that the official scrutiny of new varieties is largely of benefit to the commercial organisations, but they in turn have farmer's interests at heart, otherwise they would not be able to contract with them for their basic raw material.

In recent years the National Institute of Agricultural Botany has issued Vegetable Growers' Leaflet No. 6 entitled Varieties of Vining Peas. Varieties for quick-freezing, dehydration and canning are arranged in groups in order of maturity and the important agronomic characteristics of each cultivar are dealt with.

Farmers and growers are expected to sow several varieties in making up their acreage of vining peas and with numerous cultivars and a spread of sowing dates, the crop will mature to give a steady succession of harvests rather than the whole crop being ready at the one time. This enables the factory to freeze, dehydrate or can at the optimum maturity stage over a considerable period and prevents much of the crop getting past the correct quality stages. To illustrate this point, the acreages, varieties and sowing dates associated with an actual contract for 50 acres, grown in Scotland in 1963, are shown below in Table 8.

TABLE 8

Varieties, Acreage Sowing Dates in Respect of a 1963 Vining Pea Contract in Scotland

Meteor	Early Freezer	Kelvedon Wonder	Jade	Witham Dwarf	Dark Skinned Perfection	Johnson's Freezer
3 acres	9 acres	12 acres	2 acres	10 acres	10 acres	4 acres

"The seed shall be drilled by the Grower, as under, at the recommended 2 cwt/acre on the dates specified by the company, weather permitting":

Week commencing 18/3/63 $\begin{cases} 3 \text{ acres Meteor} \\ 5 \text{ acres Early Freezer} \end{cases}$

Week commencing 25/3/63 $\begin{cases} 4 \text{ acres Early Freezer} \\ 7 \text{ acres Kelvedon Wonder} \end{cases}$

Week commencing 1/4/63 $\begin{cases} 5 \text{ acres Kelvedon Wonder} \\ 2 \text{ acres Jade} \\ 5 \text{ acres Witham Dwarf} \end{cases}$

Week commencing 8/4/63   No sowing

Week commencing 15/4/63 $\begin{cases} 5 \text{ acres Witham Dwarf} \\ 5 \text{ acres Dark Skinned Perfection} \end{cases}$

Week commencing 22/4/63   No sowing

Week commencing 29/4/63 $\begin{cases} 5 \text{ acres Dark Skinned Perfection} \\ 4 \text{ acres Johnson's Freezer} \end{cases}$

XXI. SEED-BED PREPARATION

Part of a farmer's acreage of vining peas will have to be drilled in the early spring, thus necessitating the land to be ploughed in the previous autumn or winter. Where time permits it is advisable to have all the area for peas turned over in the previous year so that a good frost mould can be obtained through prolonged weathering. When this is done seed-bed preparation can be left until a few days before sowing and cultivation to obtain the final tilth should only be done as required since the crop will usually be drilled in batches, as indicated in the contract. Where tests have shown the soil to be acid and lime is necessary to raise the pH before pea growing commences, the appropriate quantity must be applied during the winter after ploughing for it to be incorporated by rain. If the application of lime takes place in the early spring then cultivations should follow to see that it is properly mixed into the top 4–6 inches, in order to raise the soil pH quickly and evenly.

XXII. MANURING

The levels of phosphate and potash suggested for dried peas are equally suitable for vining peas and the appropriate dressing (related to soil analyses) should be worked into the seed-bed prior to drilling. Although varieties of vining peas are just as capable of fixing atmospheric nitrogen as any others, small quantities of fertiliser nitrogen are often recommended although they may not show economic yield responses. It is often felt that the land may be completely devoid of mineralised nitrogen, particularly following a very wet winter and spring and small amounts of nitrogen applied just prior to seeding will give the crop a good start. Many farmers and growers are convinced that fertiliser nitrogen will ensure that the crop obtains and maintains a dark green colour and Reynolds (1960) reports that nitrogen has had no effect other than to improve the appearance of the crop. Excessive nitrogen applications are certain to inhibit nodulation and nitrogen fixation will therefore be reduced. Plants will have to rely on the fertiliser which has taken over the major role of "nitrogen supplier" and similar yields will usually result at additional cost. Except where the soil organic status and general level of fertility are low and following prolonged wet weather, the use of nitrogen for vining peas cannot be justified.

XXIII. SEED RATE

Field experiments carried out by the P.G.R.O. and reported by King (1966) repeatedly suggested that the optimum plant population for vining peas was in the vicinity of 11 plants/square foot. These results were obtained mainly from the varieties Dark Skinned Perfection and Kelvedon Wonder and it was thought advisable to check that the new, promising varieties emerging from Preliminary and Main variety trials would behave in a similar manner. The results obtained in 1967 from Dark Skinned Perfection and three other varieties are given in the P.G.R.O. Annual Report for that year and are reproduced in Table 9.

The data in Table 9 confirm the findings of original P.G.R.O. trials that the optimum plant population for vining peas is in the region of 10–11 plants/square foot and this holds good even when seed costs are included at £150 and £200/ton. Additional benefits to be derived from the high plant densities include shorter straw and a higher percentage of the produce in the medium size grade. Vining peas are considered small should they pass through an $\frac{11}{32}$ inch sieve, they are said to be of medium size from $\frac{11}{32}$ inch to $\frac{13}{32}$ inch, and large when they are over

TABLE 9

Plant Populations with Vining Peas. Yield and Estimated Profitability Order[a]

Variety	Actual population (plants/ sq. ft)	Seed rate (stones/ acre)	Yield (cwt/acre)	Return/ acre (£)	Cost of seed/acre (£)			Return/acre less cost of seed (£)			Order of profital		
					(i) at £100/ ton	(ii) at £150/ ton	(iii) at £200/ ton	(i)	(ii)	(iii)	(i)	(ii)	(iii)
Surprise	3·9	6·8	18·5	37	4·3	6·4	8·5	32·7	30·6	28·5	3	3	3
	5·7	9·9	33·2	68	6·2	9·3	12·4	61·8	58·7	55·6	2	2	2
	9·5	16·5	49·6	100	10·3	15·5	20·6	89·7	84·5	79·4	1	1	1
Freezer 69	5·9	10·5	48·6	101	6·6	9·8	13·1	94·4	91·2	87·9	3	3	3
	11·2	21·0	63·1	134	13·1	19·7	26·3	120·9	114·3	107·7	1	1	1
	15·7	31·5	65·1	139	19·7	29·5	39·4	119·3	109·5	99·6	2	2	2
Witham Wonder	7·1	7·8	44·5	92	4·9	7·3	9·8	87·1	84·7	82·2	3	3	3
	9·6	15·6	64·1	134	9·8	14·6	19·5	124·2	119·4	114·5	1	1	1
	15·4	23·4	58·1	122	14·6	21·9	29·3	107·4	100·1	92·7	2	2	2
Dark-skinned Perfection	5·1	9·5	45·3	81	5·9	8·9	11·9	75·1	72·1	69·1	3	3	3
	12·0	19·0	61·4	109	11·9	17·8	23·8	97·1	91·2	86·2	1	1	1
	16·8	28·5	64·3	112	17·8	26·7	35·6	94·2	85·3	76·4	2	2	2

[a] Extract from P.G.R.O.–Annual Report for 1967.

$\frac{13}{32}$ inch in diameter, and any cultural practices which can increase the percentage of those within the "medium" or "middle" grade will help the processor and in the long run will be beneficial to the grower. Many of the commercial companies use a bottom sieve size of $\frac{8}{32}$ inch and the very small peas which pass through are considered only as trash.

Seeding rates to achieve these optimum populations are governed by the germinative capacity and the size of the seed, and providing the former is satisfactory then rates between 15 and 20 stone/acre should be employed.

XXIV. SEED DRESSING

Under most circumstances, the use of thiram or captan based seed dressings can be advocated mainly as an insurance against soil borne pathogen should temperatures be low and brairding slow. As with the dried pea crop, they have greatest benefit with the exceptionally early sown peas and any which are drilled under conditions which are not ideal.

A. Drilling, Row Widths and Time of Sowing

Seed should be sown at depths similar to those advocated for dried peas, namely 1·5–2 inches below the surface and narrow row widths are preferred to wider ones since seeding rates of at least 2 cwt/acre are required and for the future these may be increased to 2·5 cwt should the results from the preliminary investigations (Table 9) be confirmed. With high seeding rates the narrow row width (4 inches) give a better spatial arrangement for each individual plant and competition from neighbours within the row is reduced.

XXV. TIME OF SOWING

It is important to harvest vining peas at the correct maturity stage; tenderometer readings 90–105 for freezing peas and 110–125 for canning. Substantial acreages of peas are grown on individual farms and to stagger the harvest a number of varieties are employed and the crop is sown in batches over a considerable period (3–6 weeks). The drilling dates are based mainly on soil temperature rise in the spring coupled with varietal maturity ratings.

A. Accumulated Heat Units (AHU)

Accumulated Heat Units are used to give a maturity classification to varieties of peas and are calculated by assessing the number of day

ownership of costly machines such as these, then it may be possible to justify the formation of a machinery syndicate and the purchase of a mobile viner. Processing companies wishing greater control of the harvest of their crops are likely to invest in these machines and general agricultural contractors may also be interested, particularly those who operate within the recognised pea-growing areas.

Figure 10 shows a close-up of the model made by Mather and Platt and Fig. 11 illustrates two of these working in a good crop of peas. An

Fig. 12. F.M.C. Mobile Pea Harvester. At the front a p.t.o. driven elevator lifts and loads the peas into the viner which has its own engine. The hydraulically operated pea collecting hopper is shown at the rear, moving to the emptying position. Photograph by courtesy of P.G.R.O. Ltd.

American machine currently in use in the East Midlands and South-east England is the F.M.C. Mobile Pea Harvester made by Scott Viner Company, Columbus, Ohio. The main features of this viner are illustrated in Fig. 12 and Figs 13 and 14 demonstrate the bulk handling of the shelled peas following harvest.

The pea haulm after passing through these mobile field viners is discharged onto the ground at the rear and is usually ploughed-in. It is seldom used for feeding or for making into silage on account of a high degree of soiling.

Fig. 13. F.M.C. Mobile Pea Harvester discharging shelled peas to bulk containers. Photograph by courtesy P.G.R.O. Ltd.

Fig. 14. Bulk loading of vined peas for dispatch to the factory. Photograph by courtesy of P.G.R.O. Ltd.

XXIX.   YIELDS   WITH   VINING   PEAS   FOR   FREEZING
AND   CANNING

The contracts for vining peas are normally placed with efficient arable farmers who are usually in possession of better than average land and yields as a result tend to be fairly high. Early varieties do not produce the same weight as the late or maincrop varieties but they are favoured in the price structure and thus the financial output per acre is not substantially lower. Low, average and high yields with vining peas for freezing are given in Table 10 and it can be assumed that the corre-

TABLE 10

Yield of Vining Peas for Freezing (cwt/acre)

	Low	Average	High
Earlies	20	30	35–40
Late or maincrop	25	35	40–50

sponding estimates for canning peas, which are taken at a higher tenderometer reading, will be higher. Berry (1966) worked out a yield–tenderometer relationship for adjusting yields to a given maturity and those engaged in experiments with peas may find this useful when the interpolation of data is necessary.

Having seen what is presently being produced per acre on a commercial scale, it is interesting to compare these production levels with experimental data associated with present variety trials and future potential yields, Table 11. Farmers and growers may argue that they

TABLE 11

Experimental and Potential Yields, Green Peas at Two Tenderometer Readings (cwt/acre)

	TR 100	TR 120
Trial yields	40–50	60–70
Potential yields	50–60	70–80

cannot emulate the yields from trials, but it must be recorded that some producers have already obtained between 3 and 4 tons of green peas/acre, not as an average, but from their highest yielding crops or fields in some years.

XXX.   ECONOMIC   ASPECTS   OF   VINING   PEAS

The contribution to the farm income from peas can be assessed through the "gross margin" technique, which takes into account the

N

variable costs in growing this crop, but does not try and allocate the fixed costs. Nix (1969) using an average yield of 35 cwt/acre and price of £43/ton estimated the gross margin as follows:

Gross Margin Calculations for Vining Peas (Freezers) (£/acre)

Production level	Low	Average	High
Variable costs			
Seed		14·5	
Fertiliser		3	
Sprays		3·5	
Total variable costs		21	
Output (excluding haulm value)	60	80	100
Gross Margin	39	54	69

It would appear that vining peas are capable of contributing twice the gross margin of barley and figures similar to those from sugar beet. Whilst it is foolish to try and justify a major farm enterprise on the value of its by-products, where appropriate they should feature in economic considerations. Pea haulm from stationary viners has to be removed and it is excellent material for the production of silage. Average pea crops (apart from the extremely short, early varieties) will produce about 6–8 tons of haulm (leaves, stalks, pods and discarded small peas)/acre and this could be included in the output. Where mobile field viners are used, then it is unlikely that the haulm can be used for feeding, but its manurial value when ploughed-in should be taken into account for the following crop.

For those readers who think that Gross Margins "only tell half the story", average enterprise costs from economic surveys in the field may be more highly thought of, but unfortunately at the time of writing a recent one could not be found. However, some useful data were collected by the Staff of the Economics Department of the Edinburgh and East of Scotland School of Agriculture and one of the main tables from their report (Wright, 1964) is now reproduced (Table 12).

In 1962, the assessed rent of £4·3/acre was more realistic than the previous year, and with average production costs amounting to about £55/acre, and yields a little below the "35 cwt/acre average", profit margins were £9 and £28/acre for canning and freezing peas respectively. Haulm valued at £1/ton would increase the profit by £6–7/acre, bringing the total profit margin to £16 and £34/acre for these two types of processed peas. Assuming that 1969–70 production costs have risen to £60/acre and 35 cwt of peas are worth £46/ton, then profit margins without and with adjustment for the haulm value would be in the region of £20 and £27/acre respectively.

TABLE 12

Extract from Edinburgh and East of Scotland School of Agriculture Economic Report No. 87. (January 1964). Peas for Processing

Average Results of Costs and Returns from an Identical Sample of 26 Whole Contracts in 1961 and 1962.

	1961	1962
	£	£
Costs per acre		
Cultivations	5·2	5·1
Miscellaneous costs	1·2	1·4
Harvesting, including depreciation of special equipment	15·9	14·6
Manures, plus and minus manurial residues	7·7	7·9
Share of crop failure	—	0·2
Seed	11·0	11·1
Rent	2·3[a]	4·3
Overheads	10·0	10·0
Total costs	52·3	54·6

	Canning	Freezing	Canning	Freezing
	£	£	£	£
Payment for peas	79·3	81·2	61·0	79·6
Payment for vining	3·2	4·3	2·8	4·0
Haulage charge	—	—	—	−0·8
Total returns	82·5	85·5	63·8	82·8
Profit	29·2	32·2	9·2	28·2
Haulm value	10·3	9·1	7·0	6·0
Adjusted profit	39·5	41·3	16·2	34·2
Number of contracts	11	15	11	15
Size of contract	42 ac.	42 ac.	58 ac.	38 ac.
Yield of peas/acre	41·2 cwt	33·7 cwt	31·8 cwt	34·0 cwt

[a] Includes many on Assessed Rental Value basis in 1961.

PART 3. PEAS FOR MARKETING GREEN—"PULLING PEAS"

XXXI. INTRODUCTION

"Peas for marketing green" is the term used to describe "Pulling Peas" in the June Statistical Returns. Before the Second World War they dominated the green pea trade in Britain but since that period there has been a marked decline as can be judged from the acreage figures reproduced in Table 13. In 1938, 87% of all peas harvested green were sold in the pod and at 62% in 1951 they still dominated the trade. However, during the next ten years the acreage dropped to less than half the pre-war figure as the vining swiftly took over the major role and by 1966 the percentage contribution from these two groups was completely reversed. The increasing standard of living was largely responsible for this major trend and as people's wages rose so they demanded higher quality vegetables such as canned, quick-frozen or dehydrated peas. There will always be a market for the fresh crop although it may decline still further in the future as the casual labour becomes harder to find.

TABLE 13

Acreages of Vining and Pulling Peas

	1938	1951	1966	1967	1968	1969	1970
Peas for marketing green	58,128	40,800	15,400	17,300	17,700	15,419	12,894
(% of total green peas)	(87%)	(62%)	(15%)	(15%)	(14%)	(12%)	(9%)
Vining peas (canning, quick-freezing and dehydration)	8731	24,900	87,100	97,200	104,500	112,974	127,500
(% of total green peas)	(13%)	(38%)	(85%)	(85%)	(86%)	(88%)	(91%)
Total Green Peas	66,859	65,700	102,500	114,500	122,200	128,393	140,394

XXXII. PRODUCTION AREAS

Peas for marketing green are produced in arable farming areas in close proximity to large urban populations and the reasons for this are three-fold. Firstly there is a local demand for the product; secondly, the large casual labour force needed for this crop can easily be recruited and transported to and from the farms; finally the peas can be marketed quickly with the minimum transport costs. Green peas which do not reach the consumer soon after they have been harvested quickly deteriorate. They lose their colour, moisture and much of their quality if not eaten fresh and this loss of quality in storage has also been responsible for much of the changing attitudes and preferences of the general public.

Statistical data concerning the acreage of pulling peas during the early 1960's showed the major production areas to be

1. West Riding of Yorkshire (South-eastern part only).
2. Essex.

3. Worcester.
4. Kent.
5. Bedfordshire.

and early crops were being produced in Somerset. In spite of the decline in the acreage since then, these are still the major production areas.

A. Size of Green Pea Crops

In 1962, when just over 30,000 acres were being grown in England and Wales, the June acreage statistics were analysed by the Ministry of Agriculture. Just over six and a half thousand farmers or growers were involved in pulling pea growing which may lead one to think that something like 5 acres were being cultivated on average by individuals. About 4000 producers had less than 1·75 acres each and those whose acreage ranged between 5 and 50 controlled more than two-thirds of all the green peas grown.

From this analysis of the holdings involved in green pea growing it is fairly clear that most of the producers are people engaged in horticulture whilst the bulk of the acreage, in the hands of a small number, is associated with agriculture.

B. Production of Green Peas

Processing companies are not involved with pulling peas and therefore the crop is not usually one grown on contract. However, in some areas it has been the custom for farmer and vegetable wholesaler to combine their efforts in producing and marketing the crop, probably halving all the costs involved and thus sharing equally any profit which may result.

Peas for marketing green are grown in a manner similar to that described for the contract vining peas and thus only notable differences will therefore receive attention.

XXXIII. SOWING DATES

The time at which green peas are sown can often be the biggest single factor in determining the profit margins. When the demand is limited and the main supply tends to flood the market over a very short period, those producers who achieve early harvests or fairly late ones usually obtain much higher prices and show larger profits.

Early	Mid	Late
February–March	April	May–June

Sowings in April will usually produce high yielding crops with the least chance of failure, but do not command the highest prices. Late February to early March sowings are a little chancy since the soil temperatures are usually low at this period and stands of peas can be lost through pests and diseases, but against this additional risk can be balanced the expectation of higher prices. Late sown peas can often be a complete failure, particularly during exceptionally wet summers, but on occasions those crops of good quality coming onto the market quite late in the summer will leave substantial margins. Where large acreages are involved, staggered sowings or the use of varieties with differing maturities can be employed to spread the harvest.

XXXIV. VARIETIES

Most of the varieties of vining peas are suitable for the green pea trade. However, they may be difficult to obtain since the processing companies are usually responsible for supplying this seed to their growers. Under these circumstances intending growers are advised to order their seed early through the trade and they may find recognised pulling pea varieties such as Onward and Thomas Laxton more readily available and equally satisfactory.

XXXV. HARVESTING AND MARKETING

Crops are ready for pulling when the majority of the pods are well filled. The end of June is considered early, July normal and August and September late. Once a crop is ready, it has to be harvested and marketed immediately and at this point it is essential to be able to call on large gangs of casual labour. As many as 100–200 pickers/acre are employed and they simply go over the crop once, stripping off all the pods into sacks or nets. This operation is often done on piece work and pickers will know before they commence how much they can expect for each 40 lb net gathered. As soon as they have been weighed up in the field, farmer's lorries or those of wholesale merchants will take them to market.

XXXVI. YIELD

In 1954 the Nottingham University Economics Department's survey of 27 growers indicated an average yield of 193 bags/acre (3·4 tons) with a range of 138–274. This would suggest the following as low, medium and high outputs.

Yield of Green Peas (Pulling Peas)

Unit	Low	Medium	High
40 lb bags/acre	120–160	170–210	220–260 or over
Tons/acre	2·1–2·8	3·0–3·7	3·9–4·6 or over

XXXVII. CROP PROTECTION

Many species of birds are responsible collectively for poor brairds of peas. The smaller species do most damage to pea crops grown in gardens and on small acreages situated to farm buildings. On a field scale the larger birds such as pigeons and corvids can play havoc with pea crops, particularly those which are sown early. One bird-scaring method on its own is not likely to be completely satisfactory. The birds usually get used to it and are soon unaffected by it. A number of bird scaring/elimination methods are listed below for the reader's consideration.

1. Carbide operated bird-scarers with time clocks and bangers (similar to fireworks).
2. Scarecrows.
3. Coloured gas-filled balloons.
4. Smoke-generators.
5. Periodic excursions to the crop with shot-guns.
6. Tin-foil scaring devices.
7. Stuffed birds of prey.
8. Long playing records of bird warning cries played periodically.
9. Erection of black cotton or black terylene above the crop.
10. Hanging up of shot birds.
11. Bird-repellant seed dressings.
12. Bird traps.

A. Narcotic Bait

During the early part of February 1969, a narcotic bait was approved for the first time by the Ministry of Agriculture for use in March 1969. Only growers of peas and/or brassicas will be able to obtain a licence to use the bait which consists of tic beans treated with alpha-chloralose at the rate of 2% by weight. It is not for widespread use for wood pigeon population control on the grounds of economics, but is simply for use by individual farmers who have bird problems on peas or brassicas. Conditions laid down in the licences will be that all affected birds shall be picked up. Pigeons are to be humanely destroyed and their carcasses burned or buried and in the case of all game birds and those which are protected by law, these must be revived and then released.

In the trials carried out in Bedfordshire in 1968, 2735 lbs of bait led to the capture of 1393 pigeons, 137 corvids (mostly rooks) and 13 protected and game birds (*Farmer and Stockbreeder*, 1969). Over a trial period of 3 years, 88·6% of all birds caught were pigeons and the approximate cost was £5/80 lbs of bait which was sufficient for about 2 acres, but it was pointed out that this need not necessarily be the present or future commercial price.

References

Bell, G. D. H. (1948). "Cultivated Plants of the Farm". Cambridge University Press, London.

Berry, G. (1966). "A Yield-tenderometer Relationship in Shelled Peas for Adjusting Yields to a Given Maturity". *J. agric. Sci.* **66**, 121–123.

Blair, I. D. and Copp, I. G. (1953). *N.Z. Journ. Sci. Tech.* Sec. A. **34**, 397–404.

Blumer, S. and Harder, A. (1953). *Landw. Jb. Schweiz.* **67**, 315–35.

Board of Agriculture. (1905). U.K. Statistics.

Cooke, G. W. (1964). "Fertilizers and Profitable Farming" (second edition). Crosby Lockwood & Son Ltd; London.

Darlington, C. D. and Wylie, A. P. (1945). "Chromosome Atlas of Flowering Plants". George Allen and Unwin Ltd, London.

Farmer and Stockbreeder (1969). 18th Feb. issue.

Fedotov, V. S. (1935). *Bull. appl. Bot., Pl.-Breed.* **9**, 165.

Fryer, J. D. and Evans, S. A. (1968). "Weed Control Handbook II. Recommendations". Issued by the British Crop Protection Council. Blackwell Scientific Publ., Oxford.

Gane, A. J. (1962). "Notes on Growing Dried Peas". P.G.R.O. Misc. Public. No. 12, II. (Dec. 1962).

Home Grown Threshed Peas Joint Committee. (1950–51). Second Progress Report on Field Experiments.

Jones, F. G. W. and Jones, Margaret G. (1964). "Pests of Field Crops". Edward Arnold Ltd, London.

King, J. M. (1966). "Row Widths and Plant Population in Vining Peas". P.G.R.O. Misc. Publ. No. 18 (March 1966).

M.A.F.F. (1957). "Pea Root Eelworm". Advisory Leaflet No. 462. Ministry of Agriculture, Fisheries and Food, London.

M.A.F.F. (1960). "Peas for Drying". Advisory Leaflet No. 357. H.M.S.O., London.

M.A.F.F. (1967). "Horticulture in Britain. Part 1—Vegetables". H.M.S.O.

M.A.F.F. (1968 and 1969). Agricultural Chemicals Approval Scheme. "Approved Products for Farmers and Growers, 1968". Plant Path. Lab., Harpenden, Herts.

M.A.F.F. (1968). "Pea Moth". Advisory Leaflet No. 334.

M.A.F.F. (1968). "Weed Control in Peas". Advisory Leaflet No. 376.

Martin, W. McK. (1937). *Canning Tr.* **59** (29). 7.

National Institute of Agricultural Botany. (1968). Vegetable Growers' Leaflet No. 6.

Nix, J. (1968, 1969). "Farm Management Pocketbook" (second edition). Dept of Agric. Econ., Wye College, University of London.

Olsen, C. (1925). "Studies on the Growth of some Danish Agricultural Plants", *Compt. Rend. Lab. Carlsberg.* (**16** (2), 1–22.)

Owers, A. C. and Yule, A. H. (1956). "Experimental Husbandry 1", pp. 58–62. H.M.S.O.

Percival, J. (1947). "Agricultural Botany" (eighth edition). Duckworth, London.

P.G.R.O. 1962. "Dried Peas in the Sixties". Miscellaneous Publication No. 12, I. P.G.R.O. Ltd, Yaxley, Peterborough.

P.G.R.O. (1963). Newsletter Number 13. Yaxley, Peterborough.

P.G.R.O. (1966). Annual Report, 1965. Pea Growing Research Organisation Ltd, Yaxley, Peterborough.

P.G.R.O. (1966). Experiments Summary—Peas 1965. The Research Station, Yaxley, Peterborough.

P.G.R.O. (1967). Annual Report 1966. Pea Growing Research Organisation Ltd, Thornhaugh, Peterborough.

P.G.R.O. (1967). Experiments Summary—Peas 1966. The Research Station, Thornhaugh, Peterborough.

P.G.R.O. (1968). Pea Growing Research Organisation Ltd. Annual Report 1967. Thornhaugh, Peterborough.

Proctor, J. M. (1957). *Farmer's Weekly.* March, 87–88.

Proctor, J. M., Reynolds, J. D. and Gregory, P. (1955). *Proc. Brit. Weed Control Conf. 1954,* 157–164.

Reynolds, J. D. (1957). *J. nat. Inst. agric. Bot.* **8,** 1, 82–95.

Reynolds, J. D. (1957). "Some Developments in Threshed Pea Growing". *Agricultural Review* **3,** 444–448.

Reynolds, J. D. (1958). *Agriculture* **65,** 1, 34–37.

Reynolds, J. D. (1960). "Manuring the Pea Crop". *Agriculture* **66,** 509–513.

Reynolds, J. D. (1960). "Right Peas for Right Conditions". *Farmer's Weekly.* 29th April, 1960.

Sansome, E. R. (1933). *Cytologia* **5,** 15.

Small, J. (1946). "pH and Plants. An Introduction for Beginners". Ballière, Tindall & Cox, London.

Trénel, M. (1927). "Die wissenschaftlichen Grundlagen der Bodensäurefrage". Berlin.

Willis, J. C. (1966). "A Dictionary of the Flowering Plants and Ferns". Cambridge University Press, London.

Wright, Edith M. (1964). "Report on Peas for Processing, 1962". Edinburgh and East of Scotland School of Agriculture. Economic Report No. 81.

Addendum

1. Pea Cyst nematode, *Heterodera gottingiana* Lieb., formerly known as pea root eelworm, is a major pest of all pea growing. Crop rotations which include peas only one year in five should be adhered to. Further information can be obtained from Advisory Leaflet No. 462 M.A.F.F. (1957).

2. Marsh Spot is caused by a deficiency of manganese and affects the mature seed. When these are split open the centre of each half exhibits brown discolorations. Preventative measures include

   (a) Application of 1–2 cwt/acre manganese sulphate to the soil or

   (b) Spraying affected crops at flowering time with 20 lb manganese sulphate/acre in 100 gallons of water.

CHAPTER 8

Forage Legumes

PART 1. LUCERNE

I.	Introduction	384
II.	Early History and Spread of Lucerne Throughout the World	385
III.	Nomenclature	386
IV.	Taxonomy of Lucerne	388
V.	Classification of Varieties or Strains of Lucerne	389
	A. Common Lucerne	389
	B. Turkestan and Caucasian Lucerne	389
	C. Variegated Lucernes	390
	D. Non-hardy Group	390
	E. Yellow Flowered Lucernes	390
	F. Named Varieties	391
VI.	British Evaluation of Varieties	391
	A. Early Type	391
	B. Mid Season Type	392
	C. Late Type	392
	D. Extra Late Type	392
VII.	Acreage and Production Areas in Britain	392
VIII.	Utilisation of the Lucerne Acreage	393
	A. Hay-making	394
	B. Artificial Drying	394
	C. Ensilage	395
	D. Grazing	396
	E. Zero-grazing or Green-soiling	397
IX.	Soil Requirements	397
	A. Texture	397
	B. Drainage	398
	C. Soil pH and Lime Requirement	398
	D. Fertility	400
X.	Climatic Requirements	400
	A. Rainfall	400
	B. Temperature	401
	C. Photoperiod	402
	D. Light Intensity	403
XI.	Lucerne Monoculture or Lucerne-grass Leys	404
	A. Summary of Benefits to be Derived from the Inclusion of a Grass Species	406
XII.	Establishment	407
	A. Lime and Fertilisers	407

          B. Time of Sowing 408
          C. Direct Seeding or Undersown 408
          D. Method of Sowing 409
   XIII. Variety and Type of Seed 412
          A. Inoculation and Treatment of Seed 414
          B. Seed Rates 416
   XIV. Management and Utilisation of the Crop 418
          A. Control of Pests 418
          B. Chemical Weed Control in Seedling Lucerne . . . 419
          C. Chemical Control of Grass Weeds in Established Lucerne Crops . 419
          D. Cultural Weed Control 420
          E. Early Management 420
          F. Manuring 421
          G. Irrigation 423
          H. Frequency of Defoliation 424
          I. Utilisation 426
    XV. Productivity 427
   XVI. Quality of Herbage 429

                        PART 2. SAINFOIN

   XVII. Introduction 431
  XVIII. Botanical Classification 432
   XIX. Types of Sainfoin 433
          A. Common or Single-cut Sainfoin 433
          B. Giant or Double-cut Sainfoin 433
    XX. Seed and Plant Morphology 433
          A. Seed 433
          B. Plant Morphology 435
   XXI. Soil and Climatic Requirements 436
  XXII. Fertilisers 436
 XXIII. Seed 437
          A. Varieties 437
          B. Sowing Rate 438
          C. Time of Sowing 439
  XXIV. Chemical Weed Control 439
   XXV. Management and Utilisation 440
  XXVI. Yield 441
          A. Forage 441
          B. Seed 442
 XXVII. Quality of Herbage 443
          References 444

PART 1. LUCERNE

I. INTRODUCTION

Lucerne, *Medicago sativa* (L.), has been cultivated as a forage crop in parts of the world for at least 2000 years and in some areas probably for a much longer period, but it is a relatively new crop compared with the cereals in British agriculture.

According to Cooper (1965), the deliberate cultivation of pasture plants is so recent compared with crop plants which means that there is an accurate historical account of the introduction and spread of many and this includes the forage legume lucerne. Bolton (1962), however, depicts lucerne as one of the oldest forage plants and as a result suggests that this distinction limits the accuracy with which its centre of origin can be deduced.

It is generally agreed that South-west Asia is the most likely centre of origin and a number of authorities suggest Iran as the country concerned. It is considered a native of the temperate regions of Eastern Europe and Western Asia having been found wild in several provinces of Matolia, to the south of the Caucasus, in parts of Persia, Beluchistan and Kashmir (De Candolle, 1919). Bolton (1962) translates these areas into the modern political divisions of Afghanistan, Iran, Iraq, Kashmir, Syria, Turkey and West Pakistan. Klinkowski (1933) suggests North-west Persia as the eastern limit and designates this area as Media and the home of lucerne.

II. EARLY HISTORY AND SPREAD OF LUCERNE THROUGHOUT THE WORLD

The earliest reference to lucerne is said to have dated back to 700 B.C. in Babylonian texts (Wheeler, 1950). It is referred to by the Greek writer Theophrastus about 300 B.C. and much is written about its virtues and requirements by the various Roman historians about 2000 years ago and of these Pliny and Columella are probably the most important as early sources of information about this legume. The former records its introduction into Greece from Media with the Persian invasion 492–490 B.C. and it was thought to have reached Italy and North Africa by 200–150 B.C. Figure 1 portrays in diagram form the development and spread of lucerne, from its centre of origin to Europe, the Americas and beyond.

In the course of spreading from Asia to the rest of the world, lucerne developed into two distinct climatic types. The first, a Southern or Mediterranean type which developed in North Africa, was introduced to Italy, France and Spain and from Spain it found its way to South America. The second type, which developed in Northern France, the Low Countries and Germany, was winter-hardy whereas the Mediterannean type evolved as cold susceptible. From Fig. 1 it will be seen that these two distinct types or races appeared in American agriculture

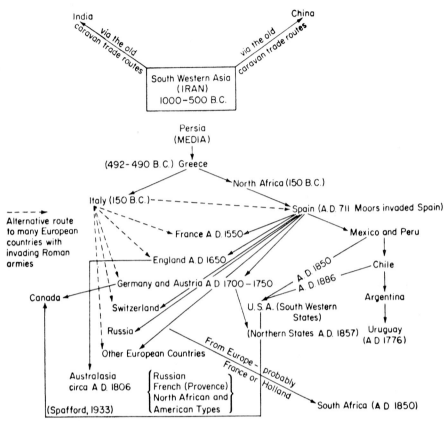

Fig. 1. Spread of lucerne throughout the world from the centre of origin, with approximate dates. Based on data collected and reviewed by Klinkowski (1933) and Bolton (1962).

directly from Europe (winter-hardy Grimm lucerne from Germany in 1857 (Klinkowski, 1933) or in the case of the Mediterranean type via South America.

Lucerne may well have come to Britain with the Roman invasion but it certainly was imported (probably from France) in the mid seventeenth century. Worlidge (1668) advocated its use alone or in mixtures with grasses for leys and he was the earliest English writer to make reference to "La Lucerne" (Davies, 1952).

III. NOMENCLATURE

The latin name *Medicago sativa* (L.) is universal, but the common name *Lucerne* is used only in Western European countries, Australia,

New Zealand and South Africa. In nearly all the other areas of the world where it is grown, the name *Alfalfa* has been adopted.

The ancient Greeks used the name medicai, the Romans medica, and both Klinkowski (1933) and Piper (1935) record that in Italy it is still known as "erba medica".

The exact philological root or origin from the old Iranian language is not known, but Piper suggests the following metamorphosis.

Possible Persian Origin		Arabic changes (with or without the prefixes el or al)		
uspust				
aspest	fisfisat	isfast	elkasab	alfafa
isfist		alfasafat	alfalfa	

Introduced to Spain as alfalfa by the Moors, this name was then transferred first to South America and then to North America where it is now used exclusively.

According to Piper the name lucerne was first used by the Frenchman Dalechamps in 1587, but there are two schools of thought on its derivation or origin.

ASSOCIATED WITH THE SWISS CANTON LUZERN

Most authorities assume that the name lucerne owes its origin to the Swiss canton and Lake Luzern, or the river valley Lucerna in Northern Italy.

(This theory is supported by Ahlgren (1949) who claimed that this legume was first cultivated in the Lake Luzern district of Switzerland in the first century A.D. and from there it spread to other parts of Europe.)

OLD SPANISH ORIGIN

Coburn (1906) claimed that lucerne was grown in France and England long before it was found at Luzern and De Candolle (1919) suggested that the old Spanish word eruye was modified by the Spanish and French to give the present name:

"eruye"	userdas
laouzerdo	luzerne

Both the common names lucerne and alfalfa are used to denote *Medicago sativa* (L.). They are synonyms, interchangeable and are not specific to any type or variety. Incidentally, alfalfa is a moorish name which means "best of fodders".

IV. TAXONOMY OF LUCERNE

Class: Dicotyledon
  Family: Leguminosae
    Sub-Family: Papilionoidaea
      Tribe: *Trifolieae*
        Genus: *Medicago* (Associated Genera—*Ononis*, *Parochetus*, *Melilotus*, *Trigonella* and *Trifolium*)
          Species: *arborea*
                   falcata
                   hemicycla
                   media
                   ovalis
                   rugosa
                   sativa

The various forms of lucerne have been listed by Darlington and Wylie (1945) in "Chromosome Atlas of Flowering Plants" and these are reproduced in Table 1.

TABLE 1

Types of Lucerne

Botanical name	Common name	Chromo-some number	Reference	Part of the world where cultivated or found wild
Medicago falcata	Yellow Lucerne	16, 32	Ledingham (1940)	Northern Temperate areas
Medicago arborea	Tree Alfalfa	32	Ghimpu (1929)	Mediterranean
Medicago hemicycla		32	Cooper (1935)	Transcaucasus
Medicago media	Sand Lucerne	32	Fryer (1930)	Temperate
Medicago ovalis		32	Tschechow (1933)	Spain
Medicago rugosa		32	Fryer (1930)	European countries
Medicago sativa	Lucerne, Alfalfa	16	Bolton and Greenshields (1950)	Cultivated forms in many parts of the world—originating in Persia
		32	Fryer (1930)	
		32, 64	Tomé (1947)	

The basic chromosome number is 8 and thus most of the species listed above are tetraploids. The two species of agricultural importance *M. falcata*, yellow lucerne and *M. sativa*, common lucerne are shown as

was fairly gradual but in the 1960's the decline was very rapid, with the present-day production area reduced to 30–40,000 acres.

When at its peak in 1954, lucerne could be found in leys in most counties of England and Wales but after the mid 1950's it became concentrated in the low rainfall areas of the East and South, i.e. East Anglia, Lincolnshire, Kent and the Downland.

Lucerne has never been of great importance in Scotland or Northern Ireland although several hundred acres are grown each year by enthusiastic farmers in Southern Scotland.

The main reasons for the decline in popularity and acreage are thought to be:

(i) It is an exacting crop in respect of soil and management both in establishment and for survival.

(ii) This legume will not tolerate serious competition from aggressive weeds or grass companion species.

(iii) The initial spring growth is late and time of first defoliation is very late when compared with early grasses.

(iv) Seed costs/acre can be fairly high.

(v) It is mainly suited to low rainfall areas.

(vi) Less suited to grazing than grasses.

(vii) Heavy leaf-drop in hay-making can result in large nutritional losses.

(viii) Although high in protein, lucerne silages are often low in energy.

(ix) There is quite a sharp decline in digestibility with advancing maturity.

Since the yield from lucerne can often be higher than from any other forage species when considering the low rainfall areas of Britain, it is still an important legume for cutting and conservation and the points listed above will receive special attention in the text in order to obtain the highest output from this potentially high yielding though somewhat special fodder crop.

VIII. UTILISATION OF THE LUCERNE ACREAGE

According to Davies (1952), the attitude in Europe, in the past, to lucerne has been to regard it almost exclusively as a forage plant for hay or ensilage

In view of the fact that it is much more suited to cutting rather than grazing and will survive longer if cut rather than grazed, it is logical to find the majority of the acres of lucerne in Britain being cut for forage.

From the June Agricultural Returns the proportion of the total British acreage cut for hay alone has been calculated. When the acreage was at its peak in 1954, just over half was cut for hay and, following the decline, the corresponding figure for the 1960's indicated that about two-thirds was being used in this manner.

It is interesting to note that during the sixties in New Zealand slightly more than 60% of the total lucerne acreage was being cut either for hay or silage (Lynch, 1967). This author goes on further to say that hay-making, the traditional use for lucerne, will probably still remain its main function in New Zealand in the future although most of the stands will be grazed at some stage in their life.

Other methods of utilisation may not be practised quite as widely as hay-making, but are important, since together they account for about one-third to one-half of the British acreage. The various ways of using the lucerne crop are listed below and although hay-making is probably still the most important, the order of placing of the remainder is alphabetic and not in any order of importance.

Hay-making.	Grazing.
Artificial drying.	Zero-grazing or green-soiling.
Ensilage.	

A. Hay-making

Lucerne is associated mainly with the low rainfall and high sunshine areas of Britain, and since these climatic conditions are the most favourable for good hay-making it follows that this conservation method should be the most important. The stems of lucerne are thick compared with those of grass and the weather must be good to satisfactorily dry them, otherwise the time to cure in the field is prolonged, resulting in reduced quality. One of the major problems in lucerne hay-making is that of leaf drop. Once the crop begins to dry out, the leaves tend to become brittle and unless the crop is turned and treated gently much of the valuable leaf will be lost.

B. Artificial Drying

Artificial drying of grass and forage crops like lucerne is once again being considered as a major farm enterprise with the introduction and use of drying plants with much larger capacities (3000 tons of dried material/season) than those previously available. Crop drying on this scale is usually associated with large commercial companies or groups of farmers getting together to form a co-operative and official approval of the latter will automatically entitle them to a one-third grant. It has been suggested that as dairy herds in Britain become larger, they will demand quality fodders, such as dried grass or lucerne in the form of cubes or cobs which can be handled and metered quickly and accurately.

One major problem of the 1960's has been the choice of suitable break crops in the cereal-dominated South and East when disease and low yields have signalled a halt to cereal monoculture or close cereal cropping. One of the best ways to rest the land is to put it into grass or perennial forage crops for several years, but the method of utilisation of the produce has often been a considerable problem. Artificial drying offers a solution to this problem of utilisation and when the end product is not required for home feeding it is in a suitable form for sale—thus making grass and lucerne into cash crops.

C. Ensilage

Silage made with fairly mature herbage from straight lucerne or grass/lucerne leys before the widespread use of forage harvesters was often over-heated and of poor quality, mainly because of difficulties of consolidation. Nowadays with machines which lacerate the crop well, there should be no problem in this connection. However, it is well known that leguminous forages tend to be high in protein and low in soluble carbohydrates and thus the fermentation process is often short of its "fuel". An additive such as molasses will help the right type of fermentation to develop by supplying the bacteria with their nutritional needs. The digestibility of lucerne stem is low particularly near the ground and when the crop is mature the energy value tends to be somewhat lower than silages made from grass. For this reason a companion grass species sown along with lucerne will help to raise the starch equivalent where ensilage is the main method of conservation.

Since digestibility drops fairly quickly with advancing maturity, the logical step is to consider ensilage of young herbage from this legume. Tristram (1956), in his account of lucerne in Northamptonshire, draws attention to difficulties experienced with silage making at an early growth stage. Very young lucerne which is leafy and succulent, packs tightly, quickly expels the air and thus limits the fermentation which can take place. The resulting silages were often under-heated and evil-smelling and part of the trouble was almost certainly due to low soluble carbohydrates in the original material. Tristram went on to say that many of these difficulties could be overcome by (a) not ensiling very wet material; (b) the liberal use of molasses (c) avoiding over consolidation, and (d) keeping a careful watch and control of temperature. Well-made lucerne silage in Northamptonshire was shown to contain up to 26% crude protein.

D. Grazing

At some time or other in the life of a lucerne ley, grazing will usually be considered. Although it has been suggested that lucerne is probably better suited to a mainly cutting rather than grazing management, it can prove to be very productive and lasting provided the animal defoliation is controlled and timely. Early spring grazing should be avoided because removal of the new shoots considerably weakens the individual plants and will lead to a reduced yearly output. Late spring and summer grazing should take place when there is a large leaf cover, it should be completed quickly and then followed by a good rest period. A system of paddock grazing or using electric fences will enable the defoliation to be effected rapidly and it must be remembered that lucerne can thrive even though heavily grazed for short periods provided it has adequate rest periods before the animals return. Continuous grazing, however light, cannot be tolerated by this legume (Tristram, 1956).

In order to enable the lucerne to withstand the winter, it must not be grazed or cut in the late summer and early autumn, otherwise the plants will endeavour to produce new growth in the autumn and so deplete their root reserves. Late autumn and winter grazing is possible, provided the land does not poach badly and in many districts lucerne or lucerne grass leys can supply useful winter grazing from foggage.

Valuable cattle grazing was obtained during the winter period mid November until late February or mid March at the Grassland Research Institute, Hurley, during the three years 1951–52, 1952–53 and 1953–54 (Pearson Hughes, 1955). With lucerne and grass sown in drills winter grazing at a stocking rate of one and a half, 18–21 months old cattle per acre was considered possible in these investigations and this was thought to be more economic than housing similar cattle inside. Larger liveweight gains during the winter were obtained from the indoor group, but this advantage soon disappeared when they were turned out to grass in the spring due to a check in growth rate.

Ruminant animals are susceptible to bloat when grazing most leguminous plants and lucerne is no exception. Farmers and graziers must be vigilant when cattle or sheep are on lucerne otherwise they may sustain considerable financial losses through reduced production or animal deaths. Animals have to be introduced gradually and restricted feeding either by limiting the grazing period or the grazing area may often be necessary. The inclusion of a grass companion species will usually prove beneficial and should sheep and cattle graze very young leafy lucerne, then access to some fibrous food such as hay or straw, at the same time, normally reduces the risk of bloat.

E.   Zero-grazing or Green-soiling

Zero-grazing is a term used to describe the feeding of animals with cut herbage from grass or forage crops. It usually infers that they are kept inside or at the farm steading and the fodder is brought to them and thus grazing of the growing plants in the fields is eliminated.

Green-soiling is the cutting of herbage from grass or forage crops to be fed outside in fields—usually worn out or unproductive grass fields.

Both of these methods of utilisation are practised widely on the European continent, particularly in France and Spain and a few farmers with large dairy herds in East Anglia have adopted zero-grazing with lucerne.

The elimination of animal grazing will increase the yearly productivity, firstly by removing spoilage from dung and urine, secondly by reducing poaching and thirdly by the introduction of a defoliation method which can be completely controlled both in time and degree. There are several other advantages when lucerne is fed as silage rather than grazed in the field. When the herbage is partly lacerated in the cutting process, the leaves and stem will be sufficiently mixed to eliminate selective feeding and thus a better utilisation of the stem is usually obtained. The incidence of bloat with lucerne cut and carted to animals is usually less than when the crop is grazed directly (Bolton, 1962).

When small quantities are involved, cutting can be done by hand or by farm mowers but if an extensive zero-grazing unit is desired then forage harvesters and self-unloading forage boxes will usually be necessary. In order to maintain animal productivity, the herbage should be cut daily and adverse weather conditions or mechanical breakdowns can prove very testing to such a system which at the outset may require considerable capital.

IX.   SOIL REQUIREMENTS

A.   Texture

Lucerne is by nature a deep-rooted plant and therefore requires deep soils in order to demonstrate its yield potential. Soil texture is not as important as fertility and water supply (Bolton, 1962), but in general lucerne often thrives best on those soils classified as light or medium textured. Root growth is extensive and unimpeded and when conditions are less than optimal on these soils the faults can usually be corrected quickly without serious losses of production. Successful lucerne cultivation can take place even on the heaviest of clays, provided other factors are in order, the most important of these being proper drainage.

Soils which are dominated by their organic content, i.e. peats and fens, are generally thought unsuitable for lucerne. The peats tend to be acid, a factor which precludes lucerne growing until the pH value is corrected and on the neutral or alkaline reclaimed fen soils, the annual release of nitrogen is less advantageous to lucerne than to associated grass species or weeds. They become more aggressive and may quickly dominate the legume. Judging by the soil pH, soils overlying and influenced by chalk or limestone would appear to be very suited to lucerne cultivation and Stewart (1951) confirms that this legume establishes well even under cereals, and will give good yields over a considerable period on a chalk soil, provided the management is right.

B. Drainage

Lucerne is a plant associated with low rainfall areas of the world, but it can be highly productive in medium to high rainfall localities provided the soils are well drained. It will not tolerate waterlogging and the land must either be free-draining or adequately drained for leys to persist (Grinsted, 1950). According to Bolton (1962) surface water can be tolerated by dormant lucerne and low temperatures delay or reduce the injurious effects. McKenzie (1951) concluded that "alfalfa" would tolerate from 10 to 21 days of flooding in the early spring, the longer periods associated with lower temperatures, but when temperatures are high and growth is rapid 1–3 days' surface flooding could cause serious injury.

In order to obtain satisfactory root development on soils which require drainage, tiles should be laid to remove surplus water from the top 2–3 feet. Graumann and Hanson (1954) suggested tile depths of 3 feet for American conditions and this figure is likely to apply also to most light to medium soils in Britain, but for heavier soils such as the clays, tile depths much in excess of 2 feet are undesirable.

C. Soil pH and Lime Requirement

Common lucerne (*medicago sativa*) is considered native to the lowland areas of Asia Minor and its natural habitat is soils which are neutral or slightly alkaline in their reaction (Iversen, 1957; White, 1967).

Acid soils, characterised by low pH values, will not allow the successful establishment and growth of lucerne. Many factors are involved and collectively they are thought of as the "soil acidity complex" (Hewitt, 1952). This complex is not fully understood but the following facts are generally agreed as being mainly responsible.

(i) Low pH values are responsible for toxic levels of manganese and aluminium in the soil and in the plant.

(ii) Under acid conditions there is seldom sufficient calcium available for legume growth.

(iii) Bacteria which are essential for nodulation and nitrogen fixation find it difficult to survive or multiply when calcium is in short supply. Bolton (1962) suggested that soils with a pH of less than 6·5 were too acid and those above pH 7·5 were usually too alkaline for the successful establishment of lucerne.

The relative yields of lucerne from a range of soils were recorded by Willard *et al.* (1934) and these are reproduced below.

| pH value | 4·7 | 5·2 | 5·9 | 6·8 | 7·4 |
| Relative yields of lucerne (%) | 4 | 12 | 41 | 100 | 94 |

The above figures indicate how intolerant lucerne is of acid conditions and Tristram (1956) was obviously correct when he insisted on a pH of between 6·8 and 7·0 for the crop in Britain. Where ground limestone is needed at 3–5 tons/acre to raise the pH to this level he suggests that lucerne is avoided for at least a year until the desired pH has been achieved through lime applications.

The beneficial effect of lime on the growth of lucerne is largely due

TABLE 3

Effect of Lime Placement on Nodulation of Lucerne Eight Weeks from Sowing

| Treatment | Percentage nodulation | Nodules/nodulated plant | |
		Crown	Lateral
4 cwt lime drilled with the seed	90·2	3·5	0·9
1 ton of lime broad-cast	68·2	2·5	0·9

After White (1965).

to the improved nodulation which automatically follows and where large quantities of lime are needed and it is cheap and easily available, it should be applied as a broadcast dressing and worked in.

In Australia on slight to moderately acid sands, lucerne is commonly established by drilling 1·5–2 cwt/acre calcium carbonate (Tiver, 1960), and in the same area where the soils are strongly acid White (1965) demonstrated that 4 cwt of lime drilled with the seed gave better nodulation than 1 ton of lime which was broadcast and ploughed in.

Hayman (1964) working with lime-pelleted inoculated lucerne seed in New Zealand was able to show immense improvement in nodulation. Only 6·1% of the plants showed nodulation from the control plots (unpelleted seed) whereas the corresponding figure for lime-pelleted seed was 90·9% nodulation and from these experiments it has been suggested that in some moderately acid soils effective nodulation of lucerne plants can be obtained using lime-pelleted inoculated seed.

Although experiments in New Zealand and Australia have demonstrated that nodulation can be obtained by drilling lime with the seed and by using lime-pelleted inoculated seed, it would be more beneficial and lasting to correct the soil pH by large lime applications prior to growing lucerne. When lime is easily available, and the cost plus application subsidised as it presently is in Britain then it would be false economy to adopt other shorter-lived techniques for establishing this legume.

D. Fertility

Lucerne demands high fertility to demonstrate its high yielding capacity as a forage plant. The organic status of the soil should be good, but not excessive and phosphate and potash levels should both be satisfactory on soil analysis. However, where deficiencies of these two major nutrients occur, they can be corrected in basal manurings prior to establishment. Davies (1952) whilst listing lime, phosphates and potash as being the major mineral demands, also records that magnesium and boron deficiencies have been recorded in Britain. Magnesian limestone use in correcting low pH values and the adoption of boronated fertilisers can eliminate these deficiencies.

X. CLIMATIC REQUIREMENTS

A. Rainfall

As a native of Asia Minor, lucerne prefers the low rainfall and high sunshine conditions, but it will establish and grow well under cool temperate conditions at medium to high rainfall provided land drainage is adequate and the right cultivars chosen.

Willard et al. (1934) demonstrated that the total root weight from lucerne in a wet year was only 55% of that produced in a dry year in the State of Ohio, U.S.A.

Davies and Tyler (1962) demonstrated the effect of growing season rainfall on yield under British conditions and their data relating to the three years 1957–59 inclusive are given on page 401.

The potential transpiration rates are included to show the level of water requirement by the legume each year and it will be noted that the dry year of 1959 gave substantially higher yields than the two previous wet years.

Most of the lucerne grown in Britain is associated with the areas which have a total annual rainfall of less than 30 inches per year and

TABLE 4

Rainfall, Potential Transpiration Rate and Yield from Lucerne

	1957	1958	1959
Potential transpiration rate			
April–September (inches)	15·66	13·71	16·96
Rainfall April–September (inches)	20·94	28·81	13.01
Average yield (lb DM/acre)	8020	9610	11,060

After Davies and Tyler (1962)

thus the amount received during the growing season (April to September) will be very close to the potential transpiration rate resulting in high productivity. This does not mean that medium to high rainfall will not support this legume. Green (1955) pointed out that successful cultivation was possible in the East Midlands Province at 750 feet above sea level with 40 inches of rain a year and even on heavy land with up to 50 inches in Wales.

B. Temperature

Wherever lucerne is grown there are seasonal variations in average monthly temperature as well as diurnal fluctuations and these are certain to affect the growth of this legume. Most species of crop plants have a threshold temperature below which growth ceases and this is usually considered to be around 5°C (41°F). As the temperature rises above the threshold value, growth rate increases up to an optimum temperature and once this is exceeded growth rate decreases. Since lucerne has been cultivated for at least 2000 years and has developed into very many distinct forms, it is not possible to give a set of temperatures which apply universally.

Where leguminous crop plants rely to a large extent on their symbiotic association with the appropriate bacteria for their nitrogen, the temperatures required by the bacteria themselves assume a greater importance. Rhizobial bacteria begin nitrogen fixation at about 10°C

(50°F) and this temperature is usually considered minimal for the growth of legumes like lucerne.

Bolton (1962) suggested that alfalfa strains have an optimum temperature range of 15–30°C (59–86°F) and that a much reduced activity is demonstrated by this crop at low temperatures 10–12°C (50–54°F) and also at high temperatures 35–40°C (95–104°F).

In reviewing the effects of environment on growth and development, Leach (1967) pointed out that although lucerne is grown at high altitude with low temperatures, higher temperatures and incidentally higher light intensities promote both reproductive and vegetative growth. Schonhorst *et al.* (1957) demonstrated that many cultivars of lucerne make less growth when temperatures exceed 77°F (25°C), which was confirmed later by Bolton (1962).

Thus as far as Britain is concerned, there are large lowland areas in the South which enjoy many summer months with temperatures between 41 and 77°F (5–25°C) and thus from the temperature aspect are very suited to lucerne cultivation.

Temperatures considerably below freezing point will usually remove above ground growth from lucerne crops and often a significant number of the plants will die through winter killing as a result of these low temperatures. Cultivars from a cold continental climate when grown in New Zealand exhibited a high degree of winter dormancy (Palmer, 1936) and the survival rate of types such as these is likely to be good. Provided the reserves of total available carbohydrate in the root are high, lucerne plants will show satisfactory recovery from freezing (Langer, 1967) and management techniques which ensure good root reserves during the winter period should be employed. These will be discussed later.

Recovery after freezing is associated with soil moisture and Calder and Jackson (1965) working with the variety Vernal, demonstrated that excessive moisture levels in the soil could be responsible for poor plant recovery. When the soil was saturated prior to freezing the lucerne plants showed a much poorer recovery than when it was at full or 25% field capacity and these experiments verify the need for adequate drainage.

C. Photoperiod

The length of daylight in a given agricultural environment is fixed and beyond the control of farmers, but it is important to see that the cultivars chosen are the most suitable.

Under a controlled environment, with day and night temperatures of 64 and 50°F (18 and 10°C) respectively, Iversen and Meijer (1967) were able to show that increasing the length of photoperiod from 8 hours to 16 hours and in some cases to 20 hours would result in higher growth rates with five lucerne varieties. This principle had been established earlier through the work of Massengale and Medler (1958) and Bula (1958) who demonstrated that dry matter yields and flowering were both promoted by lengthening the photoperiod. Coffindaffer and Burger (1958) pointed to interactions between cultivar and photoperiod and this has been verified since in some of the experiments conducted under controlled environments.

D. Light Intensity

Working at Oxford with the variety Provence, Blackman and Black (1959) calculated that for maximum relative growth rate lucerne seedlings required 2·5 times the amount of light energy they received during the summer. The light factors associated with other legumes ranged from 0·85 to 1·85, indicating that lucerne required much higher light energy values. In comparison with seedling trefoil, lucerne seedlings had a greater net assimilation rate at high but not low light intensities (Cooper, 1966) and although the effects of competition and shading are inevitable in crop plant communities, these results strongly suggest establishment without nurse crops to give the young plants the maximum amount of light.

Pritchett and Nelson (1951) grew lucerne plants in a glasshouse at various light intensities and they clearly demonstrated the need for high light intensities with this crop.

| % Glasshouse light | 100 | 27 | 14 | 9 | 5·5 |
| Relative dry weight/plant | 100 | 62 | 35 | 11 | 6 |

In these experiments plant growth was seriously reduced at the low light intensities in the glasshouse and nodule numbers were greatly reduced and it must also be recorded that significantly better growth was obtained when the plants were transferred to full daylight outside. Pritchett and Nelson also showed that shading depressed the roots more than the tops and this was confirmed several years later by Gist and Mott (1957).

The experimental work associated with light intensities quoted above has mainly been concerned with young lucerne plants in pots, under controlled environmental conditions and therefore cannot be applied widely in practice to field crops. Natural radiation levels in the field are usually higher than those in experiments as can be seen by the figures in Table 5.

TABLE 5

Experimental and Natural Light Intensities

Reference	Light intensities (foot-candles)
Gist and Mott (1957)	200, 600 and 1200
Bula *et al.* (1959)	750, 1500 and 3000
Matches *et al.* (1962)	Full sunlight, i.e. 7000 to 10,000
Steinke (1963)	2000 and 3500

Thus it is not surprising to find reductions of 25% and 46% in natural light intensity failing to reduce lucerne dry weights in the field, as indicated by Cowett and Sprague (1963) and Matches *et al.* (1962) respectively.

XI. LUCERNE MONOCULTURE OR LUCERNE-GRASS LEYS

In reviewing lucerne–grass associations, O'Connor (1967) pointed out that much of the experimental work did not include studies of lucerne grown alone but where this treatment was included no general gain could be attributed to the companion grasses in respect of total dry matter yield. These results are in agreement with the conclusion of Donald (1963) on the principles of associated growth of pairs of species when growth is competitive.

W. Davies (1952), whilst recognising that on the Continent of Europe lucerne was grown as a pure stand, pointed out that there were likely to be advantages with mixtures under many British conditions. The associated grasses, provided they did not become dominant, would be useful in suppressing weeds and could extend the season of usefulness of the ley.

In the same year, J. Davies (1952) listed the main arguments in favour of companion grasses in lucerne leys as being:

(i) The grasses help to suppress weeds.

(ii) A mixture of grass and lucerne may yield more dry matter/acre than a pure stand of lucerne.

(iii) A mixture of grass–lucerne is a better subject for hay-making or grazing than a pure stand of lucerne.

(iv) In regions of high rainfall, especially on retentive and poorly drained soils where the risk of lucerne failure is considerable, a grass companion and white clover together ensure the establishment of a ley, irrespective of the success of the lucerne.

However, the establishment trials which he conducted in the West Midlands during the period 1947–51 led him to conclude that lucerne did best as a pure stand, but where a companion grass was used, meadow fescue proved to be the most suitable.

During the period 1945 to 1952 two aspects of lucerne cultivation received attention at the Grassland Research Station, Drayton, Stratford-on-Avon. On their heavy soil derived from lower Lias clay, the use of companion species with and summer pasturage from lucerne were investigated. Green and Davis (1955) reported this work and the dry matter yields which are reproduced in Table 6 confirm Davies's early conclusions about the value of straight lucerne or lucerne–meadow fescue leys.

TABLE 6

Total Yields from a Lucerne Ley and Three Lucerne–Grass Mixtures During the First Three Harvest Years 1950–52

('00 lb DM/acre)

Lucerne alone (inoculated Provence)	Lucerne–cocksfoot (S.37)	Lucerne–timothy (S.48)	Lucerne–meadow fescue
197	182	176	202

After Green and Davis (1955).

TABLE 7

Mean First Harvest Yield and Mean Annual Yield from Lucerne and Lucerne–Grass Mixtures ('00 lb DM/acre)

Year	First harvest			Annual yields		
	Lucerne alone	Lucerne + ryegrass (S.23)	Lucerne + cocksfoot (S.37)	Lucerne alone	Lucerne + ryegrass (S.23)	Lucerne + cocksfoot (S.37)
1950	40·6	49·8	45·7	81·0	88·6	82·7
1951	25·2	30·8	29·3	57·8	55·9	55·1
1952	40·2	34·6	37·6	71·5	63·6	64·6
1953[a]	41·4	37·8	33·7	—	—	—

[a] Only one harvest in 1953.
After Davies et al. (1953), and Davies and Davies (1956).

o

Early Aberystwyth work with Hungarian White Seal lucerne showed that initially in the first harvest, lucerne plus grass outyielded lucerne alone, but this advantage disappeared with time (Table 7).

Davies and Tyler (1962) reported that lucerne plus grass companion species gave significantly higher yields in the seeding year than lucerne alone but this was reversed in the second and particularly the third harvest year. Italian ryegrass (S.22), meadow fescue (S.215) and cocksfoot (S.37) were the associated grass species and again it transpired that meadow fescue proved to be best companion. Meadow fescue had the least depressing effect on the lucerne in the seeding year and in later years, competition from this grass occurred only in the spring and thus was minimal. Italian ryegrass and cocksfoot quickly dominated the lucerne and had a severely depressive effect.

A. Summary of Benefits to be Derived from the Inclusion of a Grass Species

(i) From the experiments quoted, higher yields of herbage can be expected in the seeding year.

(ii) Earlier growth in the spring from a "cool-season active" grass species can be obtained (O'Connor, 1967).

(iii) Winter production in the form of foggage is likely when a companion grass is present.

(iv) Lucerne–grass herbage is better material for grazing—less bloat possible and may avoid luxury uptake of protein.

(v) Lucerne–grass herbage is better material for hay making—lower leaf losses and quicker drying.

(vi) Lucerne–grass herbage is better material for silage making—higher soluble carbohydrate content and thus a more satisfactory fermentation process.

(vii) The inclusion of a grass or grasses in the mixture will serve as an insurance against total or partial failure and usually minimises colonisation by weed species.

(viii) Competitive effects of aggressive species of grasses (e.g. cocksfoot) can be eliminated by growing alternate rows of lucerne and grass and thus a high yield can be obtained over a considerable period.

In order to obtain (ii), (iii) and (vii) above some loss of production from the lucerne plant is inevitable and due to competition, the period during which the legume contributes significantly to herbage yield is shortened. However, it would appear that many advantages are to be gained by including a grass companion species and on balance this should be the recommendation for most lucerne leys.

XII. ESTABLISHMENT

A. Lime and Fertilisers

Considerable attention must be given to a number of factors when establishing lucerne leys firstly to see that a proper "take" or field germination is obtained and secondly to ensure productivity over several years (usually 4–6 in practice).

Lime applications to raise the pH value of the soil solution to 6–6·5 should take place in the winter or early spring prior to seeding and some authorities even suggest a small additional application of 0·5–1 ton in the seed-bed.

Lucerne requires large quantities of phosphate and potash for successful establishment and in general terms the amount applied as a basal fertiliser dressing prior to drilling should be related to the soil levels of these two major nutrients.

TABLE 8

Recommended Levels of Phosphate and Potash in Establishing Lucerne
(units P_2O_5 and K_2O/acre)

	Level of nutrient in the soil as indicated by analyses		
	Low	Medium	High
Phosphate	80	60	40
Potash	180	120	60

When a grass companion species is included or when the soil reserves of nitrogen are considered to be very low, then a small quantity of nitrogen (20–40 units) can be added to the phosphate and potash levels recommended above.

Lucerne will not survive for any length of time if the land is in need of drainage and it is wise to check on this aspect before final decisions and sowings are made.

One other important consideration is the cleanliness of the proposed site. Lucerne is a non-aggressive species and besides being unable to compete with strong growing grasses it is equally at a disadvantage on land which carries heavy weed infestations. These must be eliminated before lucerne is sown otherwise there will be serious risks of complete failure in establishment or the stand will be so poor that it will seldom be of economic use after a year or two.

Cultivations prior to and during seed-bed preparation must be aimed at minimising weed competition and at the same time produce a fine, firm, moist seed-bed which is ideal for this forage crop.

B.   Time of Sowing

When it is necessary to obtain production from lucerne leys in the seeding year a spring sowing is essential. Better establishment, however is often gained by delaying sowing until the summer and here the loss of herbage in the seeding year has to be set against the superior "take" which in turn may mean an extra year's production.

March is normally too early for most districts. Soil temperatures are usually low and competition from annual weeds can result in many crop failures.

Early April is probably the most suitable time for sowing in the early districts of Southern and Eastern England but in other areas it is probably too soon, judging by the establishment trials in the West Midlands (J. Davies, 1952).

Late April sowings on average will result in good stands of lucerne and contribute something in the way of herbage in the seeding year. Seed-bed preparation can be thorough, but soil moisture conservation must be practised.

May is the most suitable month in late areas or in northern or western districts.

June. Sowings in June are often associated with a short fallow which may be considered necessary to clean the land prior to drilling.

July. July will usually prove the most suitable month where lucerne follows early harvested vegetable or potato crops and where the land is clean excellent lucerne leys result from July sowings.

August sowings can often result in poor establishments as the plants have barely time to grow strong enough to survive the winter (Green, 1955).

According to Grinsted (1950) autumn sowing of lucerne should never be attempted.

The above comments on March–August sowings relate to crops which are sown directly, without cereal or other nurse crops associated with them. If lucerne must be sown under cereals then this should normally take place in April, before the cover crop has grown too thickly.

C. Direct Seeding or Undersown

The lucerne establishment trials in the West Midlands during 1947–1951 referred to earlier are interesting and serve as a warning against undersowing if this can be avoided.

Undersown in spring cereals		Direct seeding	
4 trials		14 trials	
3 failed through crop competition	1 failed through non-inoculation of the seed	9 satisfactory	5 not satisfactory (either the field was abandoned altogether or it became a grass and clover ley)

The above details summarise the fate of 18 lucerne establishment trials and it must be noted that only half of these were successful and all of the successful ones were obtained from direct seedings. From lucerne trials at Colesbourne on the Cotswolds, Davies and Davis (1951) reported that a cereal nurse crop had a bad influence on the "take" and growth of this legume.

According to Green (1955) the practice of sowing without a cover crop was more common with lucerne than with any other kind of ley and on heavy land it was especially advisable to avoid cereal grain nurse crops.

Davies (1952) previously advocated that lucerne was best sown on clean land, in good heart without a cover crop during the spring or in July and that the young ley established in this manner could be grazed in less than two months.

On the other hand if lucerne has to be established below a cereal crop then the following practices will help to obtain as good a braird as possible.

(i) Choose a stiff-strawed, early maturing spring cereal.

(ii) As far as possible use low seeding rates for the grain crop.

(iii) Avoid high or even medium levels of fertiliser nitrogen.

(iv) Make sure that there is sufficient phosphate and potash available for the lucerne.

(v) Drill the legume at right angles to the cereal, 0·5–1 inch below the surface and follow with flat or Cambridge rollers.

D. Method of Sowing

(i) BROADCAST

This is the least satisfactory method of establishing lucerne or lucerne–grass leys. Without harrowing afterwards, the percentage field germination will be exceedingly low and even when this is done the seeds are at various depths, some still on the surface and some too far down, both resulting in a poor braird. It is important to use inoculated seed and the bacteria which are placed onto the seed are weakened and killed by sunlight. Thus for a successful establishment and to make sure that the bacteria survive to provide the free source of nitrogen to the legume, drilling rather than broadcasting is nearly always recommended.

(ii) DRILLING

To facilitate rapid, efficient drilling, the seed-bed should be fine, firm and flat and in order to obtain a quick establishment there should be plenty of moisture present. It is generally agreed that drilling at 0·5–1 inch below the surface under these conditions will achieve the highest field establishment, but there is no such general agreement on coulter spacings for narrow rows or row widths when grown in wide rows.

According to Grinsted (1950) the best "take" is obtained when special seed-drills with coulters spaced at 4 inches, are used and although the ordinary corn drill has a satisfactory spacing (6–7 inches usually) it is difficult to put the seed in at the optimum depth of just below 1 inch.

There have been several experiments concerned with sowing methods and row widths with lucerne and lucerne–grass mixtures during the past fifteen to twenty years. In America, Patterson and Law (1952) demonstrated lower yields from a mixed lucerne–grass seeding in 6-inch rows compared with alternate grass–lucerne rows at the same spacing whilst Kilcher and Heinrichs (1958) obtained the opposite result when using a 1 foot row spacing.

Chamblee and Lovvorn (1953) working in America with lucerne/ cocksfoot and lucerne/tall fescue, showed that alternate rows of legume and grass 6 inches apart gave lower yields than broadcast plots or when the species were mixed and sown in 6-inch rows. These findings were later confirmed by Tewari and Schmid (1960).

Mansfield (1945) suggested broadcasting lucerne as this was more likely to keep the land free from weeds for five or six years but he was aware of possible poor establishments particularly in dry climates and advocated drilling at a coulter spacing of 3 inches. Lucerne drilled in wide rows is likely to leave the land very dirty and except for special circumstances such as the production of winter fodder (Hughes, 1955), it should not be sown in this manner.

Davies (1964) working with lucerne and meadow fescue demonstrated that alternate rows of each yielded 20% less dry matter than comparable broadcast plots. In this experiment it was shown that spaced plants or plots sown in wide drills had a much lower plant population per unit area and although tillering was higher than on the broadcast areas, the actual tiller numbers remained below optimum. There is little point in sowing a non-aggressive species like meadow fescue in alternate rows with lucerne as this type of grass will not become dominant, but with an aggressive grass like cocksfoot this type of establishment could be useful. It has often been said that lucerne demands an open sward in the spring for proper development and alternate rows of this legume would appear to have distinct merit, particularly on very light sandy soils which are prone to drought. When considering winter grazing Davies (1964) suggested that alternate rows of lucerne and grass could prove advantageous both from the point of view of utilisation and possibly since it would allow differential nitrogen fertilisation.

When the Grassland Research Institute was stationed at Drayton, Hughes (1955) conducted experiments with lucerne sown in 2-foot drills with intermediate drills of leafy cocksfoot, timothy, meadow fescue or perennial ryegrass. He demonstrated that it was possible to maintain lucerne and grass together for several years by sowing in this manner. Green and Davis (1955) suggested that the balance of grass and legume was less susceptible to variations in management when the two species are separated in this way than when in closer association following a broadcast seeding. Should growth conditions be less than ideal, the adverse reaction shown by the legume is almost certainly accelerated by competition from broadcast grasses, particularly the aggressive ones.

With a poor field establishment with broadcast lucerne, but a much better yield than drilled crops on wide rows, the logical conclusion is to drill lucerne on narrow rows, in other words arriving at a broadcast plant spacing but with seeds drilled at optimum depth. Thus it would appear that a further use could be made of the new cereal drills with coulters set at 3–4 inches provided depth control and low seed rate calibrations can be effectively obtained.

Plant population clearly affects yield with lucerne/cocksfoot leys. Garner and Sanders (1940), and Jarvis (1962) experimenting with lucerne populations from 4840 up to 6·25 million per acre were able to demonstrate that the highest yields per acre were associated with the largest number of plants.

Thus in establishing lucerne or lucerne/grass leys it is imperative that a large plant number is obtained per unit area and thus wide rows, except for special circumstances like winter grazing, should be avoided.

XIII. VARIETY AND TYPE OF SEED

A large proportion of the seed which is used in Britain is purchased abroad and according to Willey and Zaleski (1955) it is mostly from Europe and the strains can be separated into two groups viz.:

Northern European group   Mediterranean group
(*Medicago media* Pers.)   (Closely resembling *Medicago sativa*)

It was further suggested that much variation could occur within these groups, but varieties or strains originating in countries with a similar climate to ours should be of immediate value to British farming.

Observation and evaluation of lucerne varieties began in 1922 at the National Institute of Agricultural Botany, Cambridge, and Parker (1931) reviewed the lucerne trials between 1925 and 1930. During the early 1950's Hawkins and Zaleski (1952) and Zaleski (1954) produced a scheme for identifying and classifying the large number of strains which were available. This work received attention earlier in the chapter, but at this stage it is worthwhile listing a number of past and present varieties associated with each of the types (Table 9).

TABLE 9

Varieties of Lucerne Grouped According to Type

Early Type	Mid season type	Late type	Extra-late type
Cardinal	Hungarian White Seal	Grimm	Argentine strains:
Chartrainvilliers	Ile de France	Ladak	Choele Choel
du Puits	Marlborough	Old Franconian	Chubut
Emeraude	Provence	Rhizoma	Hilario Ascasubi
Europe	Szarvase		La Pampa
Eynsford			Pampeana
Flamandc			Saladina
Flandria			
F.D.100			
Hybride Milfeuil			
Ormelong			
Socheville			
W.268.			

Varieties or strains classified as extra late are three to four weeks behind the early types in respect of spring growth. They are erect to semi-erect, with short, broad leaves on stems, which when mature, are thick and fibrous (Willey and Zaleski, 1955). Because of this late start they tend to be much lower yielding than the earlier strains and Willey and Zaleski report up to 60% plant mortality by the second harvest year and this level of winter-kill was six times more than in all the other strains. It is therefore unlikely that any varieties from the extra-late group would be suitable for British conditions.

The relative yields of varieties from the other three groups which currently appear in the National Institute of Agricultural Botany, Farmers' Leaflet No. 4 "Varieties of Herbage Legumes" are shown below and further details concerning these varieties can be obtained from the appropriate part of this leaflet.

TABLE 10

Relative Yields of Lucerne Varieties (N.I.A.B. data)

	Variety	Total annual yield of dry matter as percentage of control-variety du Puits
Early type	Europe	108
	Hybride Milfeuil	102
	du Puits (control)	100
	Eynsford	100
	Flamande	100
	Emeraude	98
	F.D.100	98
	Cardinal	96
Mid season	Provence	83
type	Szarvase	82
Late type	Rhizoma	87

Judging by the comparative yields in Table 10, it would appear essential for most farmers to choose a variety from the early type and when Europe is available at reasonable prices, this is the obvious choice. Differences in yield between the remaining early cultivars appear small and thus if the variety Europe is unavailable or very expensive, farmers could choose the cheapest within this group.

If mid season types like Provence or Szarvase are selected, spring growth will be later and the total annual production will be significantly lower than from the early cultivars, for example Provence has yielded only 72–83% of du Puits (N.I.A.B., 1966, 1968). In practical terms mid season varieties can only be cut two or three times during the season which means that on average there is one less cut per annum compared with the early varieties. With a growth habit which tends to be more prostrate than the early cultivars, the mid season types may prove slightly better for grazing, but it is a difficult decision to sacrifice about one-fifth of the potential yield.

The late types Rhizoma and Grimm are also low yielding and the spring growth is two to three weeks later than the early strains. Grimm was a popular variety several years ago but has been replaced by varieties from the early group. Rhizoma, also late, and prostrate in growth habit, is a bred strain which was released from Canada in 1948. According to Hawkins and Zaleski (1952) it is a hybrid between *M. falcata* (diploid) and *M. media* (tetraploid), which is reputed to produce rhizomes after the second year. These late strains are thought of as being particularly suited to a predominantly grazing management, but there is little evidence to support this in the literature.

Jekabsons (1959) working in East Anglia and considering lucerne primarily for artificial drying also indicated the yield superiority of the early varieties over the mid season types.

Vertus from Sweden, a verticillium wilt-resistant variety, looks promising for the future and also Sabilt the new cultivar from the Welsh Plant Breeding Station.

A. Inoculation and Treatment of Seed

(i) INOCULATION

The lucerne plant is able to fix large quantities of atmospheric nitrogen when associated with the bacterium *Rhizobium meliloti*. Besides supplying nearly all the needs of the legume, this symbiosis may often contribute a little nitrogen to associated grass species and once the lucerne ley is ploughed up then the following crops usually benefit as well. *Rhizobium meliloti* is effective on other Medicago species and on species of Melilotus (Sweet Clover) and Trigonella (fenugreek) but under most circumstances the seed should be inoculated with bacteria just prior to sowing. Lucerne plants which do not produce nodules in association with the rhizobia do not establish properly and can be classed as failures since they are dependent on combined nitrogen from the soil or from fertilisers for their growth. When lucerne has been cultivated fairly recently in the field which is again to be sown out with this legume then the soil will probably contain the rhizobial bacteria, but to be on the safe side and since the technique is easy and simply costs a few shillings extra per acre, it is better to inoculate the seed prior to drilling. This ensures a high concentration of bacteria in the rooting area and it means that nearly all the plant roots will be invaded by the rhizobia thus producing the symbiosis which results in nodulation.

Bacterial cultures containing *Rhizobium meliloti* can be purchased from chemical or pharmaceutical companies and it is wise to read and abide by any instructions which the manufacturers put out concerning their particular products.

In the past, in Britain, this bacterial culture has usually been transferred onto the lucerne seed in milk, but in other areas of the world, notably America, other media are suggested. Bolton (1962) recommends the application of 300–400 lb/acre of top soil from a field which supports a healthy crop of lucerne or sweet clover to be worked in with the seed. Another method suggested by the same author is the preparation of a solution of 0·5 lb of furniture glue in a gallon of water, lucerne seed to be moistened with this solution, then dusted with soil containing natural inoculum and sown immediately.

Recently cultures from the Continent have been made available in Britain which incorporate the bacteria on an inert peat carrier and the transfer to the seed is a relatively easy process. "Nodosit-lucerne" is one such preparation and brief details concerning its use are reproduced below from the distributors, instructions, Chemicovens Ltd (1967), whose full address will be found in the References. The cultures are contained in plastic bottles with sufficient material inside each to inoculate 60 lb of lucerne seed.

Instructions for inoculation

1. Remove cap and stopper.
2. Fill about four-fifths full of water.
3. Shake thoroughly.
4. Place contents in a large, clean watering can and dilute with about 6·5 pints of water.
5. Sprinkle on to the seed which has been spread out in a thin layer indoors on the floor or in the shade outside.
6. Mix the inoculum and seed thoroughly.
7. Allow time for the water to be soaked up.
8. Drill as soon as possible in the normal manner.

Receptacles containing the inoculum should be kept in a cool place which is not too dry. They should be protected from frost and sunlight and should not be opened until they are to be used. *Rhizobium meliloti* is sensitive to low and high temperatures and is killed when exposed to direct sunlight (ultraviolet rays). This should be avoided by inoculating in the shade and drilling the seed so that it is covered by about 1 inch of soil. Finally it must be stressed that the bacteria have a limited life span when transferred to the seed and it is therefore important that sowing is done immediately or soon after inoculation.

(ii) SCARIFICATION

The process of scarification, which is essentially a scratching of the surface of "hard" seeds, is employed to allow water to penetrate quickly. This results in a quick and even germination. It is normally done by special machinery associated with seed merchants and the seed has to be scratched but not cracked otherwise the germination percentage will be reduced.

(iii) FUNGICIDES

Lucerne seedlings are susceptible to soil borne fungi and fungicidal seed dressings may be useful in protecting them during the early stages of growth. Thiram based dressings can be employed against Verticillium wilt but some disinfectants and dressings containing mercury could be harmful to the rhizobial bacterial and these should be avoided when inoculation has been carried out.

(iv) DISINFECTANTS

The lucerne stem eelworm is a serious threat to lucerne growing and it can be transmitted on the seed. Fumigation with methyl bromide has been recommended and some merchants offer this service automatically to purchasers of lucerne seed. It is a cheap and successful way of avoiding the introduction of this eelworm to farms via the seed and is therefore recommended where necessary.

B. Seed Rates

Bolton (1962) recorded that alfalfa varieties in America contained approximately 220,000 seeds/pound and therefore at a seeding rate of 10 lb/acre there would be about 50 seeds sown/square foot. Seeding rates in America varied between 2 and 30 lb/acre and Bolton suggested that the lower end of the scale was most suitable for dry areas and are mainly associated with crops of lucerne destined for seed production. Where soil moisture is not the limiting factor, then higher seeding rates should be employed and when the primary purpose is the maximum production of herbage for grazing or conservation seeding rates of between 10 and 20 lb/acre should be employed.

Jarvis (1962) working at the Cambridge University farm on a light gravelly loam, demonstrated that up to 750,000 plants/acre the plant population had a marked positive effect on yield, and although further increases in numbers gave a slight increase in yield, this was of no practical significance. Rounding-off and combining the data of Bolton and Jarvis, it would appear that 4 lb of lucerne seed would be sufficient to plant an acre if there was a 100% field germination. Should one-half or one-third of the seeds sown produce plants then the theoretical optimum seeding rate should be 8–12 lb/acre respectively.

Practical recommendation prior to 1962 on the whole tended to be slightly higher than these calculated theoretical ones.

Mansfield (1945) advocated 22 lb lucerne/acre when sown alone or 20 lb with 8 lb of commercial cocksfoot and 4 lb of commercial timothy and claimed a 25% higher output of dry matter with the mixture of lucerne and grasses.

Davies (1952) suggested 15 lb of lucerne plus 2 lb of cocksfoot or 3 lb of either timothy or meadow fescue/acre.

This rate was confirmed a few years later by Tristram (1956) who advocated 14–16 lb/acre when the legume was sown alone or together with 3 lb of S.37 cocksfoot, 3 lb of S.48 timothy or 4 lb S.215 meadow fescue. This author suggested much lower seeding rates for leys where the legume and grass were to be sown in alternate rows, i.e. 6 lb lucerne with either 3 lb of cocksfoot or 4 lb of meadow fescue.

Green and Davis (1955) reported a number of experiments concerned with seeding rates for lucerne and lucerne–grass mixtures. In one trial a low seeding rate of 10 lb of lucerne plus 2·5 lb of cocksfoot was compared with 20 lb lucerne and 5 lb of cocksfoot and there was little to choose between these in terms of yield. Another experiment where lucerne was compared at 10, 15, 20 and 25 lb/acre, each accompanied by 5 lb of timothy, demonstrated that these seed rate variations had little effect on yield. Thus in the main the earlier experiments and findings are in line with the concept of plant population put forward by Jarvis (1962) referred to earlier.

Another experiment concerning seed rates and grass companion species for lucerne reported by Green and Davis (1955) is worthy of note particularly from the point of view of suitability of associated grass species. When 20 lb lucerne was sown alone or with 4 lb cocksfoot, 5 lb timothy or 8 lb of meadow fescue the lucerne monoculture produced the highest yield of crude protein, but the plots were quickly invaded by weed grasses, chiefly rough-stalked meadow grass. The lucerne–cocksfoot mixture produced the largest hay yields, but no more protein than the lucerne plus meadow fescue. The inclusion of 5 lb of timothy severely hampered lucerne aftermath growths and was clearly too high a rate for this grass species as a companion for lucerne. Finally the compatibility of lucerne and meadow fescue was clearly demonstrated and this has been verified in many experiments in Britain and other parts of the world.

Under New Zealand conditions 8 lb of lucerne with 10 lb of cocksfoot or 7 lb of *Phalaris tuberosa* (Harding grass) as suggested by Bevin

(1950) resulted in early domination of the lucerne by the grass species and later Garrett and Stewart (1952) suggested that cocksfoot if included should be at 3 lb/acre or less. Iversen and Calder (1956) showed that a lucerne–*Phalaris*–subterranean clover mixture gave higher dry matter yields than subterranean clover–grass pasture and they suggested a mixture of 8 lb lucerne, 2 lb cocksfoot, 3 lb *Phalaris tuberosa* and 4 lb subterranean clover per acre.

Recommendations regarding seed rates for lucerne or lucerne–grass leys in Britain are given in Table 11.

TABLE 11

Recommended Seed Rates and Companion Species for Lucerne under British Conditions

When grown alone	When grown in close association with grasses	When drilled in alternate, medium wide rows with a grass companion
14 lb lucerne	12 lb lucerne	6 lb lucerne
	+	+
	4 lb S.215 Meadow Fescue	4 lb S.215 Meadow Fescue
	or	*or*
	3 lb Cocksfoot (S.37, S.345 or Scotia)	3 lb Cocksfoot (light soils)
	or	
	2–3 lb Timothy (S.48, S.352 or Scots Timothy)	

XIV. MANAGEMENT AND UTILISATION OF THE CROP

A. Control of Pests

Pea and bean weevils (sitona species) will attack seedling lucerne and quite often can result in serious damage. As soon as the seedlings appear and the presence of weevils is confirmed the crop should be sprayed or dusted with DDT. The adult weevils are particularly active during April and May and this coincides with the appearance of many crops of lucerne. The following treatments are recommended for the control of this important pest (Martin, 1963).

1. DDT sprays at 10 oz active ingredient, i.e.
   (20 oz 50% wettable powder)
           or
   (2 pints emulsifiable concentrate in 20–40 gal water/acre);
2. DDT dusts at 2·5 lb active ingredient/acre, i.e.
   (50 lb/acre of 5% dust).

Pigeons may be troublesome during the early stages of growth. It is therefore important that the establishing seedlings are protected and since they are only vulnerable to significant damage during the first few weeks this will not incur much in terms of labour or cost and can mean the difference between success and partial failure.

B. Chemical Weed Control in Seedling Lucerne

Lucerne seedlings are slow to establish and can easily be dominated by broad-leaved and grass weeds and there are occasions when herbicides can be useful to assist in the production and maintenance of a successful braird. Fryer and Evans (1968) in the Weed Control Handbook recommend 2,4-DB salt and dinoseb-amine at the following rates and times of application:

2,4-DB salt	Dinoseb-amine
32 oz/acre	32 oz/acre
From the spade leaf–first trifoliate leaf stage up to fourth trifoliate leaf.	From the second to the fourth trifoliate leaf stage.

Lucerne is most susceptible to weed competition between the first and fourth trifoliate leaf stage and it is fortunate that this is the optimum time of treatment with 2,4-DB. Application after the fourth trifoliate leaf will usually result in deformity and a severe growth check. Although it will probably disappear after about two months, this is a long period for the legume to lose much of its competitive ability and spraying much after the fourth trifoliate leaf cannot be recommended.

Dinoseb-amine is a contact herbicide and can be used for weed control because lucerne exhibits considerable resistance to it. However, there is likely to be some scorch and it should only be resorted to when the weed problem is serious. The resistance shown by all legumes to dinoseb-amine is lower as the air temperatures rise and the maximum safe dosage rate for lucerne must be reduced to 24 oz/acre at 70°F (21°C) and over.

Where chickweed (*Stellaria media*) is particularly troublesome Chlorpropham at 32 oz/acre may be used once the lucerne has reached the fourth trifoliate leaf stage and this herbicide will seriously check any grasses present.

C. Chemical Control of Grass Weeds in Established Lucerne Crops

When lucerne is grown without companion grass species, weed grasses can often be a problem, since they may quickly colonize any bare ground between the lucerne plants. Both dalapon and paraquat, being selective monocot killers can be employed to eradicate these weed grasses, provided they are applied when the lucerne is dormant and when the weed grasses are actively growing. This situation occurs both in the early spring and autumn, but a spring spraying is preferable to one at the end of the growing season. According to Fryer and Evans (1968), dalapon should be used at 5 lb/acre or paraquat at 4–8 oz/acre and the latter is particularly effective in controlling meadow grasses (*Poa* spp.) and chickweed (*Stellaria media*). These authors reckon that either of the chemicals will ensure freedom from grass weeds during the following growing season, but remind lucerne growers that re-colonization by grasses or broad-leaved weeds can occur once again when the lucerne becomes dormant.

D. Cultural Weed Control

The most important and cheapest way of controlling annual weeds in establishing lucerne or lucerne–grass mixtures is by mowing or grazing and of these two, mowing is usually most satisfactory. The crop should be left until the early flowering stage wherever possible and although it may look unsightly, there are seldom any further problems with annual weeds after mowing and these will disappear provided they have not been allowed to seed. If the weed infestation is very heavy, then it may not be possible or desirable to leave the cutting until the lucerne begins to flower and under these circumstances a light topping with a mower set fairly high will be necessary at an earlier stage.

E. Early Management

The way in which lucerne leys are managed has a large influence both on productivity and longevity and in the first year it is very important to see that the plants establish well and the management must be regulated with this firmly in view.

The management in the establishment year is influenced markedly by the time of sowing and many authors have been consulted in drawing up the following suggestions for spring and summer seedlings (Table 12).

Green (1955) pointed out that lucerne tends to grow later in the autumn when young and he suggested that winter grazing should not take place until December during the early years. J. Davies (1952), confirmed the general opinion that repeated cutting of young lucerne tends to thin it out rapidly and long rest periods are necessary. Barker *et al.* (1956) working with lucerne and cocksfoot demonstrated that

TABLE 12

Establishment Year Management for Spring and Summer Sown Lucerne Leys

Spring sown (March–April)		Summer sown (July)
Ordinary conditions	Exceptionally favourable conditions	
First defoliation not later than the end of August	First defoliation in July, usually at flower bud stage.	With sowings in July, it is not usual for any herbage to be removed during active growth. The legume is left to grow and die back, but there is no reason why a grazing should not take place in the winter, say December.
September and October rest period to replenish root reserves.	Second defoliation end of September–beginning of October.	
Second defoliation, usually a grazing during November or December.	Light winter grazing or topping not later than January or February.	

frequent defoliation caused a reduction in persistency and provided there were eight-week intervals between cutting or grazing a satisfactory balance between these two species could be maintained. They also showed that early or late rest periods had little effect on the amount of lucerne present after two years if frequent defoliation was practised.

F. Manuring

Established crops of lucerne grown without grass companion species are independent of nitrogen by reason of their nodulation but lucerne–grass leys may require a little nitrogen at the beginning of the season to obtain some production from the grass component. Towards the end of the life of a lucerne–grass ley, if the contribution from the legume is very small, fairly large quantities of nitrogen will be required to obtain a satisfactory production level.

The lucerne crop requires medium to large quantities of phosphate and potash in the establishment year and where the management is mainly concerned with cutting and removing the herbage for silage, hay or for artificial drying then annual dressings to replace the amount taken away will certainly be needed. Green (1955) indicated that the usual practice was to apply fertiliser equivalent to 2–4 cwt super-phosphate and 2–6 cwt of muriate of potash/acre/annum and where magnesium deficiency was induced at high potash levels, this could be corrected by dressings of magnesium sulphate.

Jones and Dermott (1951) reported potash deficiency in lucerne and lucerne–grass leys even when the level of management was considered

TABLE 13

Fertiliser Levels for Lucerne and Lucerne–Grass Leys

	Lucerne alone (units/acre)			Lucerne plus companion grass species (units/acre)		
	N	P_2O_5	K_2O	N	P_2O_5	K_2O
Establishment year[a]	—	60	120	30	60	120
First year	—	40–60	120–180	60	40	120
Second year	—	40–60 units P_2O_5 and		60	40	120
Third year	—	120–180 units K_2O each		100	40	100
Fourth year	—	year if herbage is		150	40	60
Fifth year	—	mainly cut and removed, but smaller quantities would be sufficient if predominantly grazed.		200	40	60

[a] Phosphate and potash level for situations where soil analysis shows them at a medium level.

high and a yellowing and necrotic spotting was often followed by plant death and grass colonisation. They showed that the majority of cases occurred on low potash soils derived from chalk and under most circumstances the levels of applied potash were sub-optimal. On a potash-deficient soil in Kent overlying sedentary chalk where the levels of available lime and phosphorus were satisfactory, varying rates of muriate of potash were applied to lucerne–cocksfoot and lucerne–timothy leys. The appropriate quantities of potash were first broadcast

TABLE 14

Vegetable Characteristics of Lucerne in a Lucerne–Cocksfoot Ley and Yield from the Second Cut in 1950

	Plants/ sq. yd	Number of shoots per plant	Average height of shoot (in.)	Yield (tons green material/acre
Control (No potash)	27	7	9	1·70
60 units K_2O/acre/annum	32	8	11	2·39
120 units K_2O/acre/annum	47	13	14	2·52
180 units K_2O/acre/annum	53	15	15	3·53
240 units K_2O/acre/annum	48	19	15	3·66

After Jones and Dermott (1951).

in November (1947) and the annual dressings applied at the same rates during each subsequent autumn. During the summer of the first harvest year (1948) differences in the colour and vigour of the lucerne could be detected and trial cuts indicated substantially lower yields on the control plots which did not receive potash. Considerable reductions in plant number had occurred by 1950 and those plants remaining on the non-potash area were poor by comparison with the others (see Table 14).

G. Irrigation

Lucerne and lucerne–cocksfoot leys are thought of as two of our most drought-resisting crops and once established they can withstand considerable soil moisture deficits. From the earlier details concerning output in relation to rainfall, Davies and Tyler (1962) showed that a good yield was obtained even when rainfall during the growing season was 3·95 inches lower than the requirement as judged by the potential transpiration rate. However, it is likely that yields would have been higher if the amount of water which the crop received was nearer or equal to the potential transpiration rate and therefore in exceptionally dry years the application of irrigation water, if available, can be considered.

Very few lucerne growers in Britain practise irrigation, but in many other areas of the world it is recommended and adopted to increase productivity.

In the Canterbury region of New Zealand, Flay (1963) considered that lucerne would produce about one-third more than pasture in terms of herbage and Lobb (1967) stated that the general response of lucerne to irrigation at Winchmore had been of the order 30–40% thus indicating that efficiently irrigated lucerne could be the potentially highest yielding pasturage in this area of New Zealand. In this region Lauder (1959) demonstrated a 34% increase in total production per annum when the crop was irrigated after each cut, but there were signs that the legume lacked persistence and lost some of its vigour. Irrigated lucerne has a higher proportion of its roots in the top few inches of the soil (Lobb, 1967) and it must be remembered that an increase in the amount of soil moisture results in increased competition from associated companion grasses or weed-grasses and the latter are quick to invade thin stands.

Bolton (1962) indicated that irrigated lucerne was extremely important in the drier areas of United States, Canada, Central Asia, Near and Middle East, India, Pakistan, Australia, South Africa, Peru and Chile and he listed the following references for those who wish to consult detailed accounts of this subject; Palmer (1936), Jones and Brown (1942), Israelsen (1950), Marr (1954) and Morgan (1955).

Lucerne may require only 18–24 inches of water each growing season when only two or three cuts are taken, but with higher cutting frequencies this requirement may go up to 36–48 inches (Bolton, 1962).

In South-west France three irrigations per season are recommended and the levels and times of application are shown in Table 15.

TABLE 15

Levels and Time of Application of Irrigation for Lucerne in South-west France

| | Rate of application | |
Time of application	m^3/hectare	Approximate British equivalent (inches/acre)
At seeding time	500	2
Green bud stage	350	1·4
After flowering	300	1·2
Total season requirement	1150	4·6

($250 \ m^3$/ha $= 0·984$ inches/acre)
After Mennesson (1967).

H. Frequency of Defoliation

From the results of the early evaluation of varieties under British conditions, Hawkins and Zaleski (1952) were able to show that one additional cut could be obtained from the early strains compared with those in other groups and Zaleski and Dent (1960) suggested that in the first harvest year three cuts could be taken from the early varieties but only two from the late strains.

Most authorities on lucerne have indicated that maximum yield and persistency can be obtained where the defoliation frequency is low and adequate rest periods are allowed between each cut or grazing.

Two, four and eight cuts per season were compared at Cashmere, Canterbury, New Zealand, in respect of lucerne–grass associations where the grass component had been overdrilled and the results from two seasons appear in Table 16.

A drastic reduction in the yield of lucerne resulted from eight cuts per season and this encouraged weed growth and none of the grasses made sufficient growth to compensate for the loss in production from the lucerne. With lucerne or lucerne overdrilled with grasses, four cuts per season gave the highest output per acre.

TABLE 16

Total Herbage Yield ('00 lb DM/acre) Throughout the Growing Season from Lucerne and Lucerne–Grass Associations in New Zealand

Cutting frequency	Lucerne alone		Lucerne + timothy		Lucerne + ryegrass		Lucerne + S.170 Tall Fescue		Lucerne + cocksfoot	
	1963–1964	1964–1965	1963–1964	1964–1965	1963–1964	1964–1965	1963–1964	1964–1965	1963–1964	1964–1965
Two cuts	124	109	127	114	124	109	131	111	129	125
Four cuts	136	144	139	147	149	156	150	149	147	140
Eight cuts	91	71	89	69	92	60	91	73	92	70

After O'Connor (1967).

Davies *et al.* (1960) suggested that lucerne leys should be cut once in the seeding year and thrice in each of the following years to give maximum productivity and Jekabsons (1959) confirmed the superiority of three rather than four cuts with a large number of lucerne varieties primarily associated with grass drying.

TABLE 17

Yield of Dry Matter and Crude Protein (cwt/acre) from Lucerne Varieties over a 4-year Period in East Anglia

Cutting frequency	Yield of DM (cwt/acre)	Distribution of DM (% of total)	Crude protein yield (cwt/acre)
Three cuts/season	409·6	First cut—48·7 Second cut—31·4 Third cut—19·9	80·27
Four cuts/season	266·9	First cut—31·5 Second cut—31·2 Third cut—23·2 Fourth cut—14·2	58·59

After Jekabsons (1959).

In reviewing the results of their own work and those of previous British workers, Davies and Tyler (1962) concluded that whilst in most years in England, three cuts would produce the highest yield, they may on occasions be too severe.

Green (1955) pointed out that lucerne–cocksfoot mixtures could contribute much to summer and autumn grazing and in describing the management of this type of ley on a Derbyshire farm he indicated the value of three defoliations per season. An early cut in late May for silage or drying followed by grazing in July/August and a cut or grazing in October kept the lucerne thriving and the aggressive cocksfoot under control. Finally, on this subject of defoliation frequency, Keoghan (1967) has summarised the evidence from many parts of the world and any readers who are particularly interested in this aspect of lucerne cultivation would be advised to consult his paper and the references which it contains.

I. Utilisation

Lucerne and lucerne–grass leys may be cut or grazed and although a predominantly cutting management will usually lead to the highest output and longest productive period, at some time in the life of these leys or even perhaps every year, a grazing will be taken. They are particularly productive during the summer and autumn and many dairy farmers and graziers rely on lucerne leys during this period as the grass leys do not contribute much at this time, particularly during a very dry season. Restricting the time of grazing or the amount of herbage available will help to reduce incidences of bloat and in many parts of Southern Europe, lucerne is cut and green–soiled. Besides reducing bloat it will enable farmers to achieve a near 100% utilisation by eliminating fouling of the sward by dung and urine.

Lucerne alone is particularly useful as a green crop for drying and much of it in the drier areas of the South is used in this manner. Dried lucerne meal finds a ready market amongst the compounding companies and growers wishing to use lucerne as a cash crop can do so in this way. Protein contents are normally high and it is the level of protein which will determine the price which can be obtained. Cutting and removing the lucerne in the green state soon afterwards eliminates losses which can occur with other methods of conservation and for drying this crop really has no equal.

When the main method of conservation is hay-making, a grass companion species is usually included for a number of practical reasons. The resulting herbage dries more quickly and more easily and the losses of lucerne leaf can often be reduced. Provided the cut crop is treated gently, a fair proportion of the leaf remains in the made hay and it is certainly a nutritious and sought after winter fodder.

Silage making offers another method of utilisation, but as with hay-making there can be some practical problems. Lucerne alone is high in protein and relatively low in soluble carbohydrates and is therefore not the easiest of materials to conserve in silage form. The inclusion

of a companion grass species helps to raise the level of soluble carbohydrate in the herbage and thus a satisfactory fermentation process is
easier to achieve. At the same time additives to help the correct
fermentation to take place can be used and although they cost several
shillings per ton of green material, they will invariably improve the
silage quality. Molasses can be applied easily and quickly on the site
and this raises the level of readily available carbohydrate and the
beneficial bacteria can operate to produce low pH values which are
usually indicative of a satisfactory fermentation. Formic acid (commercial formulation Add-F, B.P. Chemicals) is another additive but
this time it is applied to the cut herbage from a forage harvester and
it immediately lowers the pH of the herbage so that the beneficial
organisms can take command of the fermentation (Figs 2 and 3).
Mineral acids are also useful in promoting the right micro-organisms
and as "AIV-acid" have been used for a considerable time in the
Scandinavian countries. AIV silage losses are lower than those experienced with other additives or when silage is made without an additive,
but the cost of this material is high. Considerable precautions have also
to be taken during application as the AIV mixture is corrosive and the
addition of sodium bicarbonate to neutralise the feed is usually practised.
For these reasons mineral acids have not been adopted by British
farmers for silage making, but they certainly are useful in promoting
the right conditions for fermentation particularly with high protein
herbage such as is obtained from lucerne.

XV. PRODUCTIVITY

Some indications of the level of productivity in terms of dry matter
and crude protein have already been referred to in the work of Jekabsons (1959). He demonstrated that lucerne for drying was capable of
yielding 5 tons of dry matter and 1 ton of crude protein/acre/annum
over 4 years when cut 3 times each year.

Green (1955) suggested a normal output of dried material of 3–5
tons/acre according to the soil fertility level and 2–3 tons of starch
equivalent/acre from ordinary leys in reasonably good vigour and
Tristram (1956) also suggested a yield of 4 tons of dry matter/acre
from 3 or 4 defoliations from lucerne in Northamptonshire.

Davies and Tyler (1962) recorded dry matter yields of 3·5, 4·3 and
4·9 tons/acre for the 3 years 1957, 1958 and 1959 respectively to show
much agreement with earlier work.

Differences between early and late varieties of lucerne in respect of

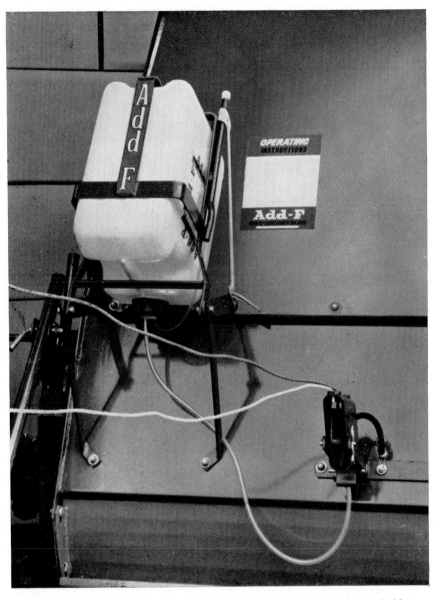

FIG. 2. Close up of "Add-F" acid applicator illustrating the bracket-held container, gravity feed tube and point of entry to the herbage and control cords which are used from the tractor seat. Photograph by courtesy of B.P. Chemicals, Agriculture Division.

dry matter and protein yields were demonstrated by Dent (1955) and these are reproduced below:

	Early varieties	Late varieties
Dry matter yield (tons/acre/annum)	5·3	3·8
Crude protein yield (cwt/acre/annum)	19·8	13·3

FIG. 3. Forage harvester picking up cut herbage and treating it with acid additive (Add-F) prior to loading and silage making. Photograph by courtesy of B.P. Chemicals, Agriculture Division.

XVI. QUALITY OF HERBAGE

Late summer–early autumn growth from a lucerne–grass ley was analysed by Bateman and Blaxter (1964) and the herbage, containing 5·2% grass on a dry weight basis, was found to have the following chemical composition (as % dry matter).

Ash	15·4	Lignin	5·4
Nitrogen × 6·25	19·2	Cellulose	23·4

Ether extract   1·0   Soluble carbohydrate   3·5
Crude Fibre   23·0   Carbon   44·1
N-free extract   41·4
(Calorific value: 4·319 kcal/g)
(Metabolisable energy: 2·81 ± 0·04 kcal/g)

The authors pointed out that the relatively low calorific value was due to the large ash (mineral) content, but that the gross energy value of the organic matter was similar to that of grasses containing similar protein levels.

Crude protein contents of lucerne dry matter are usually higher than those of grass except when heavily fertilised with nitrogen and the figure of 19·2% indicated the likely level of crude protein in herbage at the end of the growing season. Jekabsons (1959) who was particularly concerned with protein levels in lucerne for drying showed that a higher level could be maintained with more frequent cutting, but it

| | Three cuts per annum | | | Four cuts per annum | | | |
	First cut	Second cut	Third cut	First cut	Second cut	Third cut	Fourth cut
Crude protein (%)	19·0	20·1	21·8	22·8	21·6	21·6	20·4

After Jekabsons (1959).

must be remembered that a concomitant yield reduction follows very frequent defoliation.

One quality factor already referred to is a low soluble carbohydrate content and the difficulties of silage making have been discussed. Silages made from lucerne tend to be high in protein and low in energy and this is largely due to the high fibre associated with the lucerne stems.

Using the *in vitro* techniques for assessing feeding value, Terry and Tilley (1964) measured the digestibility of lucerne leaf and stem from young and mature plants and also calculated the percentage weight due to the leaf as the season progressed. A high proportion of leaf to stem has always been considered desirable and although advancing maturity will automatically give lower leaf–stem ratios, those varieties with the most leaf should be chosen for cultivation. Leaf as a percentage of the total herbage dropped from 70% during the first week of April to 35% towards the end of June (Terry and Tilley, 1964). These authors demonstrated how valuable the leaf of lucerne is when they showed the digestibility to be within the range 78–83% and the corresponding

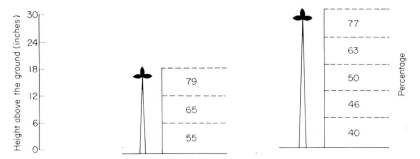

FIG. 4. *In vitro* digestibility of lucerne stem at varying heights from the ground on young and mature plants. After Terry and Tilley (1964).

digestibility figures for the stem of young and mature lucerne can be judged from Fig. 4.

PART 2. SAINFOIN

XVII. INTRODUCTION

Sainfoin, like lucerne, was being used as a forage legume during the seventeenth century. Worlidge (1668) and North writing in the middle of the eighteenth century indicated that sainfoin leys remained down from fifteen to twenty years (Davies, 1952). Jethro Tull (1733) in his classical book "The Horse-Hoing Husbandry" indicated that the name "Sain Foin" was given on account of the wholesomeness of this legume either as a green feed or dry and since he firmly stated that no description was necessary, we conclude that it must have been popular and widespread during that period.

According to Davies (1952), the specialised sowing of herbage plants has only been part of British farm practice since the middle of the seventeenth century and species like sainfoin, lucerne and red clover were introduced by the large landowners who had seen them under cultivation in nearby Continental countries.

It is difficult to assess the area under sainfoin as it has never been treated separately in the agricultural statistics, but the cultivation was fairly widespread in the eighteenth, nineteenth and early twentieth centuries. Many people associate sainfoin with either the chalk land of Southern England or the low rainfall areas of East Anglia but it has been grown in many areas of Britain. Sainfoin, known locally as French Grass, was introduced to Glamorgan about 1760 and Davies (1815) recorded that it was still being grown on a pretty large scale on the limestone soils of the Vale of Glamorgan during the early nineteenth

century. Rees (1928) traced the introduction and early spread of sainfoin in South Wales and during the late 1920's he found that twenty-
six farms in Glamorgan were growing or had recently grown this
legume.

Since the 1920's there has been a constant decline in the popularity
and acreage devoted to this legume, at first very slowly but of late very
rapidly. Hutchinson (1965) recorded that this once popular and long-
established legume has almost disappeared, since it does not respond
as well as other alternative fodder crops to the changing requirements
and circumstances of British agriculture.

A number of authorities have stated that sainfoin is indigenous in
South-east England on the chalk soils and since it has been found
growing wild the suggestion that it is native to Britain has been put
forward. Bell (1948), however, recognised the fact of it growing wild,
but pointed out that the name sainfoin is obviously of French origin,
meaning "healthy hay" and thus it is more likely to have come from
the Continent.

France and Flanders were considered to be the origin of British
sainfoin by Robinson (1937), and this author gave the date of introduction of "double-cut" sainfoin as 1834 and for "Giant" sainfoin he
suggested 1850.

XVIII. BOTANICAL CLASSIFICATION

Sainfoin is a member of the Leguminosae family belonging to the
genus *Onobrychis*. Darlington and Wylie (1955) list the various species

TABLE 18

Members of the Genus Onobrychis (Haploid number x = 7, 8)

Species	Chromosome number	Reference	Comments
O. arenaria	14	Favarger (1953)	Found in the Alps
O. caput-galli	14	Senn (1938)	Cultivated for fodder in the Mediterranean region
O. crista-galli	14	Corti (1931) and Senn (1938)	Cultivated for fodder in South-east Mediterranean region
O. pulchella	16	Senn (1938)	
O. viciifolia (*sativa*)	$\begin{cases} 14 \\ 28 \end{cases}$	Romanenko (1937) Maude (1939)	Common name sainfoin and cultivated in many areas
O. montana	28	Favarger (1953)	Found in mountains in South-east Europe

After Darlington and Wylie (1955).

belonging to this genus, and ascribe the name *Onobrychis viciifolia* (sativa) to this legume.

XIX. TYPES OF SAINFOIN

Most authorities use the name *Onobrychis sativa* Lam. and recognise two distinct forms which they term Common Sainfoin or Giant Sainfoin. Thomson (1938), studying the development of this legume in the seeding year, described these two forms in detail and it is from his work that the two types are now compared.

A. Common or Single-cut Sainfoin (*O. sativa* var. *communis* (Ahlefeld))

This type of sainfoin is long-lived and although leys of up to twenty years have been reported, it reaches a maximum yield in about the third year and it is rather unusual for it to be very prolific after seven or eight years. In the year of seeding common sainfoin remains prostrate without the formation of stems and seed. In subsequent years after the first cut, this form of sainfoin again remains prostrate in rosette form, without the formation of stems and flowers and the aftermath is normally grazed. Growth in general is less rapid and less luxuriant compared with the other type, but it remains productive for several years.

B. Giant or Double-cut Sainfoin (*O. sativa* var. *bifera* Hort.)

Giant sainfoin is a rapid growing species compared with the common form and even in the seeding year it sends up stems, forms flowers and sets seed. After cutting it again attempts sexual reproduction in the form of stem initiation and it is from this growth characteristic that it obtains the name double-cut sainfoin and two appreciable cuts can be obtained each year. Sometimes it is possible to harvest a third cut and although some authorities have suggested a third group for varieties from which three defoliations are possible, Zade (1933) pointed out that this distinction was not constant and they should be included within the giant sainfoin group.

XX. SEED AND PLANT MORPHOLOGY

A. Seed

The seeds of sainfoin are greenish-brown, bright and kidney-shaped, with prominent radicles and at 4–5 millimetres in length are considerably bigger than most of the other herbage legumes in British agriculture (Thomas and Davies, 1964). Milled seed is the true seed of sainfoin

whereas the "unmilled seed" carries with it the indehiscent pericarp
and is therefore much larger. Delayed germination in sainfoin was
first reported by Tull (1733) who found that if sowing took place with
a nurse or cover crop, some of the sainfoin seeds did not germinate
until the following spring. These findings were confirmed much later
by Rees (1931) and Thomson (1938). In experiments carried out by
the latter and sowing done in May, sainfoin seedlings continued to
appear until October in the year of seeding and they began to emerge
again from February the next year, and this delay in germination was
more abundant in the common strains.

Milled sainfoin seed has a higher germination capacity than unmilled
seed (Finlayson, 1906; Zade, 1933 and Percival, 1936). Using milled
and unmilled seed from the same bulk, Rees (1933) found that the
milled seed always showed a higher germination, a lower percentage
of hard seeds and more broken growths than the unmilled.

With unmilled seed, the husks may be empty or contain only partially
developed seed and with a fairly thick covering even the healthy germs
may find difficulty in emerging. The lower germination in respect of
unmilled seed has been ascribed to one or more of the following
factors:

(i) After the milling process some of the empty husks and small
shrivelled seeds are screened off.

(ii) Friction associated with the milling process reduces the number
of hard seeds.

(iii) The husks on the unmilled seed prevent water from readily
reaching the inner layers.

(iv) There could be mechanical difficulty in the piercing of the husk
by the emerging radicle and cotyledons.

(v) The husk itself forms a hard casing around the seed preventing
the normal swelling which occurs after water is taken up from the soil.

(vi) Within the husk there is an inhibitory substance which prevents
or retards germination.

(vii) Asphyxiation of the embryo may occur in the sealed atmosphere
of the husk.

After M.A.F.F. (1923) with additions by Thomson (1952).

Data relating to the percentage germination and proportion of hard
seeds with these two types were collected by Thomson (1952) and these
are reproduced in Table 19.

Thomson (1952) also demonstrated that unmilled seed maintained
its viability in storage better than milled seed, but from the above

TABLE 19

Germination Tests on Commercial Milled and Unmilled Sainfoin Seed

Type of seed	Number of samples tested	Average germination (%)	Average hard seeds (%)
Milled	34	71·4	1·2
Unmilled	23	61·1	7·0

After Thomson (1952).

figures in Table 19 it would be wise for growers of sainfoin to use milled seed wherever possible to obtain the best establishment.

B. Plant Morphology

Sainfoin seedlings have kidney shaped cotyledons and the leaves which appear soon after have varying numbers of leaflets. The first leaf is simple, the second and third trifoliate and all the later ones are pinnate, with 6–12 pairs of leaflets and a terminal one. Short lateral branches are formed and the plant develops into a rosette which tends to be more prostrate in the common forms compared with the giant types. Thomson (1938) studied the initial growth and demonstrated a significant difference in tillering capacity between the two main types (Table 20).

TABLE 20

Tillers in Sainfoin Seedlings

Type	Number of tillers/seedling	Standard Error
Giant	6·67	± 0·1764
Common	8·73	± 0·2482

After Thomson (1952).

Flowering in the seeding year varies with the time of sowing and variety or type of sainfoin and this was referred to earlier in the general description of the two groups.

Koreisa (1935) suggested that the difference between common and giant sainfoin was due to a variation in their thermo-stage development.

(i) SPRING PERENNIALS

These have a short thermo-phase and when they are sown in the spring, they flower the same year, e.g. Giant Sainfoin.

(ii) WINTER PERENNIALS

In comparison with the spring perennials, these have a long thermo-phase and even with spring sowings cannot be induced to flower the first year, e.g. Common Sainfoin.

XXI. SOIL AND CLIMATIC REQUIREMENTS

Sanders (1947) suggested that sainfoin was very suited to the light chalky soils and no authority would disagree with this generalisation. Mansfield (1945) drew attention to the general supposition that it was only thought suitable for dry and shallow soils on chalk or limestone formations, but was emphatic in suggesting that sainfoin would thrive under a wide range of both soil and climate. He went further, saying that sainfoin demanded a well-drained soil, with plenty of lime and a low water table.

When the Grassland Improvement Station was in Warwickshire, sainfoin was easily established on the lower lias clays, thus supporting Mansfield's theory. Green (1967) reported on sainfoin growing at the Grassland Research Institute, Hurley, and confirmed that establishment and growth was generally good on the chalky soils. It was thin or patchy on clay with flints where areas of low pH (below 6) were encountered and on the alluvial sands of the Thames Valley where pH values below 5 were recorded there were outright failures. In these areas of low pH it was possible to grow lucerne following the application of lime, but Green indicated that this remedial measure was, to date, less effective in encouraging the establishment of sainfoin.

This legume tends to produce long roots and is therefore considered a drought-resisting forage plant suitable for the low rainfall, high sunshine areas of South and Eastern Britain. In 1937 Robinson recognised these as the main areas of cultivation, but referred to claims that successful cultivation was possible as far north as the river Humber and as far west as the river Severn.

XXII. FERTILISERS

Earlier writers draw attention to the fact that sainfoin responds well to applications of farmyard manure, phosphates and potash but there is a lack of precise information concerning the amounts either applied to or needed by this crop.

Watson and More (1949) considered 3 cwt of superphosphate and 0·75 cwt of muriate of potash a good manuring for sainfoin at seeding and they suggested that further annual dressings of both these straight fertilisers would be necessary if the crop is to remain down for several years and hay crops are removed.

Provided sainfoin seedlings are able to obtain a start, using mineralised soil nitrogen, then there is no need to apply this element in fertiliser form. Experiments in Bulgaria clearly indicated that fertiliser

TABLE 21

Manuring of Sainfoin (Units/acre P_2O_5 and K_2O)

Soil levels of phosphate and potash as indicated by analyses	Seeding year		Subsequent years where first growth is cut and removed	
	P_2O_5	K_2O	P_2O_5	K_2O
Low	60	60	60	60
Medium	40	40	40	40
High	0	0	30	30

nitrogen decreased the nodule numbers on sainfoin roots and nitrogen should therefore be avoided under normal circumstances (Radomirov *et al.*, 1965).

Tentative fertiliser recommendations for sainfoin crops without companion species and based on soil analyses are given in Table 21.

Where grass companion species are included with this legume, then the annual fertiliser dressing may have to be modified to include nitrogen, but it must be recorded that sainfoin is best sown without grass companions as they quickly tend to become dominant.

XXIII. SEED

A. Varieties

There are relatively few recognised varieties available today and it is most likely that the majority of seed merchants will list common and giant sainfoin in their catalogues, in other words specifying type rather than variety.

Comparisons between and within these two groups were made by Rees (1928) who tested eleven lots of giant and twenty-one stocks of common sainfoin and small variations were recorded in time of flowering or time of first cut within each of these groups.

P

Examination of a number of stocks described as common sainfoin took place at Colesborne in 1943. Stocks originating from the Cotswolds were found to be more uniform and persistent, Hampshire stocks were lax in their mode of growth and those from the eastern counties were mixed and many contained a high proportion of stemmy, giant types (Green, 1967).

More recent examination of foreign varieties of sainfoin at the Grassland Research Institute, Hurley, Berkshire, indicated that stocks from East Germany, France, Russia and Turkey all showed the morphological characters of giant sainfoin and they appeared to be less persistent than Cotswold common (Green, 1967).

The following varieties are presently being tested by the National Institute of Agricultural Botany at their Headquarters Trial Ground, Cambridge (1969).

Eastern Counties Giant (control)	Milborne
Esparcet DSG.2002	Perly
Esparcet DSG.2004	Premier
Esparcet DSG.2005	Ternel
Fakir	

(N.I.A.B. "Guide to Varieties under Trial, Observation and Multiplication", 1969.)

During the mid 1960's, the area of Cotswold common sainfoin inspected with a view to seed certification was in the region of sixty acres and thus for an expansion in the area under this crop, it would be necessary to increase home production and at the same time use some of the foreign giant sainfoin varieties.

B. Sowing Rate

Recommended seeding rates for sainfoin in the past have always been high compared with the requirements of other herbage crops. Robinson (1937) suggested 4–5 bushels of unmilled or "cosh" seed or 50–60 lb when the husk had been removed by milling. Sanders (1947) and Watson and More (1949) make similar recommendations.

These relatively high seeding rates incur high establishment costs per acre (retail price of sainfoin seed in 1967–69 approximately £280/ton) and in the Grassland Research Institute trials which began in 1962, seed economy was thought necessary to bring the costs in line with other forage crops. A rate of 40 lb/acre milled seed was drilled in narrow rows 4 to 7 inches apart, although traditional spacing had often been of the order of 10 inches. Green (1967) reported that this seed rate gave a reasonable plant density of seedlings on the chalky soils, but commented that final establishments were too sparse and thus in further work they reverted to the traditionally high ones. It was pointed out that 56 lb milled seed/acre would provide less than 30 viable seeds/square foot (assuming average purity and germination) whereas current recommendations for other legumes and grasses allowed for much thicker stands, for example lucerne and red clover 60 viable seeds/square foot, ryegrass 80, cocksfoot and timothy 200. Cultural practices should therefore be adopted or modified to ensure the highest possible survival and establishment of seedlings as it will be certain that rates of over 56 lb/acre of milled seed cannot be justified at the present high cost. For leys sown with companion grass species 40 lb sainfoin with either 3 lb timothy or 6 lb of meadow fescue are tentatively suggested to increase the amount of available herbage and to prevent invasion of weed species.

C.   Time of Sowing

The period for sowing sainfoin is from April to July. Where the land is relatively free from weeds or the legume is to be established under cereals, then establishment will usually take place in April. The seed should be drilled about 1–1·5 inches deep in narrow rows, at right angles to the cereal, which will normally be spring barley. When under-sowing sainfoin, farmers should lower the cereal seeding rate and reduce or omit the nitrogen part of the fertiliser application to ensure the maximum chance of successful "take". Very poor establishment will result if there is any lodging in the barley nurse crops and growers should sacrifice some cereal yield as an insurance against failure.

When the spring is late or when the land needs cleaning, sainfoin can be put in without a nurse crop in May or early June. Crops establish well without cereal competition and where the ley is to be down for several years, this method is the most satisfactory.

Where rotations include crops like early peas or early potatoes or where the land can be cleared by June or July, then sowing can take place during these two months. From limited evidence and observations April and May sowings usually lead to better establishment compared with seedings in June or July and there is a chance of more herbage from cuts in the seeding year the earlier the sowing.

XXIV.   CHEMICAL WEED CONTROL

Sainfoin is best grown on narrow rows and therefore cultural weed control is restricted to the pre-sowing period. Once the crop has emerged and is in the early seedling or mature plant stage, then a few of the common herbicides may be used for the control of broad-leaved weeds and the recommendations as given in the "Weed Control Handbook" are reproduced below.

TABLE 22

Recommended Herbicides for Weed Control in Seedling and Established Sainfoin

Growth stage of sainfoin	Herbicide and dosage rate (oz active ingredient/acre)		
	MCPB-salt	2,4-DB-salt	Dinoseb-amine
1st compound leaf	32	32	Not recommended
2nd compound leaf	32	32	32
Established crops to be grazed or cut for fodder	—	—	Not recommended

After Fryer and Evans (1968).

Sainfoin is not a particularly aggressive herbage species and suffers considerably from competition from many of the fast growing broad-leaved (dicotyledonous) weeds. The above chemicals are useful in controlling many of these, but unless the crop is able to take over these areas, then the weed grasses will very quickly colonise these spaces and one weed problem is simply substituted for another.

In sainfoin at Hurley, the control of meadow grasses (*Poa* spp.) has been obtained over several years through the application of the grass killing herbicide, paraquat. This is given for information only, and does not constitute a firm recommendation, but weed-grass control was effected by the use of 3 oz paraquat/acre in the early spring of each year without apparent permanent damage to the legume.

XXV. MANAGEMENT AND UTILISATION

Common sainfoin remains prostrate in the year of seeding and it is only possible to obtain one lot of herbage from it towards the end of the growing season. In subsequent years it grows vigorously, sending up stems once only and there is likely to be a large bulk of herbage at that time which is normally cut and fed or conserved. The aftermath remains prostrate, is much lower yielding than the first growth and is normally utilised by grazing.

Giant sainfoin, which is usually termed double cut sainfoin, produces herbage twice, sufficiently bulky and upright for cutting and the after-math can be used for grazing. This type should be chosen where cutting and conservation is the primary object of sainfoin growing, but it should be borne in mind that the leys will have to be renewed much more frequently than those sown with the common form.

The first growth will usually be cut and removed from all established sainfoin leys and immediately afterwards 30–60 units each of phosphate and potash should be applied to replenish the soil levels and to promote further growth. Pure stands of sainfoin when cut for hay tend to lose a significant proportion of their leaf through shattering unless all the operations are carried out slowly and with great care. This leaf loss can be reduced by including grass companion species in the seeds mixture. Yields will often be much higher and weed invasions much reduced compared with the pure stands.

Second cuts from giant sainfoin may be taken for seed provided the initial defoliation is not a late one.

XXVI. YIELD

A. Forage

Well manured and well managed sainfoin should yield up to 3 tons of hay in respect of the common form and up to 2 tons if the giant strains are chosen for cultivation (Robinson, 1937), although Watson and More (1949) suggested that 1·5 tons would be produced under average circumstances.

Comparisons between the two forms of sainfoin have led to conflict-ing views on the superiority of either in respect of yield. Work carried out on a field scale at the beginning of the twentieth century at Woburn showed English Giant superior to English Common, but French Common was higher yielding than French Giant (Voelcker, 1902).

From small scale trials in Wales, Rees (1932) showed that total pro-duction was greater with the common varieties, but where two hay crops were required, preference was shown for the giant forms. Fagan and Rees (1930) had previously established that common sainfoin could outyield the giant forms by from 57 to 260% and Klapp (1932), working in Germany during the same era, confirmed this order of merit in the first harvest year of sainfoin.

Thomson (1951) in an exhaustive study of sainfoin in its first harvest year confirmed the findings of the majority of the earlier work that common outyielded giant forms and he demonstrated this under both pasture and hay conditions. This author also showed that common withstood more frequent cutting better than giant sainfoin. Yields are reduced as cutting frequencies are raised with most forage crops and sainfoin is no exception to this general rule, but Thomson suggested that it withstood repeated cutting better than lucerne.

At the Grassland Research Institute, Cotswold common sainfoin produced at least 2 tons of dry matter/acre in the first week in June. This yield of dry matter was similar to that produced by lucerne, but there was evidence of a higher digestibility from the sainfoin. About 1·5 tons of dry matter was obtained from two further cuts in mid July and late September and on each occasion the digestibility was similar to that of the June cut (Green, 1967).

B. Seed

In the past, seed was usually taken from leys of common sainfoin which had been down for at least 5 or 6 years, to ensure that the seed had come from long-lived plants (Robinson, 1937). Average yields of 400 lb/acre milled seed were obtained, but recent information from the Cotswold Seed Growers Association suggests the present annual expectation to be about 2 cwt/acre and this fact alone would curb any rapid expansion of sainfoin growing in Britain unless foreign seed was both satisfactory and forthcoming. Most authorities have indicated the superiority of British seed and intending growers of sainfoin are advised to obtain it in preference to foreign stocks even though they may be cheaper.

TABLE 23

Chemical Composition of the Dry Matter of the Whole Plant from Sainfoin Hay

Cutting dates	Ether extract	Crude protein	Fibre	Ash	Soluble carbohydrate
Common Sainfoin					
26th May	2·61	17·00	28·00	6·25	46·14
6th June	2·85	14·96	32·00	5·25	44·94
14th June	2·65	14·79	35·55	4·50	42·51
2nd July	2·20	12·01	37·25	4·36	44·18
Giant Sainfoin					
26th May	3.00	18·47	27·63	6·25	44·65
6th June	3·05	19·41	29·50	5·00	43·00
14th June	3·20	13·71	31·60	4·51	46·98
2nd July	2·05	15·66	33·50	6·22	42·57

After Fagan and Rees (1930).

XXVII.   QUALITY OF HERBAGE

Fagan and Rees (1930) analysed the dry matter of sainfoin hays cut at different stages and the data from their classical experiment is reproduced in part in Tables 23 and 24.

These authors separated leaves, stems and flowering head and analysed these fractions separately and it is worthwhile comparing leaves and stems from the point of view of animal nutrition, and the data in Table 24 is taken from the 6th June cut.

TABLE 24

Analysis of Leaf and Stem Dry Matter from Sainfoin Hays (% DM)

Chemical component	Common sainfoin		Giant sainfoin	
	Leaf	Stem	Leaf	Stem
Ether extract	4·10	2·10	4·60	2·45
Crude protein	21·61	8·22	26·85	10·41
Fibre	17·31	47·12	13·60	39·30
Ash	6·25	2·70	7·00	3·10
Soluble carbohydrate	50·73	39·86	47·95	44·74

After Fagan and Rees (1930).

The data given in Table 24 well illustrate the large differences in quality between leaves and stems. The leaf is much more valuable since it contains about twice the oil content, two and a half times the crude protein, only one-third of the fibre and higher ash and higher soluble carbohydrates. It is therefore of prime importance to eliminate as far as possible leaf shatter and leaf loss in conservation if a quality product is sought.

In recent years the digestibility of the organic matter of herbage has been adopted as a single measurement of quality. The 2 ton/acre yield of dry matter referred to earlier as being obtained at Hurley in the first week in June had a digestible organic matter content around 62% and was some 3–4% higher than lucerne sample under comparable conditions (Green, 1967).

Osbourn et al. (1966) compared the voluntary intake of leguminous hays using sheep and they showed that voluntary intake of sainfoin of digestibility (DM) 62·6% was 18% higher than the voluntary intake of red clover of 64·6% digestibility. This was a high voluntary intake of sainfoin hay and although the authors were not able to explain this through the chemical analyses or by current theories, it lends credence to the older concepts about the acceptability and whole-someness of sainfoin herbage and sainfoin hays.

Intake, digestibilities and chemical analyses of the sainfoin, red clover and lucerne hays are given in Table 25.

TABLE 25

The Voluntary Intake, Apparent Digestibility and Chemical Composition of Three Legume Hay (% ash, % nitrogen and % carbohydrates in the dry matter)

	Sainfoin	Red Clover	Lucerne	S.E. (Difference)
Intake				
(g/kg $W^{0.73}$/24 hours)	87·3	74·0	63·5	±2·07
Digestibility (DM %)	62·6	64·6	58·3	±0·22
Ash	7·61	9·67	7·47	—
Nitrogen	2·92	2·38	2·66	—
Water soluble CHO	10·7	11·9	7·8	—
Pentosan	8·4	7·1	9·3	—
Hexosan	22·6	25·6	31·2	—
Cellulose	28·0	29·8	34·8	—
Lignin	10·7	6·8	9·7	—
Acid detergent fibre	39·0	38·0	43·8	—
Acid pepsin soluble	33·5	36·4	—	—

After Osbourn *et al.* (1966).

References

Ahlgren, G. H. (1949). "Forage Crops". McGraw-Hill, New York, Toronto and London.

Barker, M. G., Hanley, F. and Ridgman, W. J. (1956). *J. agric. Sci. Camb.* **48,** 361–65.

Bateman, J. V. and Blaxter, K. L. (1964). *J. agric. Sci. Camb.* **63,** 129–31.

Bell, G. D. H. (1948). "Cultivated Plants of the Farm". Cambridge University Press, London.

Belov, A. I. (1932). *I.A.B. Pl. Genet. Herb. Pl. Bull.*, 6.

Bevin, R. H. (1950). *Canterbury (N.Z.) Agric. Bull.*, 247.

Blackman, G. E. and Black, J. N. (1959). *Ann. Bot. (London)* **23,** 51–63.

Bolton, J. L. (1962). "Alfalfa. Botany, Cultivation and Utilization". Leonard Hill (Books) Ltd, London; Interscience, New York.

Bolton, J. L. and Greenshields, J. E. R. (1950). *Science, N.Y.* **112,** 275.

Bula, R. J. (1958). Rep. 16th Alfalfa Improv. Conf., New York. pp. 79–81.

Bula, R. J., Rhykerd, C. L. and Langton, R. G. (1959). *Agron. J.* **49,** 33–36.

Calder, F. W. and Jackson, L. P. (1965). *Can. J. Pl. Sci.* **45,** 211–218.

Chamblee, D. S. and Lovvorn, R. L. (1953). *Agron. J.* **45,** 192–196.

Chemicovens Ltd (1967). Vintry House, Queen Street Place, London, E.C.4.

Coburn, F. D. (1906). "The Book of Alfalfa". Orange Judd, New York.

Coffindaffer, B. L. and Burger, O. J. (1958). *Agron. J.* **50,** 389–392.

Cooper, D. C. (1935). *J. Agric. Res.* **51,** 471.

Cooper, C. S. (1966). *Crop Sci.* **6,** 63–66.

Cooper, J. P. (1965). "The Evolution of Forage Grasses. Crop Plant Evolution" (Sir Joseph Hutchinson, ed.). Cambridge University Press, London.

Corti, R. N. (1931). *Nuovo. G. bot. ital.* **38,** 230.

Cowett, E. R. and Sprague, M. A. (1963). *Agron. J.* **55,** 432–435.

Darlington, C. D. and Wylie, A. P. (1945). "Chromosome Atlas of Flowering Plants", 1st edition. Allen and Unwin Ltd, London.

Darlington, C. D. and Wylie, A. P. (1955). "Chromosome Atlas of Flowering Plants", 2nd edition. Allen and Unwin Ltd, London.

Davies, J. (1952). Lucerne Establishment Trials in the West Midlands. 1947–1951. *Agriculture* **59,** 266–268.

Davies, R. O., Morgan, T. B. and Davies, W. Ellis. (1953). *J. Brit. Grassl. Soc.* **8,** 47–68.

Davies, W. (1815). "A General View of the Agriculture, etc. of South Wales (Sainfoin)". Vol. 1, pp. 592–594.

Davies, W. (1952). "The Grass Crop. Its Development, Use and Maintenance". E. and F. N. Spon, Ltd, London.

Davies, W. (1955). Prefatory note. The Grassland Research Institute, Memoir No. 1, p. 2.

Davies, W. Ellis (1964). *J. Brit. Grassl. Soc.* **19,** 263–270.

Davies, W. Ellis, and Davies, R. O. (1956). *J. Brit. Grassl. Soc.* **11,** 127–138

Davies, W. Ellis, Davies, R. O. and Harvard, A. (1960). *J. Brit. Grassl. Soc.* **15,** 106–115.

Davies, W. and Davis, A. G. (1951). *J. Brit. Grassl. Soc.* **6,** 119.

Davies, W. Ellis and Tyler, B. F. (1962). *J. Brit. Grassl. Soc.* **17,** 306–314.

De Candolle, A. (1882). "Origin of Cultivated Plants" (English edition, 1919). D. Appleton, New York and London.

Dent, J. W. (1955). *J. Brit. Grassl. Soc.* **10,** 330.

Donald, C. M. (1963). *Adv. Agron.* **15,** 1–118.

Fagan, T. W. and Rees, J. (1930). *Welsh J. Agric.* **6,** 224–237.

Favarger, C. (1953). *Bull. Soc. Neuchâtel. Sci. Nat.* **76,** 133.

Finlayson, D. (1906). "Sainfoin Seed". *J. Brit. Agric.* **13,** 147.

Flay, A. H. (1963). *Proc. N.Z. Grassl. Ass.* **24,** 104–115.

Fryer, J. D. and Evans, S. A. (1968). "Weed Control Handbook. II. Recommendations", p. 142. Blackwell Scientific Publications, Oxford.

Fryer, J. R. (1930). *Canad. J. Res.* **3,** 3.

Garner, F. H. and Sanders, H. G. (1940). *J. agric. Sci. Camb.* **30,** 182.

Garrett, H. E. and Stewart, J. D. (1952). *Canterbury (N.Z.) Agric. Bull.,* 273.

Ghimpu, V. C. R. (1929). *Cong. int. Agric., Bucarest.*

Gist, G. R. and Mott, G. O. (1957). *Agron. J.* **49,** 33–36.

Graumann, H. O. and Hanson, C. H. (1954). "Growing alfalfa". *Fmrs.' Bull. U.S. Dept. Agric.,* 1722.

Green, J. O. (1955). A Review of Practice in the Extension of Lucerne Growing. Grassland Research Institute, Hurley, Berkshire, England. Memoir No. 1.

Green, J. O. (1967). N.A.A.S. Quarterly Newsletter No. 8. G.R.I., Hurley, Berks. pp. 7–10.

Green, J. O. and Davis, A. G. (1955). "The Seasonal Use of Lucerne–Grass Mixtures for Grazing". Grassland Research Institute, Hurley, Berkshire. Memoir No. 1.

Grinsted, R. (1950). "The Cultivation of Lucerne". *Agriculture* **57,** 304–308.

Hawkins, R. P. and Zaleski, A. (1952). *J. nat. Inst. Agric. Bot.* **6,** 212–222.

Hayman, J. M. (1964). "Studies on Legume Establishment and Growth on an Acid Sulphur Deficient Soil". M. Agric. Sci. Thesis. Lincoln College, University Canterbury, New Zealand.

Hewitt, E. J. (1952). *Inter. Soc. Soil Sci.* Comm. 2 and 4, pp. 107–118.

Hughes, G. P. (1955). *J. agric. Sci. Camb.* **45,** 179–201.

Hutchinson, J. B. (1965). "Crop Plant Evolution: A General Discussion. Essays on Crop Plant Evolution". Cambridge University Press, London.

Israelson, O. W. (1950). "Irrigation Principles and Practices" (second edition). John Wiley, New York; Chapman & Hall, London.

Iversen, C. E. (1957). Linc. Coll. Tech. Publication 14, New Zealand.

Iversen, C. E. and Calder, J. W. (1956). *Proc. N.Z. Grassl. Ass.* **18,** 78.

Iversen, C. E. and Meijer, G. (1967). "Types and Varieties of Lucerne. The Lucerne Crop" (R. H. M. Langer, ed.), pp. 74–84. A. H. and A. W. Reed, Wellington, Auckland, Sydney.

Jarvis, R. H. (1962). *J. agric. Sci. Camb.* **59,** 281–286.

Jekabsons, V. (1959). *J. Brit. Grassl. Soc.* **14,** 253.

Jones, J. O. and Dermott, W. (1951). *Agriculture.* **57,** 507–509.

Jones, B. J. and Brown, J. B. (1942). "Irrigated Pastures in California". Circ. Calif. Agric. Ext. Serv., 125.

Keoghan, J. M. (1967). "Effects of Cutting Frequency and Height on Top Growth of Pure Lucerne Stands. The Lucerne Crop". A. H. and A. W. Reed, Wellington, Auckland, Sydney.

Kilcher, M. R. and Heinrichs, D. H. (1958). *Canad. J. Plant. Sci.* **38,** 252–259.

Klapp, E. (1932). "Zum Wachstumsrhythmus von *O. viciaefolia*". *Zuchter.* **4,** 280.

Klinkowski, M. (1933). "Lucerne: Its Ecological Position and Distribution in the World". *Bull. Imp. Bur. Pl. Genet. Herb. Pl.,* **12,** 61.

Koreisa, I. V. (1935). *Herb. Rev.* **3,** No. 2.

Langer, R. H. M. (1967). "Responses of Lucerne to Temperature and Light. The Lucerne Crop" (R. H. M. Langer, ed.), pp. 28–35. A. H. and A. W. Reed, Wellington, Auckland, Sydney.

Lauder, B. A. (1959). *N.Z. Jl. Agric.* **98,** 57–60.

Leach, G. J. (1967). "Growth and Development of Lucerne. The Lucerne Crop" (R. H. M. Langer, ed.), pp. 16–21. A. H. and A. W. Reed, Wellington, Auckland and Sydney.

Ledingham, G. F. (1940). *Genetics*, **25,** 1.

Lobb, W. R. (1967). "Irrigation, Management and Fertilizer Interactions. The Lucerne Crop" (R. H. M. Langer, ed.). A. H. and A. W. Reed, Wellington, Auckland and Sydney.

Lynch, P. B. (1967). "Place of Lucerne in New Zealand: Past and Present. The Lucerne Crop" (R. H. M. Langer, ed.). A. H. and A. W. Reed, Wellington, Auckland and Sydney.

M.A.F.F. (1923). "Sainfoin". *J. Min. Agric.* **30,** 426.

Mansfield, W. S. (1945). *J. Min. Agric.* **52,** 255–258.

Marr. J. C. (1954). "Grading Land for Surface Irrigation". *Circ. Calif. Agric. Exp. Sta. Ext. Serv.* 438.

Martin, H. (1963, 1965). "Insecticide and Fungicide Handbook". Blackwell Scientific Publications, Oxford and Edinburgh.

Massengale, M. A. and Medler, J. T. (1958). *Agron. J.* **50,** 377–380.

Matches, A. G., Mott, G. O. and Bula, R. J. (1962). *Agron. J.* **54,** 541–543.

Maude, P. F. (1939). *New Phytol.* **38,** 1.

McKenzie, R. E. (1951). "The Ability of Forage Plants to Survive Early Spring Flooding". *Sci. Agric.* **31,** 358–367.

Mennesson, H. (1967). "La Production de Semences Certifiées de Luzerne". I.T.C.F., F.N.A.M.S. Paris.

Morgan, A. (1955). "Irrigated Lucerne in Victoria". Reprint Aust. Dept. Agric., Victoria.

National Institute of Agricultural Botany (1966). Farmers' Leaflet No. 4. "Varieties of Lucerne". Cambridge, England.

National Institute of Agricultural Botany (1968). Farmers' Leaflet No. 4. "Varieties of Herbage Legumes". Cambridge, England.

O'Connor, K. F. (1967). "Lucerne–Grass Associations Under Different Sowing and Defoliation Systems. The Lucerne Crop". A. H. and A. W. Reed, Wellington, Auckland and Sydney.

Osbourn, D. F., Thomson, D. J. and Terry, R. A. (1966). Proc. Xth Int. Grassland Congress, Helsinki, pp. 363–367.

Palmer, A. E. (1936). "The Use of Irrigation Water on Farm Crop". Published by Canad. Dept. Agric., 509.

Parker, W. H. (1931). *J. nat. Inst. agric. Bot.* **3,** 111–126.

Patterson, J. K. and Law, A. G. (1952). *Agron. J.* **44,** 520–524.

Pearson Hughes, G. (1955). "Summer and Winter Feeding from Grass–Lucerne Drills". *Agriculture* **62,** 115–117.

Percival, J. (1936). "Agricultural Botany" (eighth edition). Duckworth, London.

Piper, C. V. (1935). "Forage Plants and their Culture", p. 671. MacMillan Co., New York.

Pritchett, W. L. and Nelson, L. B. (1951). *Agron. J.* **43,** 172–177.

Radomirov, P., Yakimova, Ya, Dzhumalieva, D. and Parushkov, G. (1965). *Rastenievud. Nauki.* **2** (1), 95–108. (Higher Agric.) Inst., Sofia.

Rees, J. (1928). *Welsh J. Agric.* **4,** 242–250.

Rees, J. (1931). *Welsh J. Agric.* **7,** 155–168.

Rees, J. (1932). "In Quest of the Best Sainfoin". *Welsh J. Agric.* **8,** 124.

Rees, J. (1933). *Welsh J. Agric.* **9,** 170.

Robinson, D. H. (1937). "Leguminous Forage Plants", pp. 74–81. Edward Arnold, London.

Romanenko, V. (1937). *J. Bot. Acad. Sci. Ukr.* **11,** 3.

Sanders, H. G. (1947). "An Outline of British Crop Husbandry". Cambridge University Press, London.

Schonhorst, H. M., Davis, H. L. and Carter, A. S. (1957). *Agron. J.* **49,** 142–143.

Senn, H. A. (1938). *Bibl. Genet.* **12,** 175.

Spafford, W. J. (1933). "Lucerne-growing in South Australia". *Bull. S. Aust. Dept. Agric.* **246,** 51.

Steinke, T. D. (1963). "A Study of Growth in Lucerne in Response to Defoliation". Ph.D. Thesis, Lincoln College, University of Canterbury, New Zealand.

Stewart, J. G. (1951). "Farming Affairs—Lucerne". *Agriculture.* **57,** 540–541.

Terry, R. A. and Tilley, J. M. A. (1964). *J. Brit. Grassl. Soc.* **19,** 363.

Tewari, G. P. and Schmid, A. R. (1960). *Agron. J.* **52,** 267–269.

Thomas, J. O. and Davies, L. J. D. (1964). "Common British Grasses and Legumes". Longmans, Green & Co., Ltd, London.

Thomson, J. R. (1938). *Ann. App. Biol.* **25,** 457.

Thomson, J. R. (1951). *J. Brit. Grassl. Soc.* **6,** 107–117.

Thomson, J. R. (1952). *J. Brit. Grassl. Soc.* **7,** 65–69.

Tiver, N. S. (1960). *Proc. 8th Int. Grassl. Cong.,* 93–98.

Tomé, G. A. (1947). *Rev. Fac. Agron. B. Aires* **11,** 299.

Tristram, J. E. (1956). "The Merits of Lucerne. Northamptonshire Experience". *Agriculture* **62,** 467–470.

Tschechow, W. (1933). *Bull. appl. Bot., Pl.-Breed. Ser.* **2,** 1, 119.

Tull, J. (1733). "The Horse-Hoing Husbandry".

Voelcker, J. A. (1902). "The Woburn Field Experiments". *J. Roy. agric. Soc. Eng.* **63,** 324.

Watson, J. A. S. and More, J. A. (1949). "Agriculture. The Science and Practice of British Farming" (ninth edition). Oliver & Boyd, London and Edinburgh.

Wheeler, W. A. (1950). "Forage and Pasture Crops", p. 752. Van Nostrand, Toronto, New York and London.

White, J. G. H. (1965). "Comparative Studies on Growth and Nodulation of Subterranean Clover and Lucerne". Ph.D. Thesis, University of Adelaide, Australia.

White, J. G. H. (1967). "Establishment of Lucerne on Acid Soils. The Lucerne Crop" (R. H. M. Langer, ed.). A. H. and A. W. Reed, Wellington, Auckland and Sydney.

Whyte, R. O., Nilsson-Leissner, G. and Trumble, H. C. (1953). "Legumes in Agriculture". *F.A.O. Agric. Series.* **21,** 367.

Willard, C. J., Thatcher, L. E. and Cutler, J. S. (1934). "Alfalfa in Ohio". *Bull. Ohio Exp. Sta.,* 540.

Willey, L. A. and Zaleski, A. (1955). "Lucerne Investigations, 1944–1953". G.R.I. Memoir No. 1. Hurley, Berks.

Worlidge, J. (1668). "Systema Agriculturae". Samuel Speed, London.

Zade, A. (1933). "Pflanzenbaulehre für Landwirte". Parey, Berlin.

Zaleski, A. (1954). *J. agric. Sci. Camb.* **44,** 199–220.

Zaleski, A. (1954). *J. nat. Inst. agric. Bot.* **7,** 131–141.

Zaleski, A. and Dent, J. W. (1960). *J. Brit. Grassl. Soc.* **15,** 21–27.

Author Index

A

Agricultural Chemicals Approval
  Scheme, 21, *58*
Ahlgren, G. H., 387, *444*
Aldrich, D. T. A., 277, 297, 298, *325*
Allen, H., 278, 279, 282, *325*
Amos, A., 234, *263*
Amos, A. J., 183, 184, *214*
Anderson, E., 218, 219, *260, 262*
Anderson, J. L., 48, *58*
Andersson, R., 132, *175*
Arable Farmer, 104, *118*
Arrhenius, O., 137, *174*, 185, *213*
Atterberg, A., 64, *118*

B

Bailey, N. T. J., 277, *324*
Barber, D., 14, *58*
Barker, W. G., 420, *444*
Bateman, J. V., 429, *444*
Beadle, G. W., 219, *260, 261*
Beaven, E. S., 64, 65, *118*
Becker, W. R., 231, *260*
Bell, G. D. H., *58*, 327, 330, *380*, 432, *444*
Bell, R. A. M., 149, 165, *174*, 188, 193, 194, *213, 214*
Belov, A. I., 389, *444*
Berger, J., 217, 222, 227, 231, *260*
Berry, G., 373, *380*
Bevin, R. H., 417, *444*
Bickmore, D. P., 138, *175*
Black, C. J., 106, *118*
Black, J. N., 403, *444*
Blackman, G. E., 234, 239, *260*, 281, *324*, 403, *444*
Blair, I. D., 340, *380*
Bland, B. F., 234, *260*
Blaxter, K. L., 429, *444*
Blumer, S., 340, *380*

Board of Agriculture, 328, *380*
Bolton, J. L., 385, 386, 388, 391, 397, 398, 399, 402, 415, 416, 423, 424, *444*
Bond, D. A., 279, 280, 281, 299, *323*
Boyce, D. S., 97, *118*
Boyd, A. G., 135, *174*
Boyd, J., 86, *118*, 145, *174*
Brown, J. B., 424, *446*
Brown, W. L., 219, *260*
Bula, R. J., 403, 404, *444, 447*
Bullen, E. R., 76, 85, *118*
Bunting, E. S., 232, 234, 235, 239, 244, 247, 248, 253, *260*
Burger, O. J., 403, *445*
Burges, H. D., 100, *118*
Burrell, N. J., 99, 100, *118*
Burtt-Davy, J., 218, *260*

C

Calder, F. W., 402, *445*
Calder, J. W., 418, *446*
Campbell, J. D., 138, *175*
Carleton, M. A., 123, *174*
Carpenter, K. J., 299, 300, *324*
Carter, A. S., 402, *448*
Castle, M. E., 234, *260*
Chamblee, D. S., 410, *445*
Chemicovens, Ltd., 415, *445*
Chesneau, J-C., 239, *260*
Cobley, L. S., 219, *260*
Coburn, F. D., 387, *445*
Coffindaffer, B. L., 403, *445*
Coffman, F. A., 122, 124, 134, *174*
Collins, G. N., 218, *260*
Cook, K. F., 138, *175*
Cooke, G. W., 140, *174*, 272, *324*, 338, *380*
Cooper, B. A., 278, 279, 282, 283, 292, 293, *324, 325*
Cooper, C. S., 403, *445*

Cooper, D. C., 388, *445*
Cooper, J. P., 385, *445*
Copp, I. G., 340, *380*
Corti, R. N., 432, *445*
Cowett, E. R., 404, *445*
Cutler, H. C., 218, 219, *260*
Cutler, J. S., 399, 400, *449*

D

Darlington, C. D., 3, *58*, 65, *118*, 124, *174*, 181, *214*, 219, *260*, 267, *324*, *325*, 330, *380*, 388, 432, *445*
Davies, J., 386, 404, 408, 420, *445*
Davies, L. J. D., 433, *448*
Davies, R. O., 405, 425, *445*
Davies, W., 389, 390, 392, 393, 400, 404, 409, 417, 431, *445*
Davies, W. Ellis, 400, 401, 405, 406, 410, 411, 423, 425, 427, *445*
Davis, A. G., 405, 409, 411, 417, *445*, *446*
Davis, H. L., 402, *448*
Daw, M. E., 76, *118*
De Candolle, A., 64, *118*, 122, *174*, 217, 218, *260*, 385, 387, *445*
Dent, J. W., 135, 136, *174*, 424, 429, *445*, *449*
Dermott, W., 421, 422, *446*
Donald, C. M., 404, *445*
Drayner, Jean M., 278, *324*
Drew, N., 200, *214*
Dunkle, P. B., 203, *215*
Dzhumalieva, D., 437, *448*

E

East Midlands N.A.A.S. Circular, 291, *324*
Eddowes, M., 232, 234, 235, 243, 247, 249, 253, 255, *260*, *261*
Eden, A., 298, *324*
Emme, H., 124, 125, *174*, 180, *214*
Emmerson, R. A., 219, *261*
Etherbridge, W. C., 124, *174*
Evans, R. E., *58*, 131, 136, *174*, 183, 184, *214*, 253, *261*, 298, *324*
Evans, S. A., 283, 284, 313, *324*, 343, 344, 345, 346, 352, 353, 366, *380*, 419, 420, 440, *445*

F

Fagan, T. W., 441, 442, 443, *445*
F.A.O., 187, *214*
Farmer and Stockbreeder, 5, *58*, 127, *174*, 380, *380*
Favarger, C., 432, *445*
Fedotov, V. S., 330, *380*
Findlay, W. M., 123, 126, 133, 134, 135, 137, 144, 145, 146, 155, *174*
Finlayson, D., 434, *445*
Finn-Kelcey, P., 97, *118*
Fisons Agricultural Technical Information, 29, *58*
Flay, A. H., 423, *445*
Foot, A. S., 234, *260*
Fraser, A. C., 219, *261*
Fraser, J. G., 64, *118*
Free, J. B., 279, 282, *324*
French, S. A. W., 95, *120*
Fryer, J. D., 283, 284, 313, *324*, 343, 344, 345, 346, 352, 353, 366, *380*, 419, 420, 440, *445*
Fryer, J. R., 388, *446*
Fyfe, J. L., 277, 279, 280, *323*, *324*

G

Gane, A. J., 310, *324*, 342, *380*
Gardner, H. W., 11, *58*
Garner, F. H., 411, *446*
Garner, H. V., 11, *58*, 188, *214*
Garrett, H. P., 418, *446*
Geering, J., 189, *214*
Gent, G. P., 308, 309, 311, 318, *324*, *380*
Ghimpu, V. C. R., 388, *446*
Gill, N. T., 65, *118*, 265, *324*
Gist, G. R., 403, 404, *446*
Gouws, J. B., 180, *214*
Graumann, H. O., 398, *446*
Green, J. O., 401, 405, 408, 409, 411, 417, 420, 421, 425, 427, 436, 438, 442, 443, *446*
Greenshields, J. E. R., 388, *444*
Greenwood, H. N., 268, 269, 273, 275, 276, 277, *324*
Greer, E. M., 3, *58*
Gregory, P., 342, *381*
Greig, D. J., 97, *118*
Grinsted, R., 398, 408, 410, *446*

H

Hall, A. D., Sir, 12, *58*, 73, *118*, 226, *261*, 310, *324*
Hanley, F., 420, *444*
Hanna, W. F., 225, *261*
Hanson, C. H., 398, *446*
Harder, A., 340, *380*
Harrison, J., 103, *119*
Harshberger, J. W., 217, 218, *261*
Haussknecht, C., 122, *175*
Hawkins, R. P., 280, *323*, 391, 392, 412, 414, 424, *446*
Hayes, H. K., 220, *261*
Hayman, J. M., 400, *446*
Heinrich's, D. H., 410, *446*
Hellstrom, Y., 132, *175*
Hewitt, E. J., 398, *446*
Hirayoshi, I., 267, *324*
Hoare, A. H., 188, *214*, 317, *324*
Hodgson, G. L., 281, *324*
Holliday, R., 27, 28, *58*, 90, 150, *175*, 190, 196, *214*
Home Grown Threshed Peas Joint Committee, 336, 340, *380*
Hughes, G. P., 410, 411, *446*
Hunter, H., 64, *119*, 149, *175*, 184, 192, 193, *214*
Huskins, C. L., 124, 125, *175*
Hutchinson, J. B., 432, *446*

I

Immer, F. R., 220, *261*
Israelson, O. W., 424, *446*
Iversen, C. E., 389, 398, 402, 418, *446*

J

Jackson, H., 86, *119*
Jackson, L. P., 402, *445*
Jarvis, R. H., 411, 416, 417, *446*
Jealott's Hill Research Station, 17, 18, *58*
Jeater, R. S. L., 17, *58*, 81, *119*
Jekabsons, V., 414, 425, 427, 430, *446*
Jenkins, M. T., 218, *261*
Johnson, C. L., 299, 300, *324*
Jones, B. J., 424, *446*
Jones, D. L., 203, *215*
Jones, F. G. W., *58*, *214*, 286, *324*, *380*

Jones, J. O., 421, 422, *446*
Jones, Margaret G., *58*, 214, 286, *324*, *380*
Jones, P. J., 79, *119*
Jugenheimer, R. W., 218, 219, *261*

K

Kalning, H., 183, *215*
Kappen, H., 230, *261*
Karpechenko, G. D., 267, *324*
Kempton, J. H., 218, *261*
Kent, N. L., *58*, 128, 132, 134, *175*, 181, 182, 184, *214*
Kent-Jones, D. W., 183, 184, *214*
Keoghan, J. M., 426, *446*
Kilcher, M. R., 410, *446*
Kimber, D. S., 196, 198, 204, 205, 206, 207, *214*
King, J. M., 362, *380*
Kinsey, C., 128, 142, 145, 150, *175*
Klages, K. W. H., 187, *214*
Klapp, E., 441, *446*
Klinkowski, M., 385, 386, 387, *446*
Koreisa, I. V., 435, *446*
Körnicke, F., 64, *119*, 122, *175*
Kostoff, D., 180, *214*
Kramer, C., 72, *119*
Kühn, H., 189, *214*
Kuleshov, N. N., 218, *261*

L

"La Culture de Mais", 232, *261*
Langer, R. H. M., 402, *447*
Langton, R. G., 404, *444*, *448*
La Potasse, 11, *58*, 186, *214*
Lauder, B. A., 423, *447*
Law, A. G., 410, *448*
Leach, G. I., 402, *447*
Ledingham, G. F., 388, *447*
Lessells, W. J., 20, *58*
Lewis, C., 138, *175*
Linnaeus, C. L., 64, *119*, 218, *261*
Linser, H., 189, *214*
Lobb, W. R., 423, *447*
Long, H. C., 188, *214*
Lovelidge, B., 97, *119*
Lovvorn, R. L., 410, *445*
Lucas, N. G., 143, *175*
Lynch, P. B., 394, *447*

M

McCance, R. A., *59*, 182, 183, *214*
McClean, S. P., 21, *59*
McConnell, P., 6, *59*
McKenzie, R. E., 398, *447*
McLeish, J., 267, *324*
Macpherson, J. F., 107, *119*
Macrae, T. F., *59*, 182, 183, *214*
M.A.F.F., 33, 36, *59*, 199, 205, 206, 207, *214*, 228, 229, 245, *261*, 285, 286, 287, 295, 296, 297, 299, 300, 303, 304, 305, 311, 315, 317, 320, *324*, *325*, 332, 333, 338, 342, 343, 344, 351, 358, 359, *381*, 434, *447*
Malzew, A. I., 122, 124, *175*
Mangelsdorf, A. J., 219, *261*
Mangelsdorf, P. C., 218, 219, *261*, *262*
Mansfield, W. S., 410, 417, 436, *447*
Marais, J. C., 203, *215*
Marr, J. C., 424, *447*
Martin, H., *59*, 247, *262*, 286, 287, 314, *325*, 418, *447*
Martin, P. C., 106, *119*
Martin, W. McK., 360, *381*
Massengale, M. A., 403, *447*
Matches, A. G., 404, *447*
Mathias, K., 48, *59*
Matsumura, M., 267, *324*
Maude, P. F., 432, *447*
M.D.A., 224, 254, *262*
Medler, J. T., 403, *447*
Meijer, G., 389, 402, *446*
Mennesson, H., 424, *447*
Milbourn, G. M., 230, 232, 234, 236, 237, 242, 244, 246, 247, 250, 251, 252, 254, *262*
Ministry of Agriculture, Fisheries and Food, *447*
Moore, F. J., 246, *262*
Moore, H. I., 182, 188, 202, 204, *214*
Moore, W. C., 246, *262*
Moran, T., *59*, 182, 183, *214*
More, J. A., 11, *59*, 188, 190, 202, *215*, 437, 438, 441, *449*
Morgan, A., 424, *447*
Morgan, T. B., 405, *445*
Mortimer, R. G., 255, *261*
Mott, G. O., 403, 404, *446*, *447*
Mundy, E. J., 21, *59*, 76, *119*

Munro, R. F., 163, *175*
Müntzing, A., 180, *215*

N

N.A.A.S., 100, *119*
Nakajima, G., 180, *215*
National Institute of Agricultural Botany, 51, 52, 54, 111, 166, 168, 192, 204, 211, *215*, 238, *262*, 270, 277, 289, 297, 303, 311, 312, *325*, 360, *381*, 413, *447*
Nelson, L. B., 403, *448*
Neumann, M. P., 183, *215*
Nicolaissen, W., von, 122, *175*
Nilsson-Leissner, G., 389, *449*
Nishiyama, I., 126, *175*
Nix, J., 254, *262*, 297, 300, *325*, 356, 374, *381*
Nixon, B. R., 106, *119*

O

O'Brien, D. G., 178, 179, 182, 183, 188, 190, 192, 202, *215*
O'Connor, K. F., 404, 406, 425, *447*
Oliver, D. L., 219, *261*
Olsen, C., 185, *215*, 333, *381*
Osbourn, D. F., 443, 444, *447*
Owers, A. C., 339, 340, *381*

P

Page, J. B., 86, *119*
Palmer, A. E., 402, 424, *447*
Parker, W. H., 412, *447*
Parushkov, G., 437, *448*
Patterson, J. K., 410, *448*
Pearson, Hughes G., 396, *448*
Percival, J., 2, *59*, 126, *175*, 178, *215*, 327, *381*, 434, *448*
Peterson, R. F., 2, *59*
P.G.R.O., 309, 311, 313, 318, *325*, 336, 338, 349, 352, 360, 362, 363, *381*
Philp, J., 125, *175*
Picard, J., 277, *325*
Piper, C. V., 387, *448*
Pizer, N. H., 228, *262*
Porter, J., 190, 202, 204, *215*
Post, J. J., 72, *119*

Price, C. D., 193, 194, *214*
Pringle, W. J. S., *59*, 182, 183, *214*
Proctor, J. M., 342, *381*
Propcorn-Users' Manual, 104, 105, *119*
Prout, J., 76, *119*
Prout, W. A., 76, *119*
Prytherch, E. I., 208, *215*

R

Radomirov, P., 437, *448*
Randolph, L. F., 217, 219, *262*
Rayns, F., 69, 71, 72, 82, 95, *119*, 188, *214*
Rees, J., 432, 434, 437, 441, 442, 443, *445, 448*
Reeves, R. G., 218, *262*
Reynolds, J. D., 336, 342, 362, *381, 382*
Rhoades, M. M., 220, *262*
Rhykerd, C. L., 404, *444, 448*
Ridgman, W. J., 420, *444*
Riedel, I. B. M., 278, 279, 281, *325*
Riley, R., 3, *59*
Rimpau, W., 64, *119*
Robinson, D. H., 11, 27, *59*, 432, 436, 438, 441, 442, *448*
Romanenko, V., 432, *448*
Rothamsted Experimental Station, 73, 74, *119*, 203, *215*
Rowland, S. J., 234, *260*
Rowlands, D. G., 277, *325*
Rybin, V. A., 267, *325*
Ryvita Company Ltd, 183, *215*

S

Salter, P. J., 228, *262*
Sanders, H. G., 190, *215*, 411, 436, 438, *446, 448*
Sansome, E. R., 330, *382*
Schleimer, A., 183, *215*
Schmid, A. R., 410, *448*
Schonhorst, H. M., 402, *448*
Schulz, A., 122, 123, *175*
Scriven, W. A., 278, 279, 282, *325*
Senn, H. A., 432, *448*
Sirks, M. J., 277, *325*
Small, J., 137, *175*, 185, 186, *215*, 270, *325*, 333, *382*

Smith, B. F., 277, 297, 298, *325*
Smith, C. E. Jr., 219, *262*
Smith, D. C., 220, *261*
Smith, L. P., 72, *119*
Soper, M. H. R., 271, 273, 274, 276, 281, 297, *325*
Spafford, W. J., 386, *448*
Spier, J. D., 124, 125, 126, *175*
Sprague, M. A., 404, *445*
Stansel, R. H., 203, *215*
Stanton, T. R., 124, *175*
Stapledon, R. G., 188, *214*
Steinke, T. D., 404, *448*
Stewart, J. D., 418, *446*
Stewart, J. G., 398, *448*
Stoner, C. R., 219, *262*
Sylvester, R., 76, *120*

T

Terry, R. A., 430, 431, 443, 444, *447, 448*
Tewari, G. P., 410, *448*
Thatcher, L. E., 399, 400, *449*
Thomas, J. O., 433, *448*
Thomas, P. T., 267, *325*
Thompson, D. J., 443, 444, *447*
Thompson, H. L., 68, *120*
Thompson, J. B., 5, *59*
Thomson, J. R., 433, 434, 435, 441, *448*
Thorne, G. N., 95, *120*
Tilley, J. M. A., 430, 431, *448*
Tiver, N. S., 399, *448*
Tomé, G. A., 388, *448*
Toynbee-Clark, Gillian, 299, *323*
Trabut, W. L., 125, *175*
Trènel, M., 137, *175*, 185, *215*, 333, *382*
Tristram, J. E., 395, 396, 399, 417, 427, *449*
Trumble, H. C., 389, *449*
Tschechow, W., 388, *449*
Tull, J., 431, 434, *449*
Turpin, H. W., 203, *215*
Tyler, B. F., 400, 401, 406, 423, 425, 427, *445*

U

Upcott, M. B., 219, *260*

V

Vavilov, N. I., 2, *59*, 64, *120*, 123, *175*, 180, *215*, 218, *262*
Vear, K. C., 65, *118*, 265, *324*
Voelcher, A., 135, *176*
Voelcker, J. A., 76, *119*, 441, *449*
Voss, A., 64, *120*

W

Wallace, T., *59*, 188, *214*, *325*
Washko, J. B., 203, *215*
Watson, D. J., 95, *120*
Watson, J. A. S., 11, *59*, 188, 190, 202, *215*, 437, 438, 441, *449*
Weatherwax, P., 217, 218, 219, *262*
Webber, J., 20, *58*
Weed Control Handbook, 29, 30, 34, *59*
Weijer, J., 220, *262*
Weinmann, W., 183, *215*
Wellhausen, E. J., 219, *263*
Werner, H., 64, *119*, 122, *175*
West of Scotland Agricultural College, 205, 207, *215*
Wheeler, W. A., 385, *449*
White, J. G. H., 398, 399, *449*
Whitehouse, R. N. H., 5, *59*

Whyte, R. O., 389, *449*
Widdowson, E. M., *59*, 182, 183, *214*
Willard, C. J., 399, 400, *449*
Willey, L. A., 412, *449*
Williams, H. T., 132, *176*
Williams, J. B., 228, *262*
Williams, J. R. Parry, 245, *263*
Willis, J. C., 330, *382*
Wilten, W., 72, *119*
Woodman, H. E., 234, *263*
Worlidge, J., 386, 431, *449*
Wort, D. A., 278, 279, 281, *325*
Wright, Edith M., 374, *382*
Wylie, A. P., 3, *58*, 65, *118*, 124, *174*, 181, *214*, 219, *260*, 267, *324*, *325*, 330, *380*, 388, 432, *445*

Y

Yakimova, Ya, 437, *448*
Yule, A. H., 339, 340, *381*

Z

Zade, A., 122, *176*, 433, 434, *449*
Zaleski, A., 391, 392, 412, 414, 424, *446*, *449*

Subject Index

A

Abyssinia, 64
Accumulated heat units (AHU), 364, 365
Acetylene gas moisture meter, 36, 37
Actinomyces, 103
Afghanistan, 180, 385
Alcohol, 67, 68
Aldrin, 21, 91, 152
Aleurone layer, 6
Alpha-chloralose, 379
Altitude, 12
Aluminium toxicity, 398
Arctic Circle, 64, 192
Argentina, 4, 186, 386, 391, 392
Armenia, 180
Artificial grass drying, 394, 395
Ascochyta, 302
Asia Minor, 123, 180
Aspergilli, 103
Atrazine, 242, 243, 244
Australia, 4, 184, 334, 386, 391, 400, 423
Austria, 391
Available water capacity (AWC), 315

B

Barban, 32, 92, 283, 284, 346
Barley
  species
    Hordeum agriocrithon, 64
    Hordeum murinum, 64
    Hordeum sativum, 65
    Hordeum spontaneum, 64
  type
    bere, 65
    four-row, 65
    six-row, 65
    two-row, 65
  variety
    Banba, 90, 117
    Beorna, 108
    Berac, 90, 112, 113, 115
    Dea, 89
    Deba Abed, 89, 90, 108, 111, 112, 113, 115
    Earl, 108
    Freja, 108
    Gerkra, 89, 90, 108, 112, 114
    Golden Promise, 90, 116
    Hassan, 90, 111, 112, 114, 115
    Herta, 108
    Imber, 89, 90, 108, 112, 114, 115
    Impala, 81, 90, 92, 113, 114
    Inis, 90
    Julia, 90, 108, 112, 113, 115
    Kenia, 79
    Lofa Abed, 90, 112, 113, 114, 115
    Maris Badger, 75
    Maris Otter, 89, 110, 111, 115
    Mentor, 90, 117
    Midas, 89, 90, 108, 112, 114, 115
    Mirra, 110
    Pallas, 89, 117
    Pioneer, 80, 89
    Plumage Archer, 75
    Proctor, 79, 89, 90, 92, 108, 111, 112, 113, 114, 115
    Rika, 81, 89
    Ruby, 90, 117
    Senta, 110
    Spratt Archer, 79, 108
    Sultan, 90, 108, 112, 113, 115
    Union, 90, 117, 118
    Vada, 89, 90, 108, 112, 113, 115
    Ymer, 90, 108, 116
    Zephyr, 89, 90, 108, 111, 112, 113, 115, 117
Basic seed, 22
Batch drier, 43
Bean mosaic, 303
Belgium, 186, 268, 269

Beluchistan, 385
Benazolin, 32, 92, 94, 154
BHC, 21, 87, 152, 236, 244, 340, 348
Birlane, 152, 245
Biscuit-making quality in wheat, 4
Black bean aphid, 291
Black bread, 179
Black grass (*Alopecurus mysosuroides*), 284
Bloat, 396, 426
Boronated fertilisers, 400
Botrytis (chocolate spot), 287, 302
Bran, 5, 6
Bread wheats, 4
British certified seed, 22, 56, 170
Bromoxynil, 32, 92, 94, 154
Bronze age, 122, 327
Brown Foot Rot, 145
Bulk or bin drying, 44
Bumble bees, 277, 278, 281
Butyl-rubber containers, 252

C

Canada, 4, 181, 184, 186, 224, 386, 390, 391, 423
Capacitance moisture meter, 37
Capillary moisture, 29
Captan, 276, 341, 364
Carbon dioxide, 95, 101
Cascade drier, 38, 39, 43
Chemical ploughing, 14, 81, 82, 83, 84
Chickweed, 420
Chile, 386, 423
China, 386
Chlorfenvinphos, 245
Chloropropham, 284, 346, 419
Christian era, 122
Cocksfoot, 405, 406, 411, 417, 418, 422, 423, 425
Cold water extract (CWE), 69, 71
Commercial seed, 22
Core beag (bristle pointed oats), 127
Cosh seed, 438
Coulter spacing, 190
Covered smut, 87, 145
Cow-grazing days, 208
Cresylic acids, 243, 284, 345
Cross pollination, 185, 213, 277, 278, 279, 281

Cuba, 217
Cycocel (CCC), 7, 189

D

2,4-D, 32, 33, 92, 94, 153, 154, 199, 200
2,4-D amine, 243
2,4-DB, 32, 92, 94, 199, 419, 440
2,4-D ester, 243
Dalapon, 420
Damping off, 247
DDT, 91, 152, 286, 348, 418
Deficiency payments, 46, 106
Demeton methyl, 292
Demeton-s-methyl, 314, 315
Derris, 292
Desiccants, 352, 353, 357
Diastase/diastatic power (DP), 67, 69, 70
Diazinon, 152
Dicamba, 32, 92, 154
Dichlorprop, 32, 92, 94, 154, 199
Dichlorvos, 314
Dieldrin, 21, 314
Dimethoate, 287, 292, 314
Dimexan, 345, 352
Dinoseb, 92, 94, 154, 199, 285, 341
  in oil, 284, 313, 345
Dinoseb acetate, 291, 313, 347, 365
Dinoseb-amine, 29, 284, 291, 313, 345, 347, 365, 419, 440
Dinoseb-ammonium, 29, 285, 347
Diquat, 243, 284, 312, 345, 352, 353
Direct drilling, 18, 81, 82, 83, 84
Distilling, 65
Disulfoton, 287, 292, 314
Diuron, 284, 346
DNOC, 29
Dormancy, 69, 95, 213
Double-flow drier, 40, 41
Downy mildew, 246
Drilling depth, 27, 85, 190, 206
Drying temperatures, 43
Dwarf beans and other types
  species
    Phaseolus coccineus, 266
    Phaseolus vulgaris, 266
    Vicia faba, 266
  type
    broad beans, 266, 267, 269
    french beans, 266, 267, 269

green pod, 303
kidney beans, 266, 267, 310
runner beans, 266, 269
scarlet runner, 267
string beans, 303
stringless beans, 303
wax pods, 303
variety
Bonvert, 311
Bush Blue Lake (274), 311
Cascade, 311
D6V111, 311
Earligreen, 311
Early Gallatin, 311
Encore, 311
Executive, 311
Gallatin (50), 311
Glamis, 311
Green pod (60209), 311
Harvester, 311, 312
I Delight, 311
Prelude, 311
Processor, 311, 312
Slenderwhite, 311
Slimgreen, 311
Sprite, 311
Tendercrop, 311
Tenderette, 311
Tenderwhite, 311

E

Egypt, 2
Egyptians (ancient), 122
Embryo, 5, 6
Endosperm, 5, 6
Ergot, 45, 209, 210
Ergotine, 209
Ergotism, 209
Extensometer curve, 5
Eye spot (*Cercosporella herpotrichoides*), 13, 18

F

Feekes–Large Scale, 7
Fenoprop, 32, 34, 92, 154
Fenugreek, 414
Fenuron, 284, 346
Field Approved Seed, 22, 56, 170

Field beans
species
Faba vulgaris, 266
Vicia faba, 266
Vicia narbonensis, 265
type
carse beans, 266, 290, 303
Horse beans, 266, 289
soya beans, 300
stockfeed beans, 265
tic beans, 266, 289
variety
Blue rock, 289, 322, 323
Daffa, 277, 321
Franks Ackerperle, 322, 323
Gartons Pedigree (Tarvin), 289, 322, 323
Gartons SQ, 277
Granton, 290
Herz Freya, 289, 303, 322, 323
Maris Bead, 289, 322
Maris Beaver, 277, 321
Mazagan, 290
Minor, 289, 322, 323
Strubes, 289
Suffolk Red, 290
Throws M. S., 277, 321
Field Certified Seed, 22
Folithion, 152
Foot rot, 247
Formic acid (add-F), 427, 428, 429
Formothion, 287, 292, 314
Four pole, 349, 351
France, 186, 224, 227, 248, 268, 269, 385, 386, 391, 424, 432
Frit fly, 13, 130, 141, 144, 245
Furfural, 133

G

Georgia, 180
Germany, 179, 181, 186, 224, 268, 269, 385, 386
Germination, 25, 68, 96, 97, 104
Gin, 181
Gluten, 4, 184
Glutenin, 184
Gramineae, 2
Grazing winter cereals, 7, 28, 150
Greece, 2, 122, 123, 386

Grey speck, 141, 153

H

Hair hygrometer, 37
Hebrews, 122
Hilum, 294
Holland, 181, 184, 186, 224, 248, 268, 269
Home-grown Cereals Authority (H-G.C.A.), 47, 96
Home-grown Threshed Peas Joint Committee, 336, 340
Honey bees, 277, 278, 282, 283, 292
Hungary, 186
Hut rack, 349, 351
Hybrid beans, 279
Hypomagnesaemic tetany, 28

I

In-bred beans, 279
In-breeding depression, 278
India, 64, 217, 390, 423
Infra-red moisture meter, 36
Inoculation (root nodule bacteria), 308, 410, 414, 415
In vitro (digestibility assessment), 430, 431
Ioxynil, 32, 92, 94, 154
Iran, 385
Iraq, 385
Iron age, 64, 122
Italy, 327, 385, 386

J

Jack Olding drier, 42

K

Kashmir, 385

L

Latvia, 186
Leaf area index (LAI), 229, 247
Leaf roll virus, 289, 303
Leaf spot (*Ascochyta* sp.), 302, 303
Leaf stripe, 87, 146
Leatherjackets, 13, 91, 141, 151, 152

Ley or lea oats, 139
Light intensity, 403, 404
Lime-induced chlorosis, 333
Lime-pelleted seed, 400
Lipase, 132
Lithuania, 184, 186
Loose smut, 21, 24, 87, 145
Lucerne
  species
    Medicago arborea, 388
    Medicago falcata, 388, 389, 390
    Medicago hemicycla, 388
    Medicago media, 388, 389, 412
    Medicago ovalis, 388
    Medicago rugosa, 388
    Medicago sativa, 387, 388, 389, 390, 398, 412
  type
    common, 389, 390, 398
    early, 390, 429
    extra late, 391
    late, 391, 429
    Mediterranean, 412
    mid-season, 391
    non-hardy, 389
    northern European, 412
    Turkestan, 389
    variegated, 389
    yellow flowered, 389
  variety
    Cardinal, 412, 413
    Chartrainvilliers, 412
    Choele Choel, 412
    Chubut, 412
    du Puits, 412, 413
    Emeraude, 412, 413
    Europe, 412, 413
    Eynsford, 413
    FD 100, 412, 413
    Flamande, 412, 413
    Grimm, 390, 412, 414
    Hilario Ascasubi, 412
    Hungarian White Seal, 412
    Hybrid Milfeuil, 412, 413
    Ile de France, 412
    Ladak, 390
    La Pampa, 412
    Marlborough, 412
    Old Franconian, 412
    Ormelong, 412

Pampeana, 412
Provence, 403, 405, 412, 413
Rhizoma, 412, 413, 414
Saladina, 412
Socheville, 412
Szarvase, 412, 413
W 268, 412
Lysine, 300

M

Machair soil, 127, 137
Magnesian limestone, 400
Maize
  species
    Zea mays amylacea, 220
    Zea mays ceratina, 220
    Zea mays everta, 220
    Zea mays indentata, 220
    Zea mays indurata, 220
    Zea mays saccharata, 220
    Zea mays tunicata, 220
  type
    dent, 220, 237
    flint, 220, 237
    Indian Corn, 217
    pod, 220, 222
    pop Corn, 220, 222
    soft, 220, 222
    sweet corn, 220, 228, 237, 238, 248
    waxy, 220, 222
  variety
    Anjou 196, 237, 238, 258
    Anjou 210, 237, 258, 260
    Asgrow 77, 237, 238, 258
    Asgrow 88, 258
    Austria 290, 237, 238, 258
    Caldera 433, 258
    Caldera 535, 237, 238, 258
    Dekalb 202, 237, 238, 258
    INRA 200, 237, 258, 260
    INRA 258, 258
    INRA 321, 237, 238, 258
    John Innes Hybrid, 238
    Kelvedon 33, 258
    Kelvedon 59, 252
    Kelvedon 59A, 237, 252, 258, 260
    Kelvedon Glory, 238
    Kingscroft KS2, 237, 258
    Northern Belle, 238

North Star, 238
Orlagold, 258
Pioneer 131, 251
Pioneer 383, 237, 238, 258
United 352, 258
White Horse Tooth, 237, 258
Maize rust, 246
Maize smut, 246
Malathion, 287, 292, 314, 315
Male-fertile beans, 280
Male-sterile beans, 280
Malting, 65, 66, 67, 68, 79, 108
Manganese deficiency, 10, 11, 153, 382
Manganese toxicity, 398
MCPA, 32, 33, 92, 94, 154, 199, 200, 243
MCPB, 32, 92, 94, 199, 347, 365, 366, 440
Meadow fescue, 405, 406, 411, 417, 418
Mecoprop and mecoprop mixtures, 32, 34, 92, 94, 154
Menazon, 287, 292
Mesopotamia, 64
Metabolisable energy, 299
Methoprotryne, 32
Metoxuron, 32, 92
Mevinphos, 287, 292, 314, 315
Mexico (city), 217, 218, 386
Middlings, 5, 6
Mildew (*Erysiphe graminis*), 26
Milling characters, 4
Milling premiums, 5
Monolinuron, 313
Multiplication seed, 22, 56, 170
Murganic RPB, 21, 87
Mycotoxins, mycotoxicosis, 103

N

Naked barley, 90
Narcotic bait, 379
Nectar (bean), 281, 282
Neolithic Age, 2
Net Blotch, 87
New Zealand, 334, 391, 394, 400, 423, 424
Nicotine, 287, 292, 314, 315
Nicotinic acid, 184
Nitrate nitrogen, 206
Nitrogen fixing, 327
Nodulation, 308

O

Oat cakes, 128
Oatmeal, 128
Oats
  species
    Avena abyssinica, 123, 124
    Avena algeriensis, 123
    Avena barbata, 124, 126
    Avena brevis, 123, 125
    Avena byzantina, 123, 124
    Avena chinensis, 125
    Avena fatua, 123, 125, 126
    Avena ludoviciana, 124, 125
    Avena nuda, 123
    Avena nudibrevis, 124
    Avena orientalis, 123
    Avena sativa, 123, 125, 126
    Avena sativa ssp., orientalis, 126
    Avena sterilis, 125, 126
    Avena strigosa, 123, 125, 126, 127
    Avena vaviloviana, 124
    Avena ventricosa, 124
    Avena westii, 124
  type
    black, 130, 146
    chinese naked, 125
    tartarian, 126
    white, 130, 146
    yellow, 130
  variety
    Angus, 149
    Astor, 147, 148, 149, 168, 170, 172
    Ayr Commando, 134, 148, 149, 150,
      170, 171
    Ayr Line, 134
    Bell, 134, 148, 149, 150, 170, 171
    Blenda, 135, 147, 148, 149, 170
    Castleton, 134, 148, 149, 172, 173
    Condor, 147, 148, 149, 168, 170,
      172, 173
    Eagle, 134
    Early Miller, 134
    Forward, 147, 148, 149, 155, 170,
      171
    Golden Rain, 134
    Gordon, 134
    Karin, 148, 170, 171
    Maelor, 148, 149, 170, 171
    Manod, 148, 149, 169
    Maris Quest, 148, 166, 167

    Marvellous, 134
    Max, 149
    Mostyn, 148, 168
    Onward, 134, 155
    Padarn, 148, 166, 167
    Pendrwm, 148, 166
    Peniarth, 148, 166
    Phoenix, 149
    Potato, 149
    Pure Line, 134
    Quality, 134
    R 30, 134, 149
    Radnorshire Sprig, 135
    Royal Scot, 134
    S.84, 134
    S.147, 143, 150
    Sandy, 134
    Selma, 148, 168, 170
    Shearer, 148, 170, 171
    Star, 134
    Stormont Iris, 148, 172, 173
    Stormont Sceptre, 148, 172, 173
    Sun, 134
    Sun II, 132, 149
    Tarpan, 148, 170, 171
    Tyrone Tawny A, 148, 172, 173
    Tyrone Tawny B, 148, 172, 173
    Victory, 134
    Vigor, 149
    Yielder, 134, 147, 148, 149, 170, 171
Offals (wheat), 6
Once grown seed, 22
On-the-floor drying, 44
Organo-mercury, 21, 87, 340
Origin of wheat, 2
Overdrilling, 424
Oxydemeton methyl, 287, 292, 314, 315

P

Pakistan, 385, 423
Paraquat, 16, 17, 243, 284, 312, 345, 420,
      440
Parathion, 152
Paris Green, 152
Pea and bean weevils, 285, 286, 348, 418
Pea aphid, 348
Pea Growing Research Organisation
      (PGRO), 308, 310, 312, 313, 318,
      336, 338, 340, 341, 343, 349, 352,

360, 362, 363, 365, 367, 368, 371, 372
Pea haulm silage, 374
Pea moth, 348, 349
Peas
  species
    Pisum abyssinicum, 330
    Pisum arvense, 330
    Pisum elatius, 330
    Pisum fulvum, 330
    Pisum sativum, 330
    Pisum sativum, arvense, 332
    Pisum sativum, hortense, 332
  type
    blue, 328, 335, 336, 337, 339, 349, 355, 356
    dried, 332, 335
    dun, 331
    field, 330, 332
    garden, 330, 332
    maple, 331
    marrowfat, 328, 335, 336, 337, 339, 349, 355, 356, 357
    processed, 333
    pulling, 329, 332, 375
    wild, 327
  variety
    Alaska, 336
    Big Ben, 335
    Canners' Perfection, 366
    Clipper, 366
    Cobri, 336
    Dark Skinned Perfection, 360, 361, 362, 363, 366
    Dik Trom, 336
    Early Bird, 366
    Early Freezer, 361
    Emigrant, 335, 337
    Faceta, 336
    Feltham First, 366
    Freezer 69, 363
    Gregory's Surprise, 360, 366
    Harrison's Glory, 334, 335
    Hylgro, 336
    Jade, 361
    Jo 06638, 336
    Johnson's Freezer, 361
    Kelvedon Wonder, 360, 361, 362, 366

Lincoln, 366
Mansholts Gek, 336, 337
Maro, 335
Meteor, 361, 366
Octavus, 336
Onward, 366, 378
Pauli, 336
Perfected Freezer, 366
Riito, 336
Rondo, 336, 337
Servo, 336, 337
Stijfstro, 336, 337
Surprise, 363
Thomas Laxton, 366, 378
Unica, 336, 337
Vedette, 336
Witham Dwarf, 361
Witham Wonder, 363
Zelka, 335, 337
No 814, 335
No 815, 336
Penicillia, 103
Pentachlorophenol, 243, 284, 345
Pericarp, 6
Permanently soluble nitrogen (PSN), 69, 70
Persia, 2, 123, 180, 385, 386
Peru, 386, 423
Phalaris tuberosa, 417, 418
Phorate, 152, 245, 287, 292
Phosphamidon, 287, 292
Photoperiod, 402, 403
Platform drier, 41
Ploughing-in beans, 275
Poland, 179, 184, 186
Polythene sacks, 160, 162
Porridge, 128
Portugal, 217
Potential transpiration rate, 401
Prebane, 29
Proboscis (bee), 281
Prometryne, 342, 346
Propachlor, 314
Propham, 344
Propionic acid, 104, 106
Protein equivalent, 299
Pulse crop, 328
Pumpernickel, 181
Pyrethrins, 292

Q

Quaker Oat Company, 133

R

Ranks–Hovis–McDougall, 5
Redland oats, 139
Resistance moisture meter, 36
Rhizobial bacteria, 270, 308, 401, 414,
    416
Rhizobium meliloti, 414, 415
Rhynchosporium, 88
Rolled oats, 128
Root rot, 247
Rumania, 184, 186
Russia, 4, 179, 184, 186, 386
Rye
  species
    Secale africanum, 180
    Secale ancestrale, 180
    Secale cereale, 180, 181
    Secale fragile, 180
    Secale kuprijanovii, 180
    Secale montanum, 179, 180, 181
    Secale vavilovii, 180
  type
    common, 192, 204
    hungarian, 180
    wild, 180
  variety
    Beka, 189
    Berna, 189
    Bernburg Fodder, 198, 204, 205, 213
    CRD, 198, 205
    Danish Brattingborg, 192
    Dominant, 194, 212
    Gartons LG, 193, 194, 198, 205
    Giant, 192, 204, 211
    Karlshulder, 189, 192
    King, 192, 193
    King II, 192, 193, 194, 205, 212
    Lovaszpatonai, 205, 213
    Mammoth White, 192
    Midsummer, 181
    Milns Grazing, 205, 213
    Pearl, 192, 193, 194, 211
    Petkus, 189, 192, 193, 194, 198, 205
    Petkus Normal Straw, 192, 198,
      205, 212
    Petkus Short Straw, 192, 198, 212

Petkus Spring, 205
Petkus Tetraploid, 192
Rheidol, 205, 213
RN 12, 193
St John's Day, 181
Star, 192, 193
Steel, 192, 193
White Russian, 192
Ryegrass, 205, 206, 209, 405, 411, 425
Rye whisky, 181

S

Sainfoin
  species
    Onobrychis arenaria, 432
    Onobrychis caput-galli, 432
    Onobrychis crista-galli, 432
    Onobrychis montana, 432
    Onobrychis pulchella, 432
    Onobrychis sativa var. *bifera*, 433
    Onobrychis sativa var. *communis*, 433
    Onobrychis viciifolia (*sativa*), 432
  type
    common, 433, 435, 436, 440, 441,
      442, 443
    Cotswold, 442
    giant, 433, 435, 436, 440, 441, 442,
      443
  variety
    Eastern Counties Giant, 438
    Esparcet DSG 2002, 438
    Esparcet DSG 2004, 438
    Esparcet DSG 2005, 438
    Fakir, 438
    Milborne, 438
    Perly, 438
    Premier, 438
    Ternel, 438
Scandinavian countries, 179, 181, 329
Scarification, 416
Schradan, 287, 292
Self pollination, 185, 277, 278, 279
Sharps, 5, 6
Simazine, 32, 242, 243, 244, 275, 283,
    284, 291
Slugs, 13
Sod-seeding, 18
Soil Acidity Complex, 398
Soil Moisture Deficit (SMD), 315

Soil pH, 10, 11, 71, 137, 141, 185, 186, 230, 231, 270, 306, 332, 333, 337, 361, 398, 399, 400, 407, 436
Somatic numbers (wheat), 2, 3
Sortex, 353, 354, 355
South Africa, 423
Sowans (sowens, flummery), 133
Spain, 186, 217, 385
Spindle tree, 286
Squirrels, 247
Stale seed-bed, 235
Starch equivalent, 71, 299
Stem Eelworm (lucerne), 416
Stem Rot (clover), 303
Sterile Guelder Rose, 286
Stocking rates, 207, 208
Stover, 218, 238, 257
Strip-grazing, 207
Subterranean clover, 418
Sulphuric acid, 243, 284, 352, 353
Sweden, 186, 391
Sweet Clover, 414
Switzerland, 123, 387
Syria, 385

T

Take-all (*Ophiobolus graminis*), 13, 14, 18, 76
Tall Fescue, 425
2,3,6-TBA, 32, 92, 154, 199, 200
TCA, 344
TCBQ, 340
Temperatures for growth, 12
Tenderometer, 358, 360, 364, 373
Testa, 6
Thermophilic fungi, 103
Thimet, 245
Thiram, 236, 276, 314, 340, 364
Threshold temperatures, 401
Timothy, 405, 411, 417, 418, 422, 425
Tri-allate, 32, 92, 283, 284, 344, 345
Tripod, 349, 351
Turkey and Turkistan, 180, 385

U

United States, 181, 186, 226, 227, 249, 329, 386, 390, 391, 400, 423
Uruguay, 386, 392

V

Vernalisation, 8, 9

W

Western Isles, 65, 126, 127
Westerwolds ryegrass, 204, 205
West Indies, 217
Wheat
  species
    Triticum aestivum, 3
    Triticum turgidum, 3
    Triticum vulgare, 3
  type
    British, 8
    Dutch, 8
    French, 8, 9
    German, 8
    Manitoba, 4
    Scandinavian, 8, 9
  variety
    Bersee, 8
    Cama, 51, 52, 53
    Cappelle-Desprez, 17, 20, 51, 53, 56, 57
    Cardinal, 54, 55
    Champlein, 51, 53, 56, 57
    Elite Lepeuple, 29
    Hybrid 46, 17, 29
    Janus, 54, 55
    Joss Cambier, 51, 52, 53, 56
    Kleiber, 54, 55
    Kloka, 56, 57
    Kolibri, 54, 55, 56, 57
    Maris Dove, 54, 55
    Maris Ensign, 54, 55
    Maris Nimrod, 51, 53
    Maris Ranger, 9, 51, 52, 53, 56, 57
    Maris Widgeon, 5, 51, 53
    N 59, 56, 57
    Peko, 5
    Professor Marchal, 29
    Rothwell Perdix, 25
    Rothwell Sprite, 54, 55
    Sirius, 54, 55
    Svenno, 5
    Troll, 54, 55
    West Desprez, 51, 53

Wheat flour types, 3
Wild oats, 29, 92, 123, 124, 125
Winter proud, 28
Wireworms, 13, 91, 151
Wood pigeon, 348

Y

Yaval oats, 139
Yellow Rust (*Puccinia striiformis*), 23, 25

Z

Zero grazing, 394, 397